Alexander King (CMG, CBE, U
Laureate) was born in 1909 in
but aspiring family. He gradu
Science, London. After a year of postgraduate chemical resear
at the University of Munich, he was appointed junior lecturer on physical chemistry at Imperial College, London, teaching and eventually forming a small research team on surface chemistry. In the 1930s he led a scientific expedition to Jan Mayen in the Arctic, and his group made a first ascent of the second peak of its striking volcano, the Beerenberg.

At the outbreak of war, Sir Henry Tizard, rector of Imperial College, invited him to forsake the laboratory to assist in scientific aspects of the war planning. King became a deputy to the Minister of Production. In this capacity he was responsible for the development of DDT and was sent to Washington to discuss it with the Americans. While there, King was instructed to act as the head of the British Mission for exchanging information with the US and he found himself Director of the British Central Scientific Office in Washington which he operated successfully for several years.

When the war ended, King accepted Tizard's invitation to work as Director of a Central Scientific Secretariat in the office of the UK cabinet and he also became scientific adviser to the Lord President of the Council. There he was involved essentially in civilian aspects of national science policy, and for the next five years operated in the centre of government with special emphasis on the problems of post-war reconstruction.

After Attlee's administration ended, King became a Chief Scientist in the Department of Scientific and Industrial Research before moving to Paris as Director of the European Productivity Agency. In 1960 he became a Director General at OECD, a position he held for several years until his retirement on age limit in 1974. During his tenure at OECD he founded, with Aurelio Peccei, the Club of Rome, an organisation of independent personalities outside government and industry with concern for the long-term problems of world society and the environment. On the death of Peccei he became President of the Club, finally retiring at the age of ninety.

Also by Alexander King

SCIENCE AND POLICY, THE INTERNATIONAL STIMULUS
(Oxford University Press, 1974)

THE STATE OF THE PLANET
(Pergamon Press, 1976)

THE FIRST GLOBAL REVOLUTION,
with Bertrand Schneider
(Simon and Schuster, 1989)

Essays and lectures on science policy,
education and technology and
three text books on chemistry

LET THE CAT TURN ROUND

One Man's Traverse of the Twentieth Century

ALEXANDER KING

CPTM

First published in Great Britain in 2006 CPTM

A CIP catalogue record for this book
is available from the British Library.

ISBN-13: 978-0-9554402-0-5
ISBN-10: 0-9554402-0-3

Typeset in Sabon by M Rules
Printed and bound in Great Britain by
The Cromwell Press, Trowbridge, Wilts

CPTM
63 Catherine Place
London, SW1E 6DY
www.cptm.org

e-mail: smart.partnership@cptm.org

'Alex King is a peace-searching, wise world citizen. A real humanist because a solid scientist, his global vision and endeavours have always been coherent with his declared values and loyal to his birth-roots. It has been and will always remain a considerable honour to being his friend and close colleague around the Club of Rome he co-founded.'

Dr Ricardo Díez-Hochleitner, Honorary President of the Club of Rome

'Dr Alex King has been on the leading edge of change for over half a century. He has been a pioneer in wartime science policy, the mobilization of science for the benefit of humanity, and he has been one of the founders of the environment movement. This book explains how and why he managed to achieve so much. *Let the Cat Turn Round* shows how one person can make a difference to world society.'

Keith Suter, Former National President, United Nations Association of Australia

'Alex and my dad (Aurelio Peccei) were an incredible pair and did much to make this world a better place. They did not get as much recognition as they deserved and I hope that Alex's book will help to rectify this.'

Roberto Peccei, Vice-Chancellor Research, UCLA

'Alex, as he is known in the Club of Rome and in the Commonwealth (by CPTM Fellows and Companions), is unique in his understanding of the limits to governance and a great innovator of new forms of co-operative governance at all levels of government. He espoused procedures combining formal and informal approaches, recognizing them as efficient, anticipatory and adaptive or spontaneous as the situation required, which is the essence of Smart Partnership today in the Commonwealth.'

Dr Mihaela Y. Smith, CPTM, Member of the Club of Rome, London

Contents

Foreword

Concerning the Ways of Cats and Men

Unlike many of my acquaintances, I am no enthusiastic cat lover. But I respect the species and am at ease in their company. As a male, I admire the ease of their independence within an accepted and valued domesticity. Cats are rather like the Japanese; they live simultaneously within two apparently irreconcilable cultures with a certain brittle comfort.

One day, many decades ago, I was sitting reading beside a cat that had spread itself out on a cushion on my knees. Instinctively I stretched out a hand and began to stroke it. There was an immediate flurry of feline irritation. I had offended the cat by stroking it the wrong way. With little effort the cat moved so that I could continue my ministrations in the right way.

This little episode appealed to me as an interesting commentary on attitudes, human and feline. The cat was able by a trivial effort, to adjust the world to its satisfaction. As for myself, was there here a possible way of life? The cat moved in response to my initiative. So let the world readjust to my aims and desires. I could certainly not expect individuals or society to change to satisfy my whims or ambitions. Nevertheless, were I to act on the assumption that they would do so, the subtle chemistry of things might tend to support my will. Here, then, was not a motto for life, but an attitude to be cultivated in order to fortify my ambition and self-esteem.

The next day, in the company of two friends I mounted the steps of the Albert Memorial and having evoked the spirit of the immortal Albert, intoned our aphorism of the month:

They say, 'You stroke the cat the wrong way.'
I say, 'Let the cat turn round.'

Abbreviations

AAPC	Anglo American Productivity Council
ACAST	Advisory Council of Applied Science and Technology
ACSP	Advisory Council on Scientific Policy
ADELA	Atlantic Development of Latin America
ADSR	Assistant Director of Scientific Research
BCSO	British Central Scientific Office – (later British Commonwealth Scientific Office)
BLS	Bureau of Labor Statistics
BSI	British Standards Institute
CERI	Centre for Educational Research and Innovation
CERN	Organisation Européenne pour la Recherche Nucléaire
CIP	Committee on Industrial Productivity
CPTM	Commonwealth Partnership for Technology Management
CSDI	Center for the Study of Democratic Institutions
CSIRO	Commonwealth Scientific and Industrial Research Organisation
DDT	Dichlor-diethyl-trichlorethane
DSC	Defence Science Committee
DSIR	Department of Scientific and Industrial Research
ECA	Economic Cooperation Administration (formerly Marshall Plan)
EEC	European Economic Commission
EIRMA	European Industrial Research Management Association
EPA	European Productivity Agency
FBI	Federation of British Industry
FID	The International Federation of Documentation

GM	General Motors
HFP	Human Factors Panel
ICI	Imperial Chemical Industries
IFIAS	International Federation of Institutes for Advanced Study
IFLA	International Federation of Library Associations
IIASA	International Institute of Applied Systems Analysis
INSEAD	Institut Européen d'Administration des Affaires
LRAM	Licentiate of the Royal Academy of Music
MIR	Military Intelligence Research
MRC	Medical Research Council
NCAR	National Conference on the Administration of Research
NGO	Non-Governmental Organisation
NIEO	New International Economic Order
NPC	National Productivity Centre
NPL	National Physical Laboratory
NRC	National Research Council
NRDC	National Research Development Corporation
NRLST	National Reference Library of Science and Technology
NSF	National Science Foundation
OAU	Organisation of African Unity
OECD	Organisation for Economic Co-operation and Development
OEEC	Organisation for European Economic Co-operation
OR	Operational Research
ORC	Operational Research Club
OSRD	Office of Scientific Research and Development
OSTP	Office of Scientific and Technical Personnel
PRA	Committee for Productivity and Applied Research
RDF	Radio Direction Finding
RIO	Restructuring the International Order
RPC	Research and Productivity Committee
SAC	Scientific Advisory Committee
SRU	Special Research Unit
TUC	Trades Union Congress
UDC	Universal Decimal Classification
UNESCO	United Nations Educational, Scientific and Cultural Organisation

Introduction

In 1974, when I retired from the OECD, I was approached by an acquaintance in one of the best-known London publishing houses. Many years earlier he had published a couple of chemistry texts for me and now, having looked me up in *Who's Who*, he wrote to suggest it was time to consider writing my memoirs. I replied, saying it was much too early. My priorities were firmly set on future activities and that 20 years hence, I hoped that I would have some life material worth recording.

The 20 years have now elapsed and, indeed, the interval has probably been the most active and enriching period of my life. I am now partly crippled and immobilised, unable to roam freely across the planet, while the passage of time has rendered me more reminiscent – an irritating trend in my thinking that I see as a creeping degeneration. In these circumstances it is not surprising that I accept it is now time to start recording at least some aspects of my life for the entertainment of my successors, family or otherwise.

But what sort of record should I leave? The history of each individual provides life material, interesting or banal, at various levels of activity, thought and emotion that might serve as the basis of a series of parallel biographies of very different and often conflicting character. In fact one seldom reads autobiographies or biographies, that portray the totality of the individual's existence, and few have the energy or temerity to delineate the several concurrent scenarios that together would integrate activity, existence and being into a coherent or discordant whole. I have no intention of undertaking such an experiment in exposing and bringing together my many lines of living. That would require a literary genius, which I am not, and would be an exposure that I am too craven to envisage. So which of my scenarios shall I cultivate? I might, for instance, write an account of my professional career, analysing the issues that have concerned me and the rationale and motivation behind the decisions I took. Or I might opt for a chronicle of family life with all its turbulent harmony, or describe my spiritual evolution.

I reject the temptation to write the traditional memoir of my professional preoccupations. They cover such a long period and involve so much detail it would be boring. Besides, with the passage of time and the freeing of official papers for public scrutiny, historians are already busy with several phases of my interests that they will doubtless illuminate far better than I could myself. They have already provided material for a flurry of Ph.D. theses. A spiritual scenario would seem too pretentious, too personal and possibly too thin to expose. A mere compilation of chatty anecdotes would be too superficial and would give little indication of what I like to think is my real self. I have decided, therefore, to steer a course between the various scenarios and attempt to write a simple account of the successive activities of my life, professional and otherwise, with only occasional glimpses of family and personal events to provide some perspective. The chief interest of the narrative will be the diversity of my activities as seen against the background of change in an eventful century.

God and the Infant

At the outset and in seeming contrast to what I have just written, it is necessary to mention a few inner experiences that have had a fundamental influence in shaping my life and behaviour as a whole.

I am continually surprised how one's memory retains, apparently indefinitely, images of certain events, often quite trivial, with a vividness they hardly seem to merit. At the same time the great mass of the minutiae of daily life that rush relentlessly through the brain are quietly eliminated. I can so easily and often recall the impressions of certain moments, events utterly unimportant in themselves, with all the details of location, landscape, light and even smells that have been commemorated for no apparent reason. I have had no say at all. So too with dreams that paint landscapes completely unknown to me in the natural world, but which persist and repeat.

Quite different is another set of memory recordings, of moments of supreme personal significance quite unlike mere meaningless flashes from the past and at a different level of consciousness; they can be conjured at will and have an indefinite validity. The first of these took place when I was nearly 3. It was undoubtedly the most vivid event I had yet experienced. I was walking beside my mother who was pushing my new sister in her pram. We were walking along a road beside Queens Park and I was aware of a line of high ornamental railings on my left. On the other side there was a steep grassy bank. Suddenly there was a blaze of intense yellow light that appeared to envelop me. I knew instantly that I was being visited by God. I cannot say how long it lasted; probably less than a minute. My mother was unaware of anything unusual happening; the baby did not cry. We continued quietly with our walk. I, of course, gave no indication that I had 'seen the light', nor did I ever mention the episode to my parents or friends. I pondered it in secret. Some 60 years after this event, I revisited the scene of my vision in a spirit of curiosity. The place was exactly as I had retained it in my memory over all these years.

As an infant I had, of course, absolutely no concept of the nature of God, as is still my situation. My conditioning to accept the traditional view of the deity must have already begun, because the word God had immediately sprung to mind as a vague someone or something important and above humanity, but connected with it. The impression of my vision was of a state beyond everyday life, yet within myself as well. For convenience I had found the name God. As I grew older, I carried the vision with me as a sort of disembodied deity, an immaterial reality unrelated to the conventional religious beliefs that I was now accepting voluntarily. Equally, I saw no contradiction to my convictions during the rationalist period that followed. I had no sense of being 'chosen' or of the vision being in any way special. I merely felt that I had the experience of an intimate and personal contact that had to be cherished.

So much, then, for God and the infant. The second seminal event was a discovery rather than an experience, a sudden realisation that had been growing in my mind. I must have been about 15 and rather studious. There was no association with any particular location. The flash of awareness was that while, as a rational being, my decisions about action or behaviour were normally built up quite logically, in practice when it came to the crunch, I frequently reacted abruptly and irrationally, in a sense opposed to what my reason had suggested. This brought about a clear realisation of the coexistence of rationality and irrationality and a need to take account of the uncontrolled power of the subconscious. Not a very profound discovery you may say, but one that transformed my attitudes to myself and others. It made me question the degree of stability of my motivations, and enhanced awareness of what was going on in my mind. I doubt if it has in any way modified the extent of my subjectivity, although a clear awareness may have provided a more reliable perspective.

The third experience that I now relate was again associated with a precise time and place. I had been to a concert at the Queen's Hall in London with my friend Normanska, where we had been exposed to an emotionally charged performance of Mahler's *Das Lied von der Erde*. At its conclusion we poured out with the audience, to find ourselves engulfed in one of London's traditional pea-soup fogs. The crowd was shocked by the sudden hostile darkness and hardly moved, despite pressure from those behind still unaware of the fog. In the confusion it took some time for us to extract ourselves and feel our way towards Oxford Street. Suddenly I had a strong feeling of certainty that I had found the central motive for my existence, namely that I had a duty to extend and deepen the zone of my consciousness. Whether the emotional force of the music or the environmental shock of the fog induced this, I simply do not know. But the image of that moment has remained vivid and determinative.

These are some of the infantile and adolescent experiences that have helped to determine my behaviour throughout life. They continue to be very real.

1

Origins

It is customary in biographies to start with a sort of pre-history, an account of some known ancestors, in the hope of providing some indication of genetic heritage. Customary certainly, but useful? When I was a student, I became fascinated by C. Lombroso's, *The Man of Genius*, which attempted to relate the phenomenon of genius to the diseases many famous people suffered from, and in particular to syphilis and tuberculosis. I became so enthusiastic with this idea that I started to write a book on genius to be called 'The Twilight of Genius', in which I attempted to differentiate the many disease-induced geniuses from the few really great people whom, I assumed or would argue, were paragons of perfect health. Why I thought it important to make this categorisation I cannot now imagine, but I mention this piece of my juvenilia because it necessitated my reading several dozen biographies and autobiographies concerning individuals who might fit into one or another of my divisions, thus learning much about the genre. Very many started with an examination of the ancestry of their subject, intended to illustrate how traits in character, behaviour and intelligence were inherited from known forebears. I accept that both heredity and childhood social environment are dominant in shaping personality, but I very much doubt if superficial observations drawn from sketchy portraits of predecessors give reliable clues to this endowment. In the future, biographies may be written using detailed DNA charts providing critical insights on genetic transmission from ancestors. At present, however, descriptions of the behaviour of ancestors do not seem to say much about inheritance, while they may yield some clues concerning the social environment which may have conditioned the next generation. Be that as it may, I have an aversion to accounts of eminent antecedents as making very boring reading, except where they are enlivened by descriptions of the antics of occasional eccentrics.

In this account of my life, therefore, I shall mention only those near relatives who were elements in the social and psychological mini-environment that helped to shape me. Rising as I do from a colourless mixture of peasantry and bourgeoisie, my knowledge of my forefathers is minimal. Nevertheless there are a few family myths worth mentioning. My father's sister, 'Aunt Kate', was very interested in the family history, obscure as it was, and did much research on the origins of the word King as a surname in Scotland, where it is rare, unlike in England or still more in Ireland; indeed in 1920 we were the only entry in the Glasgow telephone book. By tradition, the Kings are regarded as a sept of the Clan MacGregor. My family appears to have come from the countryside around the towns of Dunblane and Doune in Southern Perthshire. The family memory is that some of my ancestors played the part of one of the three Kings in the Christmas plays known to have been performed in Doune Castle in the presence of the King of Scotland. The MacGregors were proscribed on several occasions, first in 1604. Having lost the right to use their real name and to wear their tartan, my family of MacGregors chose their nickname 'King' as their official name.

There is another mystery if not a skeleton in the ancestral cupboard. Apparently, Italian blood entered on the female line a few generations ago. We even know that she was called Bellini and came from Umbria, but beyond this absolutely nothing is known about her. There were hints of the Latin in the faces of some of my relatives; two of my aunts having dark skin and pretty Italianate features. It is curious that Mediterranean blood should have mingled with that of peasants in the remote Scottish countryside. The latter would have had no opportunity to go on the 'grand tour' and return with an Italian girl, as did many of their aristocratic compatriots of the eighteenth century. My sad conclusion, therefore, is that the Bellini lady must have been a member of the Italian ice-cream community which had permeated every town and village in Scotland and was still fully operative and important during my childhood. But more of this later.

I slid onto this planet on the 26 January 1909 at 5 Battlefield Avenue, Langside, as the firstborn of James Mitchell King and Janet (Jenny) Dale. Langside was a lower-middle-class suburb of Glasgow, remembered for the battle of the same name, the last of the disastrous little battles of Mary Queen of Scots before she fled to seek the protection of Elizabeth I. I was assisted into the world by Dr Murdoch, whom I later came to know all too well when he tended me through the whole gamut of childish diseases. He was a wild highlander with fiery red hair and handlebar moustache, always wearing the kilt; so different from everyone else he appeared to belong to another species. I feared him in the sense that one is supposed to fear God, with a fear akin to awe.

My father had been first an office boy and then a clerk in a firm that manufactured wire and nails. His father had been stricken with paralysis after a stroke in his forties, unable to walk or to work. Family prospects were shattered and it was up to the children to rally for survival. The two eldest daughters set up as dressmakers, while my father had to leave school at 11 and find a job. Thus began years of penury. He was a studious boy, though, and attended evening classes for many years. Particularly interested in English literature, already a family tradition, he concentrated on practical subjects to increase his earning power and soon became proficient in bookkeeping. The family must have existed at about the poverty line during these years, but his modest earnings enabled his youngest sister Kate to go to university. She eventually became a teacher, to the benefit of the family fortunes.

At about the time of my conception, my parents were considering joining four of my mother's relatives emigrating to Canada. However, just before my birth my father was offered a job in the Nobel Explosives Company as a cashier at the princely salary of £100 per annum and decided to stay put. So I just escaped being born a Canadian. Two of my mother's siblings and their consorts did emigrate to Canada and, as we maintained close contact with them, this started for me a lifetime familiarity and close relationship with that country.

My father, as his studies continued and as he rose to modest prosperity, acquired a sizeable library. He had become a literate and cultured man. With his passion for education and his ambition, reproduced in me, it is not surprising this impoverished clerk should have ended his career as a director of ICI.

Grandfather King was an Alexander like me. He was a warm, sensitive man, self-educated and with a character not unlike that of his son James. In this time of economic deprivation, he and his wife had left the family farm in South Perthshire and, like so many Highland families, moved to Glasgow, a city of rude energies booming with the wealth and poverty that characterised the industrial revolution. He found a comfortable job and it seemed the King family had established stable conditions in that grim city. Then, tragically, he had his stroke. Fortunately, grandfather's mind and memory were unimpaired and his spirit undimmed; despite hardships, the family remained happy and cheerful. Grandfather King shuffled with difficulty on two sticks. He was a voracious reader, but he spent most of the day sitting by the window, following the kaleidoscope of the streets. From my earliest days I warmed in his company. By the time I was 8, I visited him frequently in his grey tenement flat in a working-class district; good conversation and jokes galore. On leaving, I would wave to him as he sat at his window, then turn off into the main street and disappear from his sight. I would stop for a moment to make him think I had gone for good, then peep round the corner to 'surprise' him. Waving and laughter would follow. This performance would be repeated several times with increasing moments of vanishing before I finally tore myself away. The relationship with my grandfather was one of innocence and beauty, possibly the most perfect of my life.

His wife, Margaret Mitchell, on the contrary, had little or no emotional contact with me. She was a grim woman with a smooth, unlined complexion that retained traces of classic beauty. Before their marriage she had been the cook in the 'big house' of the laird in their Perthshire village. Like so many women of the period, she had donned a black velvet bonnet at the age of 40 and dropped from active life, sitting quietly by the coal fire until death removed her in her nineties. She spoke seldom and when she did, it was usually to make derogatory and scathing remarks about some person or idea. Yet obviously she had been a very strong character, tyrannising her daughters, three of whom never dared to escape through marriage. The two eldest, Jessie and Maggie, were amiable, poor creatures who seldom left the family circle, except to go to church, but delighted in entertaining us children in their workroom. The third, Bella, ran off and married William Cook, a joiner and house builder whom the family regarded as unsuitable. Bella was very beautiful, but died of tuberculosis when I was about 5, leaving two children, William and Margaret, who were welcome playmates.

The King home was plain and poorly furnished. The kitchen was the centre of family life and dining, although there was a 'ben the hoose', a parlour for festive or formal occasions which was seldom used, except by my grandfather sitting alone for hours at the window with its view over the street. The kitchen had a black sink, called a jawbox in the vernacular. Here as in the other public rooms there was a closet with a bed in it, shut during the day by folding doors. I spent odd nights in the closet beds of the kitchen and parlour. There was no question of reading in bed as only the main rooms were lit by gas, first by fish-tail burners, and later, as technology developed, by the superior Welsbach gas mantle. There was no gas stove, cooking was done on the coal fire, no bathroom and the lavatory was a small black-painted cubicle with no light or vent. The whole house was kept meticulously clean, while its denizens likewise maintained a shiny exterior, cleansed by their ablutions at the kitchen sink. It was a typical, not too poor, Glasgow tenement at the beginning of the century.

Religion was central to the family, quite apart from its place in the spiritual life of individual members. They were 'wee frees', members of a Presbyterian sect much stricter than the established Church of Scotland, obsessed by the dogma of predestination. The Sabbath was observed in the strictest sense and alcohol was regarded as the ultimate evil. Religion to them transcended individual spiritual aspiration; it was a matter of deep, emotional, almost organic presence – a link with forebears who had suffered so much for their faith – and the basis of their cultural existence. Yet, despite the aridity of their sectarian approach, they were generously open to the world, forgiving to those who offended them and tolerant of those holding opposing views.

My mother's family was about as different as could be. They also purported to be religious, belonging to the Church of Scotland and attending regularly, but somehow automatically. I sensed quite early on that, in contrast to the Kings, the Dales carried their religion quite lightly. They observed the Sabbath strictly, it must be admitted, they were teetotal, supported the 'Band of Hope' and subscribed to African missions. Nevertheless, at New Year the occasional glass of sherry would appear. The church was the most important social and cultural focus of their lives. It was where you made friends, where the young found their mates and where, when troubles or bereavements struck, comfort, help and friendship could be relied on.

For this family, the most important thing was that they were the Dales! They knew they belonged to one of the patrician families of industrial Glasgow, and that placed them not necessarily above, but definitely apart. This was not an obsession, but a quiet and unspoken assumption. The family aura of benevolent apartness emanated from their eighteenth-century kinsman David Dale (1737–1806), who had been one of the founding fathers of the new Glasgow arising from the industrial revolution. Dale really was a remarkable man. Born at Stewarton in Ayrshire, he started life as a 'herd laddie', was then a weaver's apprentice, set up his own import business in Glasgow in 1763, and four years later married the daughter of a Royal Bank of Scotland director. He became a rich merchant, powerful banker, entrepreneurial industrialist, philanthropist, and social reformer, while he remained throughout his life a fervent evangelical who preached every Sunday in his own church until, it is reported, he became too corpulent to get into the pulpit.

Dale's memorial is the New Lanark Mills, founded in 1785, not only one of the earliest mechanised textile factories and thus a building block in the foundations of the Industrial Revolution, but also a pioneering social innovation now classified as a national monument and a popular tourist attraction.

The New Lanark experiment is commonly associated with Robert Owen, the socialist and pioneer of the cooperative movement, who purchased the mills from Dale in 1799. Owen, unlike Dale, was a flamboyant and immodest man and, while the fundamental importance of Owen's work cannot be disputed, his account of the development of New Lanark greatly understates

Dale's original achievement. Owen had, in fact, purchased New Lanark after it and another at Blantyre had been operating successfully under Dale for 13 years and was as widely known for its technology as for the enlightened working conditions. He was briefly in partnership with Arkwright during the planning phase, but was sole owner until he retired and sold it in 1799 to his son-in-law, Owen, who had married Ann Caroline Dale, David's eldest daughter. The first mill, erected in 1785, burnt down in 1788, but Dale kept his workforce fed and clothed, paying their wages while it was rebuilt. Well over 1,000 people were employed. By 1795 it had become famous and its visitors' book records the signatures of about 750 people each year from all over the world, amongst them James Watt.

The workforce of the mills, in addition to providing work and homes for evicted Highlanders after the Clearances, included a large number of children aged from 6–16, many of them foundlings or paupers. The use of child labour in Britain was still common and criticism of the practice minimal. The novel feature of the New Lanark approach was that these 'apprentices' were not only housed, fed and clothed, but were also given a sound education which prepared them for good employment at the end of their period of indenture. At one time, before Owen's takeover, there were as many as 16 teachers in the New Lanark community, who also provided education to the children of the adult workers. While there were other cases in Scotland and England of schemes to improve working conditions for children at work, it is clear that Dale had instituted a humane system which was an important step in social reform.

Dale's years of intense evangelical activity may seem to have been at odds with his position as a successful entrepreneur. As a youth he had strong religious convictions as a member of the established Presbyterian Church. Later, however, he joined with some friends to dissociate himself from Presbyterianism, apparently objecting to paternalism and what they saw as papist residuals in the Presbyterian ritual. They met in private houses and then built an 'Old Scotch Independent Church'. This appears to have been a very conservative Christian group, 'rejecting all authority, civil or ecclesiastical, other than the bible'. They disapproved of full-time salaried ministers, but had a plurality of 'elders' who could be called on to preach and who undertook the day-to-day business of the church. Dale was one of these elders. He had a talent for preaching and is said to have been the first layman to exercise the priestly function, which angered many of the good church-going citizens of Glasgow, and for a few years he was jeered and jostled in the streets. In later years he was held in high esteem by the public as a leading reformer.

During this period of intense evangelism Dale had become a successful merchant, a leading banker (he was the first Glasgow agent of the Royal Bank of Scotland) and an industrial entrepreneur. He had amassed a substantial fortune. His lifestyle had become luxurious, even ostentatious. In 1780 he bought land and erected a mansion on Charlotte Street in the centre of Glasgow, his architect and interior decorator being Robert Adam. When this house was being demolished in the 1950s, I was offered one of its beautiful Adam fireplaces. At the time I was living in a flat in Paris, so alas had to decline

It must seem perverse after what I said in the Introduction about antecedents to have devoted so many words to David Dale. When I was a child, I assumed from the talk around me, that I was a direct descendant. However, when I looked into his life story later, I found that he had no sons so I must have descended from his younger brother, which my Dale grandfather verified. So be it then, I had not inherited his genes directly. I justify this mini-biography of my eighteenth-century kinsman by the importance he played in my formative environment.

Obviously the Dales had a greatly superior lifestyle to the Kings. Indeed, I soon sensed that they felt my mother had married beneath her. They tolerated my father, were perfectly polite, there was no overt hostility, but there was no warmth, the two families had absolutely no contact during my childhood. Their house was a pleasant well-furnished villa in a neat suburb, bordering open countryside, with a garden I remember for its apple trees. My maternal grandfather was a kindly, rather simple man. As an infant, I accepted his presence much as I accepted the piano in the drawing room. He lacked the warmth and come-forwardness which enable some adults to make contact with little children, a quality grandfather King so richly possessed. In my adolescence, however, I learnt to appreciate this cheerful old man with his shock of white hair, piercing blue eyes and rosy complexion. I even discovered that he had a streak of eccentricity. Hugh Dale had owned or managed a factory for the manufacture of boots and shoes. He had retired at the age of 40 and lived happily until his mid-nineties, simply enjoying himself and indulging in good works.

Agnes Dale, my grandmother, was very different. She was a rather complex character who seemed remote, very low key and grudging. Her greatest praise for somebody or something was 'it's no bad'. One instinctively knew her for a strong character sure of what she wanted and sure to get it without fuss or conflict. As an infant I was very aware of her and slightly frightened by her aloofness. I felt no need to love her, but instinctively held her in great respect. She was a Highland woman, Gaelic speaking, intelligent and well read; she had a passion for history and it was claimed that she had read Gibbon's *Decline and Fall* seven times. Not for her the black velvet bonnet or the years' long wait by the fire for a summons to rise up to the Presbyterian Paradise. She dressed as a modern woman and lived as an active old lady well into her nineties. She had a beautiful contralto singing voice and, with her daughter Mary's soprano, sang haunting highland songs which evoked images of the mysterious world of the Celtic twilight, at once more cruel and more tender than the one into which I had been deposited.

The Dales had six children. Hugh, their eldest was an architect who built schools and churches. He married and had one daughter; my cousin Muriel, a gifted musician who later became a concert pianist, using the name Angela Dale. Mary, my eldest aunt was slightly disabled and walked with a limp, probably due to rickets in childhood, a disease that was a scourge of Glaswegians, whose supply of pure water from Loch Katrine was entirely free of minerals. Aunt Mary was an intelligent woman who never married and nor did the next daughter, Margaret, a woman of flirtatious mannerisms which I later found slightly distasteful. Then came tiny Aunt Nan, who married Archie Russell, and bore a son of the same name. They all emigrated to Canada, but later, having gained a modest prosperity, paid us occasional visits in London. My cousin Archie, whom I have frequently visited in Vancouver, became a prominent accountant. After that came my mother, Jenny, and finally David, an attractive man who was regarded as rather wild by the family. He married Jeannie Wishart who, like my father, was seen as being not quite up to the family's status. They had two daughters, born in Scotland, and a son in Canada. David became a buyer in the Hudson Bay Company and visited Europe on business every year, enabling us to maintain regular contact with our Canadian relatives and to share their problems and successes. They were important in our family.

We lived within walking distance of both sets of grandparents and I was taken to visit one or the other nearly every day. Festive occasions such as New Year or Easter demanded attendance at both houses and contrasted the atmosphere of each. It was *de rigueur* that all the children and grandchildren should gather together on these occasions. On New Year's day, for example, we would lunch with the Dales, joined by all the uncles, aunts and cousins on that side. We would have an excellent meal, but we children would be bored by the flow of grown up conversation, especially when it dragged on interminably after the distraction of eating was over. We were sometimes able to escape to another room with the cousins, but play was quite subdued and tentative, oppressed as we were by an awareness of adult nearness. But the hours passed and at last it was five o'clock and we could escape to go to the Kings for high tea with the cousins on that side of the family. Here everything was joyful and hilarious. In the maiden aunts' workroom was a long footstool which, when one end was placed on the worktable, formed an excellent slide which we all enjoyed, the laughter of the aunts mingling with our squeals. Then Grandfather King would find a large jug and give it with a shilling to my cousin Willie Cook and me to go down to the nearest Italian's to fill with ice cream. Everyone was happy and the evening all too short

One feature of the family that I have not yet mentioned was the quiet security they felt in being Scots. I doubt if any of the relatives had ever strayed south of the border by the time I was born. They accepted they were British, and were proud of the Empire with all the red areas on the atlas, but there was an inherent if low-key hostility to all things English – 'after all we fought them in bloody battles for more than 500 years', or 'the English, you can't trust them, they're no sincere'. As far as I can judge now, 80 years later, the Scots had no real resentment for past injustices; there was no violence, only a modicum of distrust and even that was shrugged off in jokes. After all, the Scots had over long periods repulsed the English armies and faced them on the battlefield, sometimes even winning. Unlike the Irish whose resentment is much deeper, the Scots never felt that they were treated as second-class citizens and, in their pride, even felt they were penetrating the English system and largely running it. The story was often told in my early days of a Scottish worthy, who had returned from a business visit to London, being asked how he found the English, 'I never met any', he said, 'I only saw the highheedjens (high head ones).' This was the society that moulded my earliest years; puritanical, narrow-minded, sometimes irritatingly self-satisfied, but warm and basically honest.

2

Glasgow at the Beginning of the Century

Glasgow is one of the oldest towns of Scotland, more ancient than its capital Edinburgh. It was founded – 100 years after St Patrick's birth at nearby Kilpatrick – in the sixth century, that obscure period following the collapse of Rome, the lost century of Arthurian legends, with Christianity still uncertainly established in the west of Scotland and struggling to reassert itself in the face of a resurgence of paganism.

The founder of Glasgow was St Kentigern, commonly called St Mungo (the Beloved), who became a bishop at the age of 25. He set up his church in a small village where the Roman road from the south reached a ford across the River Clyde. It was named 'Eglais Cu', 'the beloved church', now pronounced as Glasgow. The coat of arms includes a robin, a salmon and a ring, depicting some of the myths surrounding St Mungo's life. The city motto is his 'let Glasgow flourish by the preaching of the Word'.

The village grew in importance in medieval times, with a Gothic cathedral and a university of repute. 1707 and Union with England was a critical moment for Glasgow, then a pleasant little university and market town of 12,500 inhabitants. The Union had given Scotland the right to enter England's colonial markets, transforming the country's economic prospects. Glasgow, with its access to the Atlantic, quickly sought a share of the markets of North America and the West Indies, becoming a major importer of tobacco and raw cotton. Soon a widespread cottage industry arose that, by the time of Dale, had begun to consolidate and mechanise through the use of waterpower in large, efficient cotton mills.

The population of Glasgow had risen to about 45,000 and

was increasing rapidly. It was at the head of an estuary from which its products could be exported all over the world, in a region rich in coal and iron and was about to become one of the cradles of the Industrial Revolution. Most important were its excellent educational facilities and the highly developed entrepreneurial talents of some citizens. Amongst them was James Watt, whose steam engine was the motive force for the construction of the new era of industry. Amongst the immediate local effects of Watt's engine was the draining of water from the coal pits to provide ample reducing material for the smelting of metals, and the dredging of the Clyde to make it navigable, making possible the shipbuilding that dominated the region's economy for a century. By 1800 the population had reached 83,000. It had become the commercial centre of the new Scotland, with a vast industrial hinterland. When I was born 100 years later, its population neared one million and it had become known as the Empire's second city.

This phenomenal population growth transformed Glasgow into a large, rough and unplanned 'frontier' city, with most of its inhabitants recent immigrants. The eighteenth century was a period of demographic growth for the kingdom overall, but natural increase only accounted for a fraction of population growth in Glasgow. The Agricultural Revolution just before the Industrial Revolution led to a great reduction in the farming population all over Britain that was quickly absorbed by emerging industry. In Scotland, however, agriculture had been so primitive that, following the Union, improved methods of cultivation actually increased agricultural employment. There the two revolutions were simultaneous and socially symbiotic.

Extensive poverty in both Scotland and Ireland, had, for centuries, driven emigration and inevitably the promise of opportunity that Glasgow and Clydeside offered was an irresistible magnet. For many, alas, it merely resulted in an exchange of rural for urban poverty. The movement of people from the Highlands to the city was dramatic. I remember as a child being taken on a Saturday evening to visit the 'Highlander's umbrella'. Large numbers of Highlanders congregated regularly on a stretch of road under the central railway station sheltered from the rain, to meet friends, speak the Gaelic, swap stories and evoke the homeland.

Still more of the workers for the new industries were Irish. Emigration from the poverty there to the United States was already massive; now they made the easier step a couple of hours eastwards to the opportunities of Clydeside in their thousands. Thus industrialised Glasgow, not unlike Ulster, had a mixed population predominantly of Presbyterian Scots and Catholic Irish. Frictions and rivalries between these two communities were never dangerously violent, because, unlike in Northern Ireland, both were poor and equally oppressed. Rivalries persist today, but largely played out as a ritual on the football field between the two prominent Scottish league clubs, Catholic Celtic and Protestant Rangers, in games charged not only with emotion, but also with history and its bitterness.

Unprecedented urban growth and lack of planning had shaped the Glasgow of my birthtime into a conglomerate of districts and suburbs of very diverse character that gave a general impression of dirt and disorder. This image was somewhat misleading, because the city did have many noble buildings, whose beauty is now recognised, but still had some industrial plants close to the centre, noticeably an ironworks. I remember sudden bursts lighting the sky from blast furnaces on dark winter evenings, a phenomenon known to all Glaswegians as 'Dixon's Blazes', as familiar in the streets as the noise coming from the shipyards or the fog-horns of ships on the Clyde.

There were huge areas of grey tenements with few amenities, most notoriously the Gorbals. These housed large numbers of the poor and were amongst the worst slums of Europe, centres of crime, lawlessness and drunkenness. These were the Glasgow equivalent of Blake's 'dark satanic mills', the black face of the Industrial Revolution. The streets were thronged by 'shawlies', poor women who clasped their infants and wrapped themselves in large tartan plaids. They were even to be seen on the fashionable shopping streets in the centre of town. On the other hand, there were pleasant districts such as Kelvinside and Pollockshields that provided luxurious living for the few. The city was set in beautiful country just south of the Highland Line, so near that it was even possible to reach Loch Lomond by tramcar. To the south there were green hills, wild moors and plenty of unspoiled countryside.

The Glasgow accent, considered the ugliest in the spectrum of so-called English speech by many, is not really a single phenomenon, there were several spoken within the city. At one extreme, the Keelie accent using more or less English words rendered guttural and jerked through a drink-sodden throat; from a distance it sounds rather like Arabic. At the other extreme, pure 'Kelvinside', very precise in a drawling, lilting voice, can sound ludicrously precious. In between there are many others such as mine with its mixed vocabulary of English and vernacular phrases. I still find myself using colourful Lowland Scots expressions as more apt than the equivalent English. However, at about the age of 10, my parents insisted I take elocution lessons to improve my accent and enunciation. I submitted reluctantly to what I thought very sissy, but later was very thankful when, in the cruel atmosphere of an English public school, I was derided for my slight Scots accent.

An obsessive feature of Glasgow life was alcohol and attitudes to it. I understood very early that everyone was either a drunk or a teetotaller. There seemed nothing in between; moderate drinking was unheard of. As a small child I was often taken to visit friends on the other side of the city, going by tramcar and changing cars in the middle of town. On Saturday evenings, when walking from one tram stop to another, we had to thread our way through drunks every few yards. I clung to my parents, disgusted but not frightened. The drunks were noisy and often maudlin, but never menacing. I remember my embarrassment on one occasion when one of them tried to press a bawbee (sixpenny piece) into my palm. The teetotallers seemed to be equally but oppositely obsessed by alcohol. They talked about it as the great evil, signed pledges, attended the 'band of hope', and listened to endless sermons about it on Sundays. In our family abstention was inevitably total.

Alcoholism seems to be a difficult issue in many Northern, cold countries such as Russia, Finland and Canada. I remember well my first visit to Toronto during World War II. I had a meeting there on the Monday and profited from a free weekend to visit Niagara Falls. I caught a train for Toronto on the Sunday evening to find it packed with drunks singing or vandalising all the way to the city; Ontario enforcing Prohibition, trainloads of men repaired to Buffalo in neighbouring New York State each weekend to drink themselves silly.

In many countries there is rivalry, at times bitter, between the two largest or most important cities; Berlin and Munich, Milan and Rome, Moscow and St Petersburg. Scotland is no exception. Edinburgh possesses the aura of a capital city, although it had, in my youth, only a quarter of the number of inhabitants of Glasgow and, in comparison, was always a refined city with refined citizens enjoying an intense cultural and intellectual life. The distinction was neatly expressed by referring to 'the Glesca folk' and the 'Edinburgh people'. Nevertheless, there has always been a degree of mutual respect and sense of complementarity between the two cities.

I have only a few flashes of memory from my first years; one being a clear recollection of learning to walk. My mother placed a piece of chocolate on a chair and I struggled to my feet to retrieve it. This could be dismissed as false memory, as a mere retelling, assuming my parents told me about it, but I am convinced it is a direct image registered by my infant mind. Another early memory is of standing in the street watching our furniture being loaded onto a removal van to be taken to a new home in a red tenement in a better area called Crosmoloof. I must have been about 4 years old. About this time also I was beginning to read and I remember looking for hours at an atlas and imagining the delights and curiosities of all sorts of exotic places. This is a love that has lasted my lifetime. By the age of 5 I knew the names of all the main countries of the world and their capitals. We were undoubtedly still quite poor then and my father still had obligations towards his parents. We ate very simply; much porridge, fresh herrings and potatoes with coarse oatmeal. Meat was a rare luxury. We were never hungry, but I was frequently ill. I contracted every possible childish infection, including a bad bout of diphtheria, attended by the wild Dr Murdoch.

We stayed only a couple of years in the new dwelling, moving to a modern bungalow in Dumbreck, a very good suburb, where we lived for seven years until moving to London. Until Dumbreck, play had been in the streets. Little girls chalked out squares on the pavement and hopped about playing peever or skipped interminable singing rhymes, often with primeval words passed down the ages only through children's games. Little boys fooled around, fought, shouted and made general nuisances of themselves. The scene was undoubtedly one of crude happiness;

there was no poetry to be sensed. Even the children's songs were banal:

Annie Rooney had a wee wain.
Someone hit it wi' a stane.
The neebors said it wus a shame
To hurt pair Annie Roonie's wain.

Dumbreck opened up exciting new perspectives; my lifestyle changed completely. Here was space and mature trees. Our house backed on to a large public park with birds and squirrels and, as I lay abed on summer evenings, the shrill cry of the peacocks was pleasurably disturbing. Soon I was able to climb over the railings and roam around the park at will after it was closed for the day. Later on I ventured further afield with friends, often ranging around the Pollock Estate after hours. This was a large, well-maintained gentlefolks' park with grassy paths, vistas of bluebells and banks of rhododendrons. It was a sheer joy to be there and I started a collection of wild flowers, an enthusiasm which lasted for years and was one of the real pleasures of my childhood.

I had already begun going to kindergarten in the basement of Langside Hall. I remember being collected each morning by a 'big girl' who was probably 8 or 10, who ushered me across the traffic of the main road. My parents had made friends with her father, the local fishmonger, through the church. When we moved to Dumbreck, I was enrolled in the lowest class at Albert Road Academy in Pollockshields, toddling off each morning in MacGregor kilt to catch a tram that took me to school in 20 minutes. I started in the middle of the academic year and as the holidays approached, the teacher read out a list of those pupils promoted to the next grade in the autumn. Alas my name was not on the list, meaning I would have to repeat the first class for a further year. My pride devastated by this blow, I burst into tears and howled. The wretched teacher paused, and then added my name to the list of promotions. This was my first experience of the cat-turning-round phenomenon. The next academic year passed; at its end I received a certificate to say that I was 'Dux of Class 1a', but that war shortages meant no Dux medal that year.

I remember quite clearly World War I breaking out when I

was five and a half. None of my relatives suffered directly, but several family friends were killed or wounded. My father's firm was the main manufacturer of explosives in Britain and would obviously have to expand for the production of munitions. He was directed to stay with the company; although, influenced by the famous Kitchener poster, he had volunteered. The most serious impact of the war was that I seldom saw my father, who had long and irregular hours of work. In 1915 he had a car crash when driving home in the wee hours of the morning after a prolonged planning session. My chief recollection is of him lying very still in bed with a block of ice atop his stomach. He recovered quickly.

Of course, I remember the blackout and my amazement at seeing the windows of the trams obscured by dark blue paint. We were warned about the Zeppelin menace and there were some futile raids near London, but it was too distant a threat to strike fear in us. More exciting was the frequent sight on cold, dark winter nights of the multi-coloured flickerings of the Aurora Borealis flapping across the sky. After leaving Glasgow, I never witnessed it again, until many years later I saw a spectacular display of the Southern Aurora from Hobart in Tasmania.

One day, a couple of Australian cousins turned up on leave. They arrived in uniform with their large Anzac hats and I felt very proud to be associated with them. We took them to see a pantomime. This was the first time I had been in a theatre so it added to the exotic effect of the Australian visit.

Ration books seemed to make adequate amounts of food available. We never felt deprived, but shopping was difficult, with long queues at the butchers, and I remember my mother's indignation at having to pay an exorbitant 10 pence a pound for bacon. As for shops, there were some department stores in the city centre; notably the luxury houses of Sauchiehall Street and the rather down at heel Anderson's Royal Polytechnic. Supermarkets were far in the future. Few goods were prepackaged; sugar, tea, flour, all were scooped into little paper packets, generally dark blue or plum coloured, and weighed on scales. It took so long and was tiresome for little boys accompanying grown-ups. Our grocer, like many others, proclaimed himself grandly as an 'Italian Warehouseman'. As one entered, a bell alerted the salesman and one noticed on the left a barrel made of light unpainted wood, containing white Spanish grapes packed in cork dust.

This leads me to the phenomenon of the 'Eyetalian' ice-cream culture in Scotland, already hinted at and probably of marginal interest to other than small boys. Ice-cream shops run by Italian families were everywhere, in cities, towns and villages as well as the itinerant barrows, some of which came to the school playground during breaks. There were three or four of these shops near us and we divided our patronage between them. For a penny, you could buy a slider (ice cream between two wafers) or a poke (cone) for a ha'penny, or you could sit down inside and eat a dish of the same ice cream – only vanilla. Often, when my parents were flush, I would be sent out with a jug to be filled with scoops. Occasionally, I was allowed a McCallum, a prototype of the sundae, a scoop of the same vanilla ice covered with raspberry sauce. Their chilly delights were served summer and winter, augmented during cold months by hot green peas with vinegar. In my view, this Eytalian system represented an exotic and highly desirable sub-culture with many associated events and anecdotes.

I remember a day some 35 years later, when strolling with a friend beside Loch Lomond, we spied in front a couple with a small boy who was obviously pestering. Suddenly, with no obvious provocation, the father turned to the boy and shouted in a loud voice, 'but there are no Eyetalians here'. My English friend could not understand how the absence of Italians in this idyllic Scottish scene should give rise to such an angry outburst – but I knew!

Finally, a small coda to this Italian Symphony. Many years later, travelling from Paris with my wife on the Rome Express, we decided to have dinner. Entering the dining car, we were astonished to find it nearly filled by people laughing and joking in broad Glasgow accents. In the morning as we approached Turin, we returned to the dining car for breakfast to find the same people still bantering loudly, this time in fluent Italian. Here was obviously a group of the Glasgow ice-cream fraternity returning home for a vacation.

Returning to my childhood; when I was about 7, my Aunt Kate began to sometimes take me to places of interest such as the cathedral and the botanic gardens, which I enjoyed enormously. One weekend we went as far as Stirling, where I was thrilled to visit the castle on its rock. From there we went to the Wallace Monument to the famous Scottish patriot, a few miles away on top of a hill at Causewayhead. We were directed to a tram that would take us there and I assumed it would be a normal electric vehicle, like in Glasgow. To my great delight it was horse-drawn. I was proud to be within an anachronism, but I had a shiver of intimation of my own transient place in history.

This period, from the dawning of consciousness until 12, was one of the happiest of my life. Family life was harmonious, my parents had a very happy marriage. Family quarrels were rare and short-lived. Their very rarity made the few quite disturbing. I remember a very few such occasions when there was a strange silence between my parents that cast a gloom and even a vague fear in me, a mystery that I was unable to comprehend. My father was a generous, kind man, a little self-satisfied with regard to trivial actions, but who, although conditioned to orthodoxy was still surprisingly open-minded. My mother was intelligent and sensitive, always inquiring about the larger questions of life, death and purpose. I suspected later that my father was highly sexed and probably too demanding.

Nowadays it is generally supposed that in those early days, the macho image of the father prohibited him from helping in 'women's work'. This was far from the case in our family or in other households we frequented. My father was a willing partner in domestic chores. He was active in the kitchen and I was expected to participate from the time my interventions could be helpful rather than delaying. I accepted this as normal and the tasks were quite light, but very soon a country girl came as a live-in maid.

In our family, the church relationship was much less heavy than in the previous generation. Activity on the Sabbath was still restricted, but much less so than with the grandparents and aunts. We still went to church on Sundays, generally morning and evening and I attended Sunday school in the afternoon, but our social life was no longer dominated by relationships formed within the congregation. I did not find it too heavy a burden; I accepted it as a fact of life and found some of the sermons quite interesting. On one occasion, I was taken to a service in Glasgow cathedral. The minister was Rev. Laughlan McLean Watt, famous for his poetry and fiery Highland oratory. The sermon was indeed one of Hell fire, but not in the traditional sense of souls condemned for venial sins, but rather a lament for man's betrayal of the potentialities of human nature. People around me were in tears and when the church 'scaled', the congregation slid away in guilty silence. By this time my father had many personal friendships with business colleagues and did much entertaining. The social level of his friendships also rose, as he became trusted and *persona grata* with managers and directors of his own and other companies.

Holidays were a joyful feature of my childhood. In the earliest years they were spent in simple accommodation in farms and village houses not far afield. I had the excitement of leaving for the railway station in a cab or, later on, by taxi. We had holidays at Dunblane, at coastal resorts on the estuary of the Clyde, in nearby Argyllshire and the Highlands. Of the many images which remain from this period I recount a few: long runs in a horse and carriage through the beautiful Dulveen Pass to Nithsdale and Dumfries; being engulfed in a swirling mass of people descending from the train at Wemyss Bay and proceeding under the watchful eye of the stationmaster, dressed in a morning coat and black top hat, towards the paddle steamer for Rothesay on the Isle of Bute; climbing Ben Lawers from Loch Tay to identify a rare flower found only near its summit; admiring displays of minerals, ranging from gold nuggets to galina in show-cases in the windows of miners' cottages at Leadhills; shivering with horror in the same village to see a metal worker immerse his bare arm in a bucket of water before plunging it deep into a cauldron of molten lead and withdrawing it unscathed.

As the years passed, my education proceeded normally. I enjoyed most of the lessons, and remember my excitement when we moved from arithmetic to algebra. I have no detailed images of classes or teachers, but preserve painful recollections of itchy, burning chilblains as I cowered near the hot pipes to keep warm in icy classrooms in winter. But I have vivid memories of a score of fellow pupils, with a few of whom I used to roam the moors and oil shale deposits outside the city at weekends. The school was co-educational and there were a few pretty girls in our class. I remember one of these who was pert and flirtatious. She lived quite near the school and was often followed home by a flock of admiring boys, whose cheeky homage she took for granted. I somehow disliked her and kept well away. But I did greatly admire another, quiet girl with long, flowing hair of fiery auburn. I don't think I ever actually spoke to her; I had, as yet, little contact with the opposite sex. My parents had never given me even a whisper of the facts of life. I had, of course, assimilated a mass of myth, fantasy and fact on the subject from other boys, but it was not until I was in hospital at the age of 12 that my education in these matters made realistic progress, thanks to the ministrations of a young nurse.

A landmark of my last year in Glasgow was a visit to Ardeer, the Nobel dynamite factory on the Clyde estuary, on the invitation of the chief chemist, a friend of my father. The high point was an inspection of the nitrating plant in the production of dynamite. It was located in a wooden hut in a hollow in the sand dunes. At the centre of the hut, was a large circular bath with a charge of glycerine. A monster thermometer was immersed in the liquid. It had very bold graduations, especially one very prominent red marking. Nitric acid trickled into the glycerine with mechanical stirring. The one attendant sat with his eye glued to the red mark. If the mercury rose as far as the red mark, he had to pull a lever so that the whole room would be plunged in cold water to prevent an explosion. He sat on a one-legged stool, so that should he fall asleep, he would tumble to consciousness in time to extinguish the reaction. All this fired my imagination concerning the mystery and utility of chemistry. Then the research director of Ardeer gave me a simple textbook of organic chemistry. I read it avidly with only partial understanding, but I dreamed of the linking of atoms and the possibility of designing structured compounds of desirable properties through a sort of molecular architecture. I decided then and there that I was going to be a chemist when I grew up. I announced this decision to my father, who was not displeased but damped my enthusiasm somewhat by telling me I should not specialise in science until I had acquired a good grounding in history, geography, languages and literature to avoid being as one-sided as he found some of his scientific colleagues to be.

By 1921 my father was secretary of Nobel's Explosives Company which then amalgamated with some other firms to form Nobel Industries Ltd. This then joined with Brunner Mond and United Alkali to form Imperial Chemical Industries. Sir Harry McGowan, of Nobel's, whom my father knew through church connections, became managing director of the huge new combine. Sir Josiah Stamp, the economist, became company secretary and my father was appointed his deputy. This meant that we had to move to London. Our house in Dumbreck was sold and a new one bought at Highgate.

The move was scheduled for early summer so I could finish my school term and we were busy packing when, ten days before departure, I went down with a heavy infection, diagnosed as scarlet fever. I was sent to an isolation hospital and was very ill. There was some hope I might be free of infection just in time to accompany the family to London, and I was beginning to recover and having the skin rubbed off the soles of my feet to remove contagion when I relapsed in a second deep fever. This time it really was scarlet fever, caught in the isolation ward. The doctors conferred later and decided that my first fever must have been Malta fever, whatever that is. Being already in a weak state, the illness was quite dangerous, and it was decided I would have to stay on in hospital when the family went south. My father visited me in the evening before they left. He was dressed in white protective clothing and, in the dim light of the ward, seemed quite ghostly. He was very solemn saying goodbye and I felt I might never see him again. The door shut behind him and I shut my eyes, feeling abandoned and alone.

3

Descent amongst the Sassenachs

Finally, I was well enough to travel. Aunt Kate came to the hospital and accompanied me to London. We arrived at the new house in Highgate at about 6 p.m. on a beautiful June evening. My relief and comfort at being back in the family was complemented, if not overwhelmed, by the excitement of having arrived in the great city for the first time. After the jubilation of reunion, inspection of the new house and a simple supper were over, I insisted on going into town that very evening. After protests that I was too tired from the journey, that I was crazy, that I should wait till tomorrow, I pushed until they accepted. I was soon seated on the open top deck of what was to me an exotic red London bus. It was sheer heaven. Our adventure only went as far as the edge of the centre and then back again, but it was a symbol of a major transition in my life.

The summer of 1921 was, for southern England, perhaps the most beautiful of the century, warm, but not too hot, dry, but with the occasional refreshing shower. The good weather persisted from June until well into October when it broke with an enormous thunderstorm. I had an agreeably false impression of the English climate in contrast to that of Scotland. It was a celebration of my arrival amongst the Sassenachs and I certainly made the most of it. Our house was pleasant enough and conveniently located for easy access by bus or tube. Each day I ventured into the City, Westminster or the East End, exploring the sights; cathedrals, churches, parks, palaces, museums and monuments, absorbing the history of a millennium. In no time, I gained a working knowledge of the London streets and could find my way anywhere. We had no family holiday that year, the move to London was novelty enough and the weekends were spent exploring Kew, Hampton Court, Epping Forest and St Albans.

There was, however a vague shadow thrown over that delicious summer, a foreboding about my forthcoming immersion in an English public school. All our contacts with the English since we had arrived in London seemed to confirm my Scottish prejudice: they were a different and difficult race, their aloofness was a masked hostility and their understatement, a barely concealed contempt. My parents tried to enrol me as a day boy in Highgate School, quite near our new home, a minor public school founded in 1565 and well thought of for its science teaching. Unfortunately the day boy quota was already filled and I would have to be a weekly boarder for at least a year, until space was found in one of the day 'houses'. This was very troubling; I was to be incarcerated during the week with a bunch of hostile English boys and unable to visit my home a mere five minutes away.

The dreaded moment arrived and I was duly delivered by my mother to Cordell House and into the hands of Mrs Sylvanus, House mistress and wife of the teacher of class 1a to which I was allocated, at the very foot of the academic ladder. I hated the whole atmosphere, sleeping in a dormitory with six other boys, forced to eat unfamiliar and hence unpalatable food, having my every moment dictated and monitored, my accent unacceptable and communication difficult. It was every bit as ugly as I had feared. The maleness of the school appalled me. In my academy in Glasgow, the presence of girls had never been obtrusive, but had somehow softened the atmosphere. At Highgate, values were different and it seemed less human. In Cordell House we were a motley lot and much less English than I had expected. It included a Pole, known as Podzymack, an Icelander, Olafson, a Belgian, Vandam, a remote Persian with the Imperial name of Rieza and a 17-year-old English lout, the house bully, who terrorised all the quiet little boys. No attempt was made to bully me, however, and I even managed to stand up for some of the most vulnerable. What alienated me most was the power of the dominant value of the school – success in sport. Lip service was paid to academic distinction as evidenced by lists of names of scholarship winners to Oxford and Cambridge colleges emblazoned in gold lettering on the walls of 'big school'. But it was really only sporting achievement that counted. Recently, a former Highgate boy had won a medal at the Olympic games and had thus become a role model, overheating the sports cult, to my mind intolerably.

I was placed in the lowest form and found no challenge in the curriculum, most of which I had already covered in Glasgow. The school day started with a short service in the rather beautiful gothic chapel, in which the poet Coleridge was buried. I found it automatic, repetitive and empty of any religious feeling. Little did I think that later, when I had become a senior boy, I would on occasion have to don a surplice and read the lesson. Teaching was very uninspired and boring, especially the long hours of Latin grammar. Prep in Cordell House in the evening completed each dreary day. One success enlivened this, my most wretched year. There was a Cholmondly Masonic Lodge geography prize offered each year on the basis of an examination paper common to all classes. To my astonishment I won it from the lowest class and was awarded five volumes of outdated travel narratives, bound in sumptuous red leather. I am told I came first in this examination in every subsequent year I was there, but fortunately was never awarded the prize ever again.

My second year, should have begun well. I became a dayboy so could share fully in the social life of the family and balance this with the idiosyncratic mores of the school. However, the reality turned out to be even more dismal than ever. The teacher's ideas were banal, his voice slurred and he was haunted by the suspicion that his pupils were cheating him, as they no doubt were. The boys seemed to have neither the will nor the ability to learn anything. They were amused only by opportunities to outwit a despised master and his stupid vigilance. This seemed to me to be the slum class of the school, where a few hours of boredom had to be endured before the boys emerged to enjoy the realities of school life – football or cricket.

Having survived this grey zone, school life became more interesting year by year. I had come to terms with the domination of sport. I was reasonably proficient at football; good enough to ward off contempt, insufficiently proficient to tempt the toughs to draw me into the sporting elite. An unexpected triumph came with a success in the annual cross-country race earning me house colours. I was allowed to wear a blue blazer with red trimmings, a passport of sporting respectability for the rest of my time at school and a guarantee of a quiet life. The classroom was no longer boring. The teaching of history was brilliant. In Glasgow it had been restricted to Scottish history, so I had to catch up on English history before the Union of the Crowns. This was not just ‘1066 and all that’ and good and bad kings, but serious stuff such as Magna Carta and the prehistory of democracy. For the first time I sensed the flavour of social and political history. Later we did a course on medieval European history and one on Poland. Mathematics was brilliantly taught by one master and in a lacklustre manner by the others. Then science teaching commenced. Initially it was entirely descriptive, with lots of bangs and stinks that little boys revel in. As yet it held no inkling of the meaning of science, but was enough to stimulate me to set up a primitive laboratory in a shed in our garden, to the utter disgust of my mother.

Meanwhile, family life flourished. Many of my father’s Glasgow colleagues had established their families in various London suburbs and soon a ritual cycle of visiting was established. My parents joined a Congregational church in East Finchley and made many new friends; mainly London Scots, kind people, more urbane and cultivated than the church people I had known in Glasgow. We had very few essentially English connections. My parents were voracious readers. We all read the same books and there were many lively discussions around the dining table at weekends. I was at ease in this kindly bourgeois environment and even submitted, protestingly, when my father insisted on buying me a bowler hat! I wore it once, but it gave me a headache so I never wore it again.

I had made few friends at school who I would have liked to invite home, but there was one exception. In the Cordell House days I had protected one timid little boy who constantly attracted the attention of the bully. In later years he would accompany me on long runs over Hampstead Heath. Although somewhat puny, he had considerable staying power, and we both cultivated this sport of cross-country as a means of escape from the local environment of school. I discovered gradually that he had a penetrating knowledge of literature and a facility for creative writing. I was invited to his home, which was very near the school. There I met the Carters, my first real English family, whom I naturally assumed must be typical. The father was a tight-lipped accountant, pleasant enough, but uninteresting. The mother was rather intense, chatty and gossipy; her very ordinary conversation was interspersed by occasional rather bizarre remarks. There was also a sister who appeared to me a pleasant, normal little girl. It seemed odd that they kept the telephone on the floor, under the grand piano covered with a green baize cloth. When it rang, there was a flurry of activity as one or more of the family got down on their knees to answer. Conversation at table was quite banal, about school life, politics and family trivia. Then, suddenly, the mother looked at me and said, 'I think you are quite handsome and you will look quite lovely in your coffin.' Nobody blinked an eyelid. I knew immediately that I would have to modify my ideas about the typical English family.

I had known my friend by his school designation as Carter I, but I soon realised that, for some obscure reason, he was also known as Reubke, while I discovered that his Christian name was Geoffrey. I should mention that in school tradition only surnames were used. At Highgate I never knew the first names of any of my schoolmates. I was merely listed as King, until, near the end of my time there, a small boy with the same name entered and I automatically became King I and he King II.

1925 was an important year. I passed the London University matriculation examination and the family agreed that I should prepare for university entrance. My father was very happy about this, being determined that I should have all the educational opportunities he had been denied. As a reward for my exam success, I was given a holiday abroad, a first-time event for the family. It was arranged that I should go on a package tour to Switzerland and Paris with my Dale grandfather and my lame Aunt Mary. This included a week at Grindelwald in the Bernese Oberland, followed by a week in Paris. The long night run from Calais was so exiting I could not sleep, but had to get up and pace the platforms to sense the foreign sounds and smells of the various French provincial towns at which the train stopped. We reached Basel just as dawn was breaking and changed to another train to Interlaken. Here was another experience, sitting unslept in the station buffet, breakfasting on delicious rolls and coffee as dawn gradually transformed dream to reality. Then came the first sight of the glorious snow-capped Alps. Every day of that week discovered fresh glories and I was sad to leave the Alpine scene. Paris was, of course, another world that I skirmished as a relentless tourist.

This expedition was most memorable, however, for the extraordinary behaviour of my grandfather who, as soon as we had left behind the white cliffs of Dover, seemed to drop his usual quiet domestic self to become rather wild and unpredictable. When the ferry berthed at Calais, he was nowhere to be found with our train due to leave in a few minutes. Eventually he was discovered in the engine room, talking vivaciously to the ship's engineer and furious that we should be worried he would miss the train. Then, a few days later, in the course of a walk in the high Alpine meadows, he discerned some berries he thought resembled Scottish blaeberries. Eating some, he was horribly sick for a couple of days. The final straw was when we took the mountain railway high up to the enchanting snowy wonderland of Jungfraujoch. Unnoticed, he wandered off by himself across the snow towards the peak. Worried by his absence and alarmed for his safety, we were still searching for him when the last train of the day was about to descend. Poor Aunt Mary was distraught. He strode dramatically into the station with a few seconds to spare, again jeering at our anxiety.

In Paris he behaved normally, but I was amused to notice how calmly he accepted the proffered nudity at a performance at the Moulin Rouge. One day a couple of people from our party were trying to persuade me to go off for the weekend with a girl from the perfumerie they had patronised – for my education of course. Aunt Mary was shocked and I was terrified, but grandfather encouraged the idea, knowing perfectly well that it would not happen. Then, on the last day in Paris after visiting the Palais de Justice, grandfather made a flamboyant speech in broad Scots thanking the French judge who had explained their system of justice to us, much to the amazement of the other tourists and the horror of Aunt Mary and myself. Once back in Britain, he relapsed into his old quiet style. He had never before given the family the faintest inkling of this wilder aspect of his personality, but I often thought that it must have manifested itself during his long retirement. I also wondered at times if I might have inherited a streak of his sporadic bouts of irresponsibility.

A taste for foreign travel had begun its germination within me, but it was a couple of years before I could indulge it again. Meanwhile the quality of life at home expanded. My parents took me to the theatre occasionally, generally to light drawing-room comedies or fashionable musicals such as *No, No Nanette* and *Rose Marie*. I quite enjoyed these but was not enamoured with them. One Saturday evening my father proposed going to see some popular show in the theatre in nearby Golders Green. Fortunately, as it turned out, the theatre was sold out. At a loose end, we wondered what to do. Passing back through Hampstead, my father remarked that there was a theatre there called the Everyman. We had no idea what was playing, but decided to go anyway. There were some seats free, we entered the small auditorium and the curtain rose to reveal a Victorian interior with a woman dressed in black velvet standing front stage. This was Laura Cowie as Hedda Gabler. One look and I was trapped for life by serious theatre. It was a superb performance and I knew I was experiencing something real. A couple of weeks later in a West End theatre I saw Robert Lorraine in Strindberg's *The Father*, that insane beautiful play by a creative, insane dramatist. Then I discovered a set of volumes of Ibsen in the school library and devoured them rapidly.

I had broken into a new world where the aims and reasons for existence, belief and behaviour were questioned, so different from my conformist upbringing, and with it a sense of freedom. This inevitably led to devouring the works of Nietzsche and Schopenhauer, Marx and Freud. I tried to share my discoveries with my parents, who were guardedly sympathetic and placed no obstacles in my way. I should mention that I had already concluded that traditional religion's image of the world did not measure up to what I, in my arrogance, saw as reality, despite the validity of the beautiful myths and symbols of the bible. Nevertheless, whatever God – if any – existed, I had a conviction of the power of the spiritual and the moral.

At school I plunged into serious study. In understanding chemistry I had crossed the line from the merely descriptive to the scientific, demanding an effort of reorientation. Being now in the science side of the school, I found myself one of a very different type of pupil, mostly boys with a modicum of real intellectual curiosity and capacity for understanding. Dr Stevens, a new young chemistry master who had recently taken his Ph.D. at

Cambridge, appeared and his live contact with research brought a new vitality to its teaching and an affirmation of its relevance both to learning and to the practical matters of everyday life.

But Dr Stevens turned out to be no narrow specialist; he had a strong interest in cultural life generally and a special love of the theatre. A few months after his arrival, he initiated a series of play readings to which a few of his interested pupils were invited. Soon a group of seven or eight of us went regularly each Sunday evening of term to the house where he lodged. We would read a play that we had studied the previous week, then discuss it as we toasted crumpets over the fire to continue philosophising until midnight. This procedure persisted for over two years, by which time we had become familiar with a broad selection of British and European drama augmented by many visits to the live theatre. I was able to see practically all of Ibsen's plays, a rare privilege, uniquely possible because of an outburst of popularity on the occasion of his centenary. The Stevens seances were an extremely important part of my education. The fact that they were initiated by a teacher of the natural sciences fitted in with my father's advice that I should relate my science education to the wider humanistic issues. From this, my lifelong interest in interdisciplinary approaches, the unity of the sciences and a holistic approach to life and learning was born.

Around this time my father became the proud possessor of a car, a very flimsy Standard 10. As soon as I reached the age of 16, I was allowed to drive. It was not yet necessary to pass a driving test to obtain a licence. I learned easily, but, fortunately for me, my father did not and it soon appeared he was a danger on the road to the extent that my mother refused to allow him to take the wheel. This meant that, in effect, the car was mine. We spent many weekends exploring the countryside all round London. The car was greatly underpowered and often the passengers had to push it up quite moderate hills. One visit to Devon had to be abandoned because, even with pushpower, it simply couldn't manage the hills. A couple of years later it was replaced by a sporty Riley 9, which became my personal car during the heady years of college, giving me a high degree of mobility and many advantages (e.g., with girls). In return I had to do a great deal of parent ferrying, but that too was positive as it kept me close to the family during my flighty years.

By the summer of 1927 I had the urge to visit the continent again. I had never possessed a bicycle, but it seemed that cycling and youth hostels might be the most practical way for a person of modest means to explore Europe. The son of a colleague of my father who lived near us at Highgate, Jimmy Laing, agreed to join me in a cycling tour of Germany. The bicycle was delivered to me the day before departure and I taught myself to ride it that evening and on the way to Victoria Station in the morning.

We proceeded to Brussels by boat and train, cycled across the city from the Nord to the Midi station and then on to Cologne. There we spent the first night in the somewhat sordid youth hostel in a tower of the famous Rhein bridge, encountering for the first time a boisterous group of German boys and girls. I was happy to find my schoolboy German was sufficient for everyday communication. I had a warm feeling of being an independent individual abroad, never experienced during the previous holiday with my grandfather. Next morning we set off on our bicycles up the Rhein and continued thus until the end of the trip, doing as much as 120 miles a day. At the Swiss frontier we turned east towards Bavaria, crossed over the mountains into the Tirol, before, at Innsbruch, crossing back to Munich. Then we cycled north to Nurenberg and on to Turingia, before heading west to end up in Amsterdam. Jimmy and I got on reasonably well, but had no common interests or compatibility and never travelled together again. I much enjoyed seeing the diversity of the German landscape with its mountains and long flat plains, but there was little of human interest in this journey, little to remember, indeed my most vivid recollection was astonishment at being able to buy cups of coffee on the Amsterdam trams. However, the physical accomplishment of this strenuous journey fired my ambition for more adventure in the future.

That autumn was the beginning of my last year of serious study at school, leading to the London University intermediate exam at its end. Throughout my years at Highgate I had shown little interest in national and political affairs; indeed, I cared very little about what was going on in the world. I was in such an apparently safe cocoon that I was desensitized to the possibility of any external threats. I realised vaguely that there was a great deal of social unrest. From the novels of Dickens and other fiction, I was aware that poverty, exploitation and crime had been rife in Victorian times, but I assumed the social reforms they evoked had been accomplished and that we were living in enlightened and equitable times. The general strike had made a big impression on me, but I learned about it mainly in anecdotal terms through one of my schoolmasters who drove a London bus as part of the general bourgeois strikebreaking effort. I never thought to question the reasons behind the dispute, assuming as did all those around me that the strikers were in some way subversive and had to be forced back to work for the common good. My advance towards the questioning society had obviously not yet shaped my world view.

I did have some feeling about the general economic climate and was exposed to government exhortations to save in order to provide capital for industry. The consumer as an economic force had not yet been invented and the credit card was still many decades in the future; political correctness led straight to the savings bank. Amongst the middle classes debt was regarded with opprobrium, goods were only purchased when one had the necessary funds, and hire purchase was seen as a new and somewhat immoral device. I remember advertisements for the first pay-by-instalments firm I became aware of, guaranteeing that the furniture you purchased would be delivered in a 'plain van', to prevent you losing the respect of your neighbours. How different from the consumption-driven society of today with its debt burdened multitudes and its deterrent to saving. Perhaps there is still something to be said in favour of thrift after all.

The end of term came and I left Highgate for good. I remember feeling that I was at the high point of adolescence. I had the naive idea that, elevated to university life, I would become more responsible and be metamorphosed into a progressive young adult. So far, so good; I had broken the cast of my conditioning; I had enjoyed and suffered the pangs of romantic love without making too much of a fool of myself; I was on the threshold of a new world.

In the previous two years I had become friendly with another science student and member of our play reading group named Dan Piggott with whom I shared many interests. I had been invited to his home, the second English home into which I had been welcomed. Would it be as typical as the first? Dan was an only child. His father, headmaster of a secondary school, was a dapper man, precise and friendly, but a cerebrally determined puritan. His mother was a true bluestocking, a mathematician who had been the first woman to reach the summit of the Matterhorn. She was kind and had a good sense of humour. In later years I became very fond of her. Dan and I did a few short cycle/camping trips in the south of England and we proved to be compatible. His only mannerism that jarred on my hedonistic instincts was that even at the end of a strenuous day's cycling, he would never admit to being hungry, although he might agree on occasion that he was 'weak for want of food'. I found also that he had never had a hot breakfast, never tasted bacon and eggs; this seemed to me to be strange in an Englishman, however puritanical. Finally, I managed to persuade him to join me in celebrating our freedom from school by setting out on a cycling expedition to Istanbul, or Constantinople as I preferred to call it. Both sets of parents bravely agreed to the plan. To cover expenses my father, with what seemed profligate generosity, gave me a banker's letter of credit for £60. This triggered us to begin having the necessary injections and collecting visas for the many countries we expected to traverse. In this post-World War I period, visas were required for entry into most European countries, including Germany. Getting them was quite a task, therefore, the most difficult being Albania, but I finally cornered an Albanian consul in a basement flat in Bayswater.

4

The Globetrotters

My obsession with travel – to explore the world in its magnificent diversity, to visit other peoples and possibly understand them – has taken me to all the continents and innumerable countries. Curiosity, at the root of this urge, is a common property of humans and many other species. Clearly an important mechanism of evolution, built into the genes, essential for survival in foreseeing dangers and avoiding predators, it is also the origin of innovation and change. Science can be seen as rationalised curiosity and wanderlust is one of its less significant manifestations.

My type of wandering was determined by the conditions of the period and the possibilities open to a student lacking financial means. At the beginning of the century, trains and ships were the main means of transportation; there were few cars, roads, especially in Eastern Europe, were still primitive, hitchhiking was many decades in the future. A further constraint was that I had various competitive ambitions, especially a strong professional drive that kept me at home most of the year, thus travel was limited to short periods not too expensively far abroad. Transportation had to be mainly self-powered, by bicycle on excruciating roads and also, as I soon found, by canoe on rivers, linked by train and ship at the least cost. I was extremely fortunate with Dan Piggott, a kindred spirit who had a much better ability than I for practical detail; also his puritanical way of life was useful in difficult situations.

This chosen means of travel made for much discomfort *en route*, huge physical effort, extreme fatigue, constant attack by insects, especially mosquitoes, difficult sleeping in inadequate tents, deluges and sunburn, inadequate food with little variety. Much to my surprise, these unpleasantnesses also had their satisfactions, especially in retrospect, and I recognised a trace of masochism in my wanderlust after our first few expeditions. The road to Constantinople was not a road to Damascus, but it was the pursuit of happiness through tolerance of physical suffering.

We set off from London one morning in early July in a heavy thunderstorm, our bicycles hopelessly overloaded, absurdly wearing 'Bombay bowlers', pith-lined tropical helmets, and inciting jibes in plenty as we rode to Victoria. At the station, people stared at us with ill-disguised derision. But this is not a travel book. My intention is to convey impressions and personal encounters as the itinerary unfolds not to describe scenery or historical sights. We crossed to Calais and took a train to Passau, near the source of the Danube. Entering Austria we cycled through beautiful countryside to Vienna, passing through the Wachau and crossing over the Wienerwald. Straightaway the first mistake of our preparations was evident – cycling in shorts. After a few days my thighs were red and swollen with sunburn, making progress painful and raising my temperature by night. Too late to change now though.

Arrival in Vienna was a shock after the sun-drenched roads of Austria. It was quite different from other great cities I had visited, somewhat artificial as if it had been left stranded in time, the capital of a great empire that had faded away, left isolated without a country to dominate or on which to feed. The classical baroque of its great buildings seemed to me heavy and irrelevant, monuments of a defunct empire. The people seemed confused and politically uncertain. Social conflict was in the air and attempts at 'Gemutlichkeit' by the bitter and defensive Viennese seemed contrived. It was only after the trials of the Nazi unification, World War II and the Soviet occupation that a new self-confidence emerged to support and reinstate the true Vienna. Of course we also had glimpses of the other side, the lights, the busy shops, the opera, the elegantly dressed women. This aspect seemed to me more Johann Strauss than Mozart or Freud.

One day we visited an old gentleman acquaintance of the Piggotts. He had been a colonel in the defeated army of the Empire and was now one of the leaders of a Royalist party, plotting with very little hope for restoration. He described the current state of Austrian politics in an obviously partisan way; how the socialists dominating the city were in conflict with the great mass of peasants outside, who were well disciplined, armed and threatened to march on the city. Our host had been thrown into prison several times by the detested socialists and was at present under house arrest. His villa in an eastern suburb backed on to the battlefield where the Polish King Sobiesky had defeated the Turks who had subdued the Balkans, overrun Hungary and threatened the rest of Christendom until they were halted in his backgarden.

Soon we were on our bicycles again, heading for Budapest. East of Vienna, there is a sense of flatness and aridity; the villages of Hungary with their single streets of long featureless houses seemed hardly European. There was no butter because there were no refrigerators and in cafés cups of (Turkish) coffee were invariably accompanied by little glasses of cold water, at first single glasses then two or three per cup as we journeyed eastwards. It was in one of these villages that we first encountered 'globetrotting', a strange phenomenon that we met all over Eastern Europe. We cycled into its one dusty street of seemingly windowless houses and stopped outside the inn, where we immediately noticed a small crowd clustered around a red-bearded man who was drawing one of the old village houses. Not far away, another crowd had gathered around a second redhead who was sketching a similar house. At this moment a vehicle with a large sail moved in from the east to be greeted by calls of 'globetrotter'and the crowds around the two artists dispersed. The owner of the road yacht, a sturdy flaxen-haired fellow dressed in the minimum of clothes, started to demonstrate his contraption. He was soon brandishing some cards, bearing a badly printed picture of himself standing beside his apparatus, that he proceeded to sell cheaply to the villagers.

One of the artists moved over to him and they began to speak in German. The road yachtsman was invited to inspect the conveyances of the others and they moved off, followed by the villagers, who had in the meantime discovered us, with fresh waves of 'globetrotter' uttered in a shrill note of surprise. The artists were also cyclists, but much more sophisticated than us. Behind normal machines, they towed large, egg-shaped boxes on wheels. By pulling a rope, the top part of the egg lifted to reveal a bed of sorts, protected by the suspended top.

In the evening we joined the other globetrotters in the inn. We were welcomed as brothers and they told us in detail about their wanderings. They were all making extensive tours of Europe and beyond, and had already been on the road for months. None of them had any money except from selling the cards about themselves that they carried in many languages, or else they sketched features of villages to sell as they passed through. The arrival of us globetrotters created great excitement in the village. Never before had so many foreigners congregated at one time. The room in the inn was so crowded that it was difficult even to lift an arm to drink, while the window was crowded by many others outside, craning to catch a glimpse of the strange intruders.

We encountered many other globetrotters on our journey and were greeted as such wherever we passed. They were all young and inevitably poor men, who, suffering from an excess of wanderlust, travelled for months or maybe years from country to country. They seemed to have little curiosity about 'sights' or local customs, their satisfaction appearing to derive from globetrotting itself. Nationality had little meaning to them, having rendered themselves virtually stateless. They were very different from contemporary tourists whose affluence permits a certain comfort and who proclaim their nationality noisily wherever they go. They were the antithesis rather than the forerunners of the jet set, nor, with their lack of ideology had they any kinship with hippies. We never encountered any women globetrotters. Perhaps the most extraordinary of them all was a Dutchman we met in a village in Transylvania. He was standing beside what appeared to be a large white egg on wheels. This was his mobile home. Lined with cherry-coloured velvet and womblike, he could lie in it, secure and comfortable, peeping out the front through a spyhole and propelling himself by moving levers backwards and forwards using either arms or legs. On top was an aerial, connected to a radio within, keeping him in contact with the world as he dawdled along. Here again was a professional globetrotter, with cards of his bizarre machine on sale to pay his way. He was on a trip around the world in his cocoon, apparently happy despite many hardships including nights spent in prison in Turkey as a vagrant without visible means of support.

Often when locals greeted us as globetrotters, we were interrogated about our journeys and origins, to great astonishment that we were British. The universal view of the British was of dignified and reserved conservatism that would not allow for the existence of riffraff such as Dan and I. Once the novelty of 'globetrottery' and its fraternity had worn off, we began to tire of the attention as it greatly curtailed our privacy and the possibility of exploring new places.

I doubt that globetrotting in this form still exists in Eastern Europe. My friends with their leisurely, crazy means of transport are an extinct species. The automotive age with its fast, crowded roads leaves no space for eccentrics, but it does meet the needs of new generations of travel-hungry young people, both boys and girls, through hitch-hiking. Increased affluence as well as travel concessions make it possible for them to satisfy their wanderlust, travelling farther than in my days and with minimal physical effort. Not long ago, I flew from Moscow to Novosibirsk. At the back of the plane was a group of young people from Russia and various Western European countries, lolling about on their sleeping bags as they were carried to a holiday in Siberia. What a contrast with my own early experiences. The ease and superficiality of their adventure shocked my puritanical mind. I must still be a subconscious masochist .

To return to our itinerary where I left off, approaching Budapest. We found it one of the most beautiful cities of Europe, brilliantly situated on the banks of the Danube, just as it turns south for its monotonous run through the great plains to Belgrade. The river makes the city and the city adorns the river in perfect symbiosis. What a difference a great river makes to a city; it is impossible to imagine Paris without the Seine or London without the Thames. Vienna, which has so many of the characteristics of a great city, ignores its river, which flows, unexploited and unnoticed at a considerable distance from its centre.

In Budapest, the sweeping elegance of the Margerit and Ezebet bridges give unity, dimension and distinction to the two contrasting cities of Hungary's capital. Buda, on the right bank, rises steeply from the river to a high ridge of exciting, historical buildings, giving a fabulous skyline, while the low-lying Pest has a remarkable river front of striking edifices, behind which is a busy, very ordinary city. Climbing to the heights of Buda, amongst the palaces and cathedrals, the view across the city from the Fischer's Bastion is truly magnificent, with the plains beyond fading blue to the horizon. In Pest we visited the parliament building, a nineteenth-century construction in mock gothic that I thought dramatically wonderful without questioning its taste. I had no idea that my next visit there would be 60 years later, paying a courtesy visit to the President of the Magyar Republic.

We found the Hungarians charming and friendly. Despite our ignorance of their difficult language, communication was easy, as most of them spoke German. They seemed embittered by the loss of Hungarian territory in the 1918 treaty of the Trianon and mourned especially losing Transylvania, engendering deep hatred of Romania, its new master. Nevertheless there was a spirit of solidarity in the air, tinged with an almost fanatical patriotism following from the break up of the Hapsburg Empire; a new form of national identity that had survived the horrors of the communist rebellion of Bela Kun.

From Budapest we took the train across the flat Hungarian pustas to the Romanian frontier and crossed into Transylvania at Arad, where the transition into another culture again seemed abrupt. At first sight the new environment seemed alien, very different from Western or even Central Europe. The customs people, with armed soldiers looking on, meticulously examined every item of our miserable baggage, seeking in vain for any books, letters or newspapers in Hungarian. They were courteous enough, but impounded our bicycles until certain formalities had been completed that they suggested might take several days. In our naivety we were at a loss how to react, but soon learnt that a small 'sweetener' was essential to almost any transaction in Romania.

We felt we must find lodging close to the bicycles, so left the station that was clearly some distance from the town, to find a simple inn. We entered a large, dusty, unpaved smelly square, bordered by low mud and wood buildings. At one corner was a market. Men clad in white pyjamas so voluminous they looked like skirts and wearing wide hats, sat cross-legged behind piles of melons and gourds. Women dressed in black costumes scurried here and there. Other white-pyjama'd men lounged in groups smoking cigars, while a horseman in a long, white felt cloak, embroidered with brightly coloured wools moved slowly across the square. There were no obvious inns, but eventually we entered a low wooden building that we thought might be one – and it was. A bright-faced peasant girl greeted us in Hungarian, went off to fetch someone who could speak a few words of German and soon we were settled in a clean, bare bedroom. An hour or so later a man was summoned to enquire about us and to help in filling out the official forms. He was an elderly Magyar, who talked with us for a long time, complaining about the treatment he received from Romanian officials with their superior ways and inferior laws. He spoke in a low voice for fear of being overheard, and looked forward to a day when the Hungarian army would return and bring civilisation back to the town.

Finally, we recovered the bicycles and began to explore the town. We found it much pleasanter than its outskirts had suggested. There was a broad street, a few buildings that might claim architectural merit and some quite decent shops. It even boasted a steam tramway that rumbled and bellowed along the main street. People thronged the street, many dressed in normal European clothes, including quite a few well-dressed women, along with many gold-braided officers with waxed moustaches and rouged cheeks. Rouging in the Romanian army was limited apparently to officers of the rank of major and above. If that was so, a surprising proportion of high-ranking officers peopled the Arad streets. Romanian officials excepted, the population was almost wholly Magyar. The fashionable street life was, however, only part of the picture; mingling with the chic ladies and effeminate officers were multitudes of peasants in white pyjamas and many poor conscripts in tattered uniforms. Life is sad and unreasonable!

I had never given any serious thought to Transylvania either as a country or as a concept. Dimly in my imagination it was an imaginary place of dark mystery, a shadow within the prehistory of Europe, concurrent perhaps with the *Nieberlungenlied* or the *Kalevala*. It was the home of Count Dracula; it had been colonised by the children of Hamelin, spirited there by the Pied Piper. Arad proved to have been a sardonic Ruritanian prelude to the stark reality of Transylvania; a land like other lands, beset by ethnic, economic and political problems. These were some of my meanderings as we started on the long and painful ride through Transylvania, over to Brasov, across the Carpathians and down to Bucharest many days away. The mid-August sun burned and blistered our bodies and turned our protective helmets into ovens; insects stung, our throats were continually parched and drinking water hard to come by.

The road was abominable, rutted and often consisting of rocks and pebbles like the beach at Brighton, but with the interstices filled with fine white sand that we churned into clouds so thick they obscured our view. Closer to the mountains, there was a thunderstorm every evening and, although the cold rain soothed burning skin, wet camping hardly added to our comfort. Food was difficult; there were no shops in the villages, and meals were monotonous it usually being only possible to buy dark bread at the millers and to entice a peasant to sell a couple of eggs – and well water for the tea. A blessing were the plums, dripping ripe from trees along the roadside, cultivated for distillation into slivovitz. Without them, I doubt it would have been possible to continue.

Then, by chance, we stumbled on a gastronomic oasis. One evening, exhausted by a particularly bad stretch of road, we stopped early and began scouting for a suitable site to pitch our tent for the night. Suddenly we were greeted by a voice speaking French. We had come upon a summer camp run by the Red Cross for refugee children. The leader, a pale-faced, hollow-cheeked vivacious Hungarian woman named Faber Emma, invited us to dinner, an invitation too enticing to decline. Luckily we had arrived on a gala occasion celebrating the visit of the regional President of the Red Cross, who turned out to be a glamorous Russian Princess, the owner of the French-speaking voice that had greeted us. She was a refugee too.

There were some 60 children at the camp, of various ages and nationalities, principally Hungarian, Romanian, Saxon and Russian, with a goodly number of grown ups in addition. What a wonderful meal it was after days of frugality. It culminated with a roast of venison shot by the local count whose shooting lodge was in the hills behind the camp. We were waited on by Justinian and Leucretia, both aged 12. The children stood at our elbows like slaves, silent and expressionless, replenishing our plates and filling up our glasses. Conversation at the table was in

French, German and English, amusing and urbane. At last, after a large bowl of brambles and cream, the whole company moved away into the woods where the campfire had been lit, scenting the air with its aromatic perfume and revealing a wide circle of people.

An old fiddler struck up a folk tune and for a long time the children danced round the fire, executing complicated steps and singing until suddenly there was silence. Then one by one, children stepped out of the darkness into the brilliant firelight to sing songs or recite poems. It was during this that the Princess told us her story; the death of her brother at the hands of the Bolsheviks, her escape from Petersburg with her fortune, her association with the Budapest opera. At last the long evening came to its end. We had not erected our tent when we arrived and it would have been difficult to do so in the dark. Faber Emma suggested that we might pass the night in their hay shed, an offer we accepted with alacrity. We dossed down comfortably on the hay with the songs, the languages and the crackling of the fire dancing in our minds until we slept. In the middle of the night I was awakened from deep sleep by a heavy object falling on my body followed by the scampering away of an animal. A large rat had fallen on me from the rafters. Alarming enough anyway, but when I remembered that we were in Dracula country, positively terrifying.

As we proceeded towards the mountains the road became progressively worse, but we pressed on and eventually reached the first town since leaving Arad. Sibiu is its official name, but to its German-speaking inhabitants it is Hermannstadt, and Nagyszeben to the Hungarian minority. This is a main town of Siebenburgen, the German area of Transylvania that, according to legend, was where the children, whisked away by the Pied Piper of Hamelin finally emerged. We found it uninteresting. My chief recollection is of sitting in a quiet Gothic Lutheran church that seemed strangely out of place in this sun-scorched plateau. We found plenty of people to converse with in German. As in Arad there was a sprinkling of elegant and over-decorated Romanian officers. After the steep climb up from Brasov and over the Carpathians, we descended to the plains via Sinaia, the beautiful summer resort of the Romanian kings. I have a vivid memory of its magnificent trees, last remnants of the primeval

European forests. It is somewhat sinister in its foreboding grandeur, reminiscent of Scene 1 of the Valkyrie. To escape the ominous oppression of the forest, we decided to seek an open space and finally set up tent in a meadow above the trees. Alas, the worst of our nightly series of thunderstorms crashed down and our camp was washed away. What a contrast that night was from my next visit to Sinaia 65 years later, when my wife and I were lodged in the royal suite of the castle and visited by the President of Romania.

Entry into Bucharest, as to most great cities, was through sordid streets and industrial slums, but as we approached the centre there was a certain grandeur in its wide tree-shaded avenues. We cycled past the famous Arc de Triomphe, a replica of the Etoile in Paris. Romania at the beginning of the century was proud of its Latin blood and language, surrounded as it was by 'barbaric' Slavic countries. Bucharest aspired to be the Paris of the East, so what could be more appropriate than to erect its very own Arc de Triomphe. This was duly built and paid for, but at the inauguration, when the scaffolding and sheets were removed it was found to be not marble, but papier maché. I can vouch for the reality of the papier maché, the holes in which were crudely repaired by pasting over with old newspapers.

Arrival in the city was difficult. We had to find a hotel, but where? We tried to stop and ask the way to the railway station, around which cheap hotels flourish in most cities. However, when we stopped, a crowd of 50 to 100 people gathered immediately, shouting, 'globetrotters', making it impossible to make verbal contacts. We could not understand why, considering that Bucharest was a city of many bicycles. I suppose that with our heavy packs and pith helmets we stood out as freaks. Be that as it may, the situation was embarrassing and frustrating. The street gathering soon began to impede the traffic, but was soon broken up by the peremptory advance of a bell-clanging horse-drawn tram that barely managed to stop at the edge of the crowd that was quickly dispersed by the police. Apparently there had been a press photographer in our midst, and next day we discovered that news of our arrival was acclaimed by all the Bucharest newspapers and little vignettes of us appeared with photographs, even in papers published as far off as Moldavia. We cycled away quickly to avoid the police, before stopping again in an attempt to communicate, but masses of people surrounded us at every attempt. After four or five of these receptions, we did at length manage to make contact with a workman who spoke French and offered to be our guide. This was Jacob Ghorghe we discovered, a foreman at a construction site. He led us through a maze of mean streets to a hotel and installed us in a not too clean room, but it was a relief to have found a resting-place and to be free of the bicycles. Jacob Gheorghe, although a Romanian, had once worked for six months in Belgium whose wonders he extolled. 'In Romania,' he said, 'the people are stupid and have no culture. In Belgium everything is different, even the workmen are intelligent and everyone delights in conversation.' This view of Belgium surprised me as being at odds with my own ignorant image of that country derived mainly from reading Baudelaire.

We spent the rest of the day with Jacob, walking the streets of Bucharest, enlightened on every trivial detail by his endless talk. Poverty and misery were everywhere; large numbers of peasants in their white costumes mingled with the townsfolk; there had been a disastrous crop failure following a drought, causing a rural exodus and mass unemployment. After dinner we saw Jacob off in a tram to his home in a distant suburb and returned to the hotel in the expectation of a quiet night. This was not to be. Our room was on a balcony and only partially shut off, making us fear for the safety of our small resources. Our neighbours were very noisy, with much laughter, screaming and unidentifiable noises. Next to our hotel was a large beer hall whose band blazoned the night through. It was not long before we realised that our hotel was also a brothel. The situation was simultaneously terribly exciting and horrifying for two naive British boys.

The next two days were spent in a bureaucratic nightmare. We had been assured, on entering Romania, that no further formalities were required during our stay. Now, on preparing to leave the country, we found that we should have reported to the police within ten days. Failure to do so merited a large fine or even imprisonment. Our bicycles should have been registered and here the penalty was confiscation or another fine. Jacob assured us that the passing of quite small sums to the right people would easily resolve our difficulties, but this quickly proved to be simplistic. How small could the grubby notes be and into whose palms had they to be pressed? We immediately found ourselves caught in an unfamiliar web of corruption and exploitation, which it took two days of hard work to extricate ourselves from; we had to visit six police commissariats, immigration buildings and lawyers' offices before we were able to sign a notarised declaration apologising for our ignorance of the regulations. After a night in a police station we were free and permitted to leave the country – with our bikes.

On the last evening we were invited to dinner by Jacob. He had a beautiful young wife who cooked a gastronomic meal and the wine was excellent. Conversation was, as we had feared, mainly a comparison of the splendours of Belgium with the squalor of Bucharest. Jacob seemed to us a good and kind man, but in the end pathetic. While his vision had been lifted by a brief glimpse of the outside world, it had frustrated his life with unattainable aspirations.

We reached Constantinople, the object of our journey by ship from the Black Sea to the Bosphorus. The city with its domes and minarets was even more beautiful than in our dreams. We cycled out of the customs house at a moment of symbolic import. There was a large clock the Arabic numerals on which were being painted out to be replaced by Roman ones. The Attaturk reforms were still in implementation. Through introduction by a family friend working at Robert College, we were found a place in a boy's camp on the shores of the Sea of Marmara overlooking the Princes Islands. From there, the sights of the city were in easy reach. At the camp I met a Greek boy called Voticas whom I ran into by chance years later in the streets of Narvik in Norway. One day we crossed on the ferry to Scutari on the Asian side of the Bosphorus and climbed a hill beyond, from which we had a superb view of hills and valleys stretching temptingly into the infinity of Asia. Constantinople was, then, but a lure, my travels had only just begun.

After a delirious time in Constantinople we sailed to Greece, landing at Piraeus, and cycled up to Athens. We spent some days there visiting sites around the capital, then to Delphi and the near Peloponnesus, before leaving Patras by sea for Corfu. We had a deeply happy time in Greece. Arriving at Corfu, we cycled over to Paleocastritza to appraise the subject of Boeklin's remarkable picture, the *Toteninsel*. Our tent was perched on a high cliff overlooking the island with its foreboding cypresses and white marble ruins. I realised how close high romanticism can come to realism when the reality is inescapably romantic. I am no lover of the grandiose and dramatic in painting, but I can only praise Boeklin's genius in evoking with such realistic precision, the quality of beauty of this mysterious Isle of the Dead.

That year, 1928, had seen a severe epidemic of dengue throughout the Eastern Mediterranean. The armies of both Greece and Turkey had been decimated by it and it was rampant in the cities. Dan and I had both been stung by mosquitoes in our hotel room in Athens and dreaded that we would develop the disease. I woke up in our camp overlooking the 'Isle of the Dead', with a high fever, feeling like death itself. Dan fortunately was not infected. I spent all that day half asleep and wretched. The *Toteninsel* penetrating my delirium.

We were due to leave by ship to Albania. When morning came, I was just able to mount my bike and pedal back to the port on the other side of the island. In fact, I slipped off the bicycle about every 20 minutes, fell asleep, woke up ten minutes later and repeated the performance, finally reaching embarkation by uneasy instalments. We decided, in view of my illness, that it would be unwise to tempt fate by entering backward and uncertain Albania, so we paid for a journey to Kotor, the first Yugoslav port of call. My recollection of that cruise up the Albanian coast is dim. Lying on the deck, I was vaguely aware of blue seas and sunshine and of stopping in the port of Dures (Durazzo). I knew that we were loading a cargo of skins and hides, but my memory of this town was of an Albanian peasant monotonously repeating the words 'Durazzo schön, Durazzo schön'. Lying, as I was semi-conscious on the deck, I was an obvious object of suspicion to the ship's officers who were contemplating the welcome we could expect at Kotor. The whole coastline was alerted to the danger of the dengue epidemic spreading to the west and here was I an obvious carrier, a target for vigilant quarantine staff. They therefore decided to off-load me before we reached Yugoslavia. With our bicycles, we were placed in a small boat, rowed to a bleak beach and unwillingly dumped in Albania, near a small town, Saint Johann di Medua. We managed to pedal to a village not too far from the coast and pitch our tent just off the village square where I lay for a couple of days, taking nothing except lemon juice for the vitamin C until the fever subsided. I soon began to eat a little and it was not long before we were able to pedal gently northward to Skhodar (Scutari) at the end of the lake of the same name, across which runs the Albanian/Montenegrin border. Here at last I was able to enjoy the dubious comfort of a bed in a bare room in an inn. In a week, I was back to normal. Dengue isn't really dangerous, but it is debilitating. As with that other mosquito-transmitted disease, malaria, it recurs from time to time without reinfection. I had an acute recurrence of dengue a couple of years later in Helsinki.

Albania, the ancient Illyria, was and remains a very unformed state. Liberated from Ottoman rule only a few decades earlier, it was sadly lacking in institutions and infrastructure, and heavily under Italian influence. Shortly before our arrival its king, Zog, had ascended a very shaky throne. Arriving by sea, he had been unable to land. As a result of vendetta, assassins were waiting to get him as he landed and he was forced to spend some time camped on a barge off the coast. He managed one foggy night to elude his foes, proceed to the capital Tirana and seize the throne.

The people had no confidence in the new Albanian currency and shops would not accept it. Its only use was to buy stamps at the post office to send postcards home. We badly needed to change some money to buy food. After a good deal of difficulty we found a Greek moneychanger willing to change our Greek drachmas into the *de facto* local monetary system. This proved to be a mixed coinage in which small monetary units of many European countries, such as the British shilling, French franc, German mark and Yugoslav dinar all had the same value, despite their differences in international exchange. The Albanians, even more isolated then than now, were unaccustomed to meeting foreigners, so naturally they were curious about us, but surprisingly open and friendly. Happily they had not yet learned the term 'globtrotter'. From our limited view of the Albanian scene, the most striking feature was the colour and variety of the native costumes. It appeared that every village had different designs and colours. But this was not all; in some villages there were three sharply different traditions of dress. The predominant one was Moslem, easily identified as the women wore half-veils and baggy trousers. Then there was the different style with flared skirts of the Orthodox Christians and a third for the few Catholics. The result was a kaleidoscope of colour and design unique in Europe.

Eventually we left on a small steamboat across Lake Scutari for its northern end in Montenegro, in company with an uxorious and loquacious Moslem with four wives. After a modicum of trouble with Albanian immigration there being no stamp in our passports recording our entry into their country, we entered Montenegro in the evening and camped on low marshy land at the end of the lake, overlooked by stark limestone mountains. We found later that this was a sort of no-man's land between the two not very friendly countries, commonly used by the young men of both who enjoyed stalking each other in the dark. Unaware of this we soon settled to sleep, but were rudely awakened by rifle shots in the vicinity. However, our camp was neither approached nor endangered. The next morning I felt a renewal of energy, fortunate because of the steep road up which we had to push our bicycles to reach Cetinje, the former capital, built in a saucer high in the mountains. It was difficult to believe this village was the capital of a European sovereign state; the Royal Palace, a single-storied building, could have been the house of a well-to-do peasant. We also visited a nearby market town, Podgaritza, later to be renamed Titograd as the capital of the People's Republic of Montenegro. A strange country lacking in natural resources, it is as poor as Albania. The coastal region, dominated by the Karst, has little land capable of producing food. We caught occasional glimpses of small fields in precipitous depressions in the limestone rocks, but otherwise nothing green. The landscape is one of indescribable desolation. There are, of course, more fertile lands behind the coastal mountains, but this rugged land symbolises the character and strength of the people. The inhabitants, dressed uniformly in a rather sombre national costume, seemed to be well above average stature and fiercely proud and independent like most mountain peoples – only more so. They retained their freedom over the centuries, repulsing the Turks and remaining a Christian enclave within the vast Moslem area of the European Ottoman Empire. At one time the population apparently fell to a few hundred, but they survived and even rejected Napoleon's expansionist ambitions.

We left Montenegro with regret, descending from Cetinje down the steep zigzag road to Kotor, thence by ship to Split and, finally, Venice. Entrance to the 'magnificent' city was inauspicious – we felt silly pushing the overloaded bikes up and down the steep bridges over the canals on the way to depositing them in the railway station consigne. We were cheered on by hoards of small boys as numerous as our tormentors of Bucharest, shouting as ever 'globetrotter, globetrotter', but with a pretty Italian accent.

5

They Called it Higher Education

Returning from Constantinople, I found that my parents had sold the house at Highgate and moved to another at Hatch End, a prosperous outer suburb bordering on unspoiled countryside. Another step up the social ladder for them, and for me a longer journey to town, to college and to be with my friends.

I started immediately at the Royal College of Science and, having passed the London University intermediate exam, was exempted from the first year of the degree course. The freedom of college life, of being accepted as a young adult, was exciting at first, but the work was not as different from school as I expected. It seemed easier, or perhaps there was just less pressure. I found an easy camaraderie with my fellow students but few real friendships. The familiar Reubke was already there, pretending to be a scientist, although his attitudes were essentially literary and subjective. He had linked up with an extraordinary young man, Norman Stuart, for some reason known as 'the Normanska', a brilliant musician who had also drifted erroneously into chemistry as a career. I was quickly incorporated into an ill-assorted trio with these two, who were clearly misfits in a science college. This placed me in a somewhat distorted position in the array of student personalities. In no way a misfit, I still shared many ideas and delights with Reubke and Normanska. I soon realised Normanska was a homosexual, although he was then completely unaware of this and had several frustrating relationships with young women in the years I knew him. Reubke was uninterested, while I liked women.

Our trio shared many experiences, plans and conspiracies.

Reubke just scraped through in the end. He became a painstaking scientific journalist and died of lung cancer in his mid-thirties. Normanska did somewhat better and worked as a chemistry lecturer at a polytechnic. Our friendships continued after college days. I even went abroad with Normanska a couple of times, on travels the antithesis of those I had made with Dan. We met with enthusiasm from time to time over many years until, driven by manic depression, he set fire to himself in 1962. They were unfulfilled people of quality, but lacking any clear objectives or capacity for survival. I grieve for them still and for their wasted talent. I was a serious student of chemistry, but the slow tempo and lack of challenge made it all too easy to defect. We received good, solid if uninspiring teaching that seemed somehow unrelated to life. There was little sense of wonderment and little to incite curiosity. Our teaching provided a firm basis of classical chemistry up to the Bohr atom; we were well trained, but there was little real education involved. Despite it being a revolutionary period in the development of science – relativity, quantum theory, uncertainty principle, wave mechanics – we were given little hint of it. Most students seemed unaware of, or unperturbed by, any shortcomings in attitudes and curricula and showed a lack of interest in almost everything else. This was specialisation rampant. I remember an incident when we were invited to a social occasion at the home of one of our professors. He asked one of us what his main interest was. 'Physical chemistry', he replied. 'Good', said the professor, 'but I meant what are your hobbies?' 'Chemical physics', was the instant reply. He passed out easily head of the class and was never heard of again. It was the idea as well as the reality of accepted specialisation that incited rebellion in our trio and led us to indulge extravagantly in non-scientific pursuits. In fact, our lives became so rich in so many ways that this chapter could give the impression of unrestrained dilettantism. But, before launching into our extracurricular activities, I must mention one area of light and humanity in our rather sterile academy.

To qualify for the degree, chemistry students had to take physics or botany as a second science subject. I had already passed the intermediate physics exam while at school so chose botany, a happy choice that greatly added to my education. The head of the department, Sir John Farmer, immediately named

Farmer John, was a portly, urbane man who gave a series of basic lectures. He might have been an Edwardian gentleman farmer, inspired by the ideas of Darwin and Huxley to become a teacher. It quickly became obvious that his course was not simply about botany, but biological principles in general, as illustrated by the life of plants. A first impression was that his lectures would be banal and uninspiring. They were loaded with clichés, worn out proverbs and Latin tags. His favourite saying appeared to be 'half a loaf is better than no bread'. Yet everything he said had a measured importance. There were no declamations; everything he said seemed directed personally to us and we were unwillingly captivated. My curiosity was stimulated by his discussions of the Darwinian and Lamarckian approaches to evolution. Soon we were involved in the complexities of heredity and the uncertain mix of hereditary and environmental elements in human as well as plant behaviour. Everything he said appeared related to life, and even to my life.

Under Farmer a group of young lecturers supervised our practical work, giving opportunity for discussions and personal conversations lacking in chemistry instruction. We were given a new picture of the pulsating nature of life and its struggle through the aeons. Inevitably we were confronted by the inconceivable, but inescapable, concept of First Cause or of the mindless genius of evolution having stumbled into the creation of man with his wonderful and terrible world. These occasions led to late-night adolescent gropings about the meaning and end of existence, much like my earlier experiences in Dr Stevens' house; equally futile, but at a higher level.

We were surprised to find that the College offered an annual prize for an essay on a non-scientific subject and decided to compete for it. I was awarded first prize for an essay on Ibsen, while Reubke came second with one on Wagner. The modest amount of prize money started my now considerable library. Like other colleges, we had a magazine, 'the Phoenix', named appropriately as it died frequently only to be resuscitated by new cohorts of students. It had a certain glory: its first editor had been H. G. Wells. Having read a couple of numbers, our trio dismissed it as hopelessly boring and unworthy. At first we tried to shame it by submitting parodies and caricatures, but our irony was unnoticed, so we decided to launch a competitor. I named it *The*

Muckrake after a plate *The man with a Muckrake* in a wonderful old edition of *Pilgrims Progress* in my father's library. This, we hoped, would be spicy, provocative and debunking, but not so disrespectful as to invoke the ire of the authorities. We wrote the first number almost entirely, the stencils typed laboriously by me, run off and then distributed at a penny a copy. It was a great success and we soon attracted sufficient copy for subsequent numbers. It got me into some trouble, but not too serious.

The Muckrake became so popular the *Phoenix* finally capitulated, inviting me to become editor of the new *Phoenix*, which I accepted. It started out by being very bright, but after a few issues had appeared, the burden of its production with exams approaching dimmed my enthusiasm – it probably became just as boring as its predecessor. I did have my moment of glory at a college dinner a year or so after I became editor; the guest of honour was H. G. Wells. As his successor editor of the Phoenix, I sat beside him throughout the meal and had the fun of coping with his conversation, part iconoclastic, part visionary, delivered in his squeaky voice. My conversation was probably pretty inane.

During my last year we published a review named *Monochrome* that had serious pretensions and carried some reasonably good essays and poems, but did not survive our departure. And then there was the Literary and Debating Society, moribund and ripe for a take-over, which we quickly accomplished. As its new President I was able to attract sizeable audiences for the first time. We had some primitive, but amusing debates with students of a London theological college. Then I invited a number of well-known writers to talk to us. Most declined, I had a typically ironic postcard from Bernard Shaw. Vita Sackville West came and gave a good conventional talk about her art. I had a pleasant conversation over tea afterwards. She impressed me as a serious elderly lady at peace with herself and I would never have guessed her passion and non-conformity. The high point of my success was a talk by Edith and Osbert Sitwell. Already notorious and known to everybody from popular press reports of their extravagant statements, an audience of several hundred turned up. The college toughs occupied the front rows, their rugger scarves mockingly tied in bows, in parody of the 'aesthetic'. I expected trouble, but there was none. Edith charmed them with her extravagance and Osbert taunted them with a malevolent kindness that quickly disarmed them.

Edith invited Reubke and I to her Saturday afternoon salons in her flat behind Whitleys in Bayswater. There, over tea and bath buns, we met many contemporary writers, or at least those on speaking terms with the Sitwells. Edith orchestrated the bizarre conversations – denouncing writers and critics whom she detested with a malice that would have been mean had it not been so grotesque. I was already familiar with her exploitation of the sounds of words and the shadows of those sounds in the Russian Ballet type of unreality her poems created, so it was fascinating to follow the showers of sounds she produced in actual conversation. She insisted that she was forming a society for the propagation of inhibitions and I guess that is what she had always been doing in relation to her own life. Behind the carefully constructed facade of eccentricity she presented to the world, one could sense a vibrant and concerned essence. Over these weeks of contact I acquired a warm affection for her. It was through these salon occasions that I started to read *Ulysses*, which I felt must be significant because of the weight of Edith's deprecation. Enjoying the mocking fantasy spun by her, it is not surprising that we should be attracted by the Russian ballets that were being performed at the time. Diaghilev was at his peak and we were captivated by the whole series and in particular with the symbolism and ambiguity of Petroushka.

Meanwhile, my interest in the theatre and in classical music continued. I felt fortunate as a student in London, able to enjoy the immense range of arts that only that great city can offer. How poor, in comparison, are the cultural amenities that the provincial universities, even Oxford and Cambridge, can provide. Yet I had murmurs of puritanical guilt. Was I overindulging in the ambrosia of a contemporary but ephemeral culture? Could it be that these external stimuli, delivered too early to the immature individual have a power of conditioning, compulsive at the age of awakening, and possibly inhibitive of further development? It is all too easy to have one's attitudes and beliefs stamped by the alien agency of the theatre, rather than gradually through the sweet and bitter real experiences of life. But all this introspection didn't worry me too much.

The most poignant experience of the theatre at this phase of my understanding was of Ernest Milton in Pirandello's *Henry IV*, with its magical interweaving of reality and illusion. Normanska joined me in my devotion to the Everyman theatre which Hedda Gabler had engendered. This theatre was one of the world's first Kammerspielhauser, valiantly experimenting in the new theatre of the twentieth century. We became acquainted with its director Malcolm Morley and were able to make suggestions for new productions. We became familiar backstage and able to argue with the actors who largely shared the vision of the director. Here we experienced more Ibsen, also Strindberg, Gogol, J. J.Bernard the dramatist of the unspoken word, Lenormond, Georg Keiser and the German Impressionists; these and many other masters of drama at the Everyman and Gate theatres helped me sense the Zeitgeist of life and language. Thinking myself destined a playwright I even had a brief try at acting, or rather overacting, in an attempt to get a feel for the stage. Reubke and I wrote *The Wicked Duke of Mantua*, a parody of Elizabethan tragedy, performed at the College and later produced by me in Munich. In music my taste had moved from Wagner to Bruckner, Mahler and Sibelius and was drifting towards Bartok. Normanska, who played the organ, proclaimed Karg Elert, an Austrian composer, a genius. He came to London and we rallied round him with suitable devotion. Then there was endless reading of novels; we followed the Virginia Woolf vogue and liked especially the subtlety of *Mrs Dalloway* and *To the Lighthouse*. Aldous Huxley was particularly significant to us; erudite, stylistic, honest and amusing, we found *Point Counterpoint* a telling statement of the brittle shallowness of our age and our own behaviour.

The wanderlust lived on, although in hibernation. Here I have to introduce a new member of our dramatis personae, Petroushka. Dan and I noted that many of our European counterparts had taken to the rivers in canoes; a promising way of exploring areas of interest, remote from roads and traffic, but with the possibility of reaching historic cities on their banks. We decided to try it out.

Dan, after a year at Cambridge working for his tripos, joined me *en route* for Rosenheim on the River Inn. There, at the Klepperboat factory, Petroushka awaited us, a cleverly designed collapsible canoe we found robust and safe. When we first met, she was a collection of sticks and struts and a reinforced rubber skin in three canvas bags. They could be erected easily and quickly into a shapely canoe for two, with enclosed stem and stern and perfect for double paddling. There were fixtures for two masts and we had three sails, a mainsail, a mizzen and a jib, and side-boards to maintain stability when sailing. A waterproof cover with two holes allowed our bodies to project and prevented the craft from flooding should we overturn. There was also an axle with two wheels that allowed dry-land transportation either erected or in its bags. Herr Klepper assisted us in erecting Petroushka and, with the masts and sails stowed away, we launched into the wild waters of the Inn. It felt wonderful, if a little precarious, but we soon mastered the art of paddling and steering. In no time we reached the confluence with the Danube and began the easy and picturesque passage of Europe's greatest river, parallel to the road on which we had cycled a couple of years earlier on the way to Constantinople. This time it was effortless bliss to be carried along by the current with only minimal effort required to maintain a straight course.

This delightful and uneventful journey was essentially to try out Petroushka for more serious future work. We gradually identified her strengths and weaknesses, experiencing her tantrums and subtleties that had to be understood if she was to give her best. Once at Vienna, Petroushka was dismantled, packed into her bags, wheeled to the station and taken by train to Budejovice in Czechoslovakia. We hardly bothered to look at this historic town, which I have cause to remember from events much later during the 'Springtime of Prague'. We launched immediately into the River Vlatva (Moldau), where we had trouble navigating over the shallows, but were soon on our way to Prague and on through 'Saxon' Switzerland, back to Germany and the Elbe. At a place with a railway station, we packed everything on a train to Berlin. Just north of that city we paddled and sailed on the Mecklenburg Lakes for two idyllic weeks through lakes, rivers and canals, eventually reaching Lubeck, near the Baltic coast. Petroushka had behaved beautifully.

The next year passed quietly, with study becoming more difficult and time consuming, curtailing our external activities as final exams approached. Between written papers and practical tests, exams devoured two weeks of precious life. Was it worth it? I was not quite sure. I had recently won an essay prize of tickets for really good seats at Covent Garden for the whole of Wagner's Ring. The first three operas were magnificent, but *Gotterdammerung* was to take place as usual at 4 p.m. on the day of the practical exam in organic chemistry. Getting there in time would mean leaving the college an hour and a half before the end of the exam and risking losing critical marks. Should I miss this golden opportunity to experience the opera or possibly miss a first-class degree. *Sein oder nicht sein; das ist hier die Frage*. Should I be bravely stupid or cowardly wise. I settled for the former and left the exam hall at 3.30. The performance was terrific and I did not allow my conscience to interfere with it. In the event, I doubt if it made much difference to my career having never been asked to present my degree certificate when transferring to a new job. Amongst the 50 students of my cohort, only one gained a first-class degree, the boy whose hobby was chemical physics! I came next as top of the seconds.

This hurdle stumbled over, time now to decide on a career. One of my father's friends was Professor Raistricht, a biochemist whose work fascinated me. I had some thoughts of launching into industrial research, on biochemical products. Professor Baker, head of the chemistry department interviewed me. I mentioned that I was considering going into industry. He was quite shocked. 'You have done too well in your exams', he said, 'to be allowed to be lost to industry.' I listened to his blatantly ivory tower arguments and decided to try for an academic career, which I thought might yield space for wider activities. Having been exempted from the first undergraduate year, university rules stipulated that I had to put in a further year before my degree could be validated. However, this could be devoted to research towards a Masters. I was found a place in Professor Baker's laboratory. He was one of the last survivors of the Edwardian school of British chemists. He had made the discovery that if the reactive gases oxygen and hydrogen are intensively dried over phosphorus pentoxide and then allowed to come together and subjected to an electric spark, they would not detonate as would normally be the case. This surprising, but not too fundamental, discovery caused him to be known everywhere as 'Baker-dry'. As one of his research students I was subjected to working under his eccentric ideas of apprenticeship as the best method of learning how to do research. Professor Baker disapproved of 'higher degrees' which he dismissed as frills and especially the Ph.D., which he denounced as a new-fangled and unnecessary German invention.

The best way to become a good researcher was by trial and error and the mastery of laboratory techniques; wisdom would follow by the doing of it. I was given the daunting task of measuring the effect of light on the electrical conductivity of very thin films of intensively dried atomic weight pure silver chloride. He then refused to see me for a couple of months; that is, before I had acquired some of the very difficult techniques involved. I set to in dazed hopelessness. The laboratory had very little, mostly primitive, laboratory equipment; there were no research funds. I had to start by learning to be a competent glass blower, which stood me in good stead later. I had even to make my own glass filter pumps that by then were standard in every laboratory. That Christmas all my friends received little glass animals as presents. The preparation of silver iodide to the required degree of purity was extremely exigent but feasible, as was the technique of sputtering thin films on glass. I struggled hard to become a good laboratory technician, but there was no intellectual challenge.

I realised with anxiety that during the previous five years, half of Baker's students had abandoned their tasks and left before the end of the year. It was of little comfort to be told that those who had persisted ended up in good jobs, academic and industrial. Eventually Baker deigned to see me again and I asked about the conductivity measurement progress. He showed little interest and merely told me that we did not have the necessary equipment, nor the funds to buy it. Nevertheless he seemed quite pleased with what I had done and explained gently that he had just been testing me out and would now give me a new and quite feasible task, investigating the influence of his intensive drying technique on the chlorine/ammonia reaction. This occupied the rest of my research year and gave me the basis for an M.Sc. that I wrote up successfully the following year. I did not regret this year of harsh training and began to feel quite kindly towards my professor. Outside the lab he was quite friendly. I visited him at home in Gerrards Cross and met his wife and daughter and discovered his real interest in carving beautiful objects in wood. He even invited me to the theatre with his family. It was a Gilbert and Sullivan opera, which on my own or with Normanska I would have been too superior to witness. Which one it was escapes me, but I enjoyed it despite myself. In the interval he introduced me to Arnold Bennett, the novelist, who was in the audience

As the next step towards an academic career, the professor advised me to spend a year doing research in a foreign laboratory, under a scientist of international reputation. After an interview I was awarded an Anglo-German exchange fellowship to work in Munich with Professor Kasimir Fajans.

Then came the long vacation once more. I set off with Dan on Petroushka's greatest adventure. We joined a ship at Hull, with Petroushka in her bags, and sailed via Copenhagen to Helsinki. A couple of friendly girls on board helped to pass the voyage pleasantly and we impressed them on the last day by erecting Petroushka on deck, ready to launch into the harbour at Helsinki. As we approached the city a naked girl stood on a rocky island welcoming us (and the other passengers) to Suomi or Finland, the country of innumerable lakes and islands. Some travel brochures claim that the country possesses 70,000 lakes, others 130,000.

Soon we were sailing the canoe eastwards from island to island, along the Gulf of Finland in beautiful weather and midsummer light. The sea was moderate, with enough wind to give us some speed. The evenings were especially beautiful with the sun shining until about 11 p.m., the sea pea-green and the western horizon bright orange. Most of the islands were quite small and uninhabited, camp sites were easy to find but very uncomfortable as there were few flat places and then very rocky with little possibility of securely sinking tent pegs. After about ten days we packed up and moved into the interior to join up with the system of lakes. We spent several weeks in this delightful countryside, sailing in lakes, paddling in rivers and occasionally crossing one of the larger lakes on one of the innumerable speedy little steamers that seemed to be the country's main means of transportation. Finally, a long train journey took us to Turku, the second city of Finland which its Swedish citizens call Åbo, situated in the southwest corner of the country.

We set off through moderate seas on a long sail from island to island of the Turku archipelago until eventually we reached the Åland Islands, a Swedish-speaking province of Finland. The capital Mariehamn was charmingly pastoral with friendly people anxious to talk with us. This was the take-off point for our crossing the 60 miles of open sea which separates Finland from Sweden, the most stupidly dangerous but vastly exhilarating exploit of my life. Looking westwards everything seemed serene, but the huge empty horizon did cause some trepidation. We surveyed sea and weather for a couple of days to gain some sense of the meteorological environment and then, one evening when everything seemed to be set fair, we launched off into the open waters with all sails set. Although it was already late summer, the evenings were still light. As we passed the ultimate islands, we cooked ourselves a dish of scrambled eggs on our small primus stove and then made sure that our cover was snugly fixed around us for what turned out to be a perilous journey.

The sea was almost calm as we set out and the sky clear, but when we were still in sight of land, clouds began to appear and the wind freshened. As darkness fell, the sky became black and the wind howled. We were soon facing a wild sea with waves over a metre high; at one moment we were on the crest of one, then tobogganed down its side, to be plunged beneath the next, submerging us for an instant until we bobbed up again to its crest. For hour after hour we plunged through this tumultuous sea. The physical effort needed to keep us on course brought us near to exhaustion. Occasionally we saw the lights of a ship steaming up the Gulf of Bothnia. It was daylight before the wind fell and the waves became less formidable. With only a small compass for navigation, we had somehow managed to keep approximately in the right direction.

At last we saw, with relief, the distant outline of the outermost islands of the Stockholm archipelago. Eventually we reached the first Swedish island, landed and, completely exhausted, slept for hours. Towards the end of that day we paddled into Stockholm, sodden and bedraggled, but triumphant. Our foolhardy voyage was over. Petroushka had crossed the Baltic. Time to return to London, try to be a responsible grown up and start a career.

6

Munich in the Hitler Dawn

Why go to Germany when advised to spend a year doing research abroad and, amongst all their research centres, why Munich? First, Germany was the country *par excellence* for scientific research. The drain of Europe's scientific talent to the United States was still many decades away. Americans were still coming to Europe and especially to Germany for research experience. Munich University held a high postion for research with several Nobel laureates there. Importantly, too, Munich is close to the Bavarian Alps and renowned for its opera and music, so I could hope to spend a useful research year there surrounded with cultural amenities. Like many British students, I had considerable sympathy for the Germans as having been given unfair treatment at the Treaty of Versailles. They had suffered the disaster of hyperinflation and were, after all, good, serious, hard-working chaps. There was little hatred for Germans after World War I. It had been a senseless, non-ideological struggle instigated by a power-crazy Kaiser whose soldiers had suffered much the same as ours. We knew about the communist revolution in Russia and the tensions it had generated in other countries, including Germany. Hitler's name was appearing more and more often in the English newspapers, but I had no understanding of how frail the Weimar Republic was.

So, at the end of October 1929, I left London for Munich, looking forward to a profitable and pleasurable year. A senior German student had been sent to meet me at the station and I was conducted to the *Studentenheim* in the Turkenstrasse, where most of the foreign students lodged and where I was to reside for the year. My fellowship assured tuition, lodging, breakfast and a small stipend. This was reinforced by a small allowance from my father, who wanted to ensure that I had sufficient means to make full use of this opportunity. In addition I soon learned how to earn small sums by teaching English and translating brochures and manuals into English for German companies.

I was welcomed in the *Studentenheim* by the *Hausmutter*, allocated a pleasant, simply furnished bedroom/study and soon became acquainted with its other mainly foreign but some German residents. There was an unlikely Armenian from Erivan, a cheerful Catalan nationalist, a Bulgarian biochemist, a couple of cheeky and often inebriate Irishmen, one of whom later found fame as a poet, a handsome Indian and quite a few British and Americans. Among these was Howard Robinson from Schenectady, N.Y. with whom I was to make many excursions and with whom I am still in contact 65 years later. I realised that this was the first time I had enjoyed the camaraderie of a student community. The contrast from my coterie life in London was considerable. We got along well together apart from the inevitable occasional tiffs and misunderstandings, mainly about girls.

Reporting to the university, I first had to have a medical examination including a chest X-ray. The doctor was horrified that my lungs were scarred from tuberculosis. He warned I should take care not to exert myself, learn to walk slowly and not indulge in sports. This was a shock at first, especially in view of the intensity of physical effort of cycling and canoeing in the previous five years, and I determined to ignore the warnings and continue to live as I felt I should. Next came a visit to the university registrar. I was registered as a research student, but had the option of also signing up for undergraduate courses. I had no intention of spending these rare days in Munich listening to lectures in German, but I did register for a series of lectures in photochemistry and one on *Tristan and Isolde*. Then came my first interview with Professor Kasimir Fajans, director of the university Institute of Physical Chemistry, very formal and correct. He cross-questioned me on my research experience, but obviously disbelieved my account of work in the London laboratory. He insisted that I, in common with the other aspirants for research in his institute, should first submit to a test of skill by undertaking a series of set experiments. This proved to be no obstacle, but it did consume some ten days before I could get down to serious work.

Work in Fajans institute was as orderly as Baker's had been disorderly. There were a few assistant professors with specialised lines of research such as photochemistry or chemical kinetics, but most activity contributed to the professor's programme. This was divided into a few distinct elements, each under a *Privatdozent* who supervised his students' day-to-day work. I was allocated to a Dr Wulff who was responsible for studies of the mutual distortion of ions in the crystal lattice, for the measurement of which he had developed an interferometric method of extreme precision. It took me weeks to master this technique and be capable of using it with certainty on a series of alkali metal halides.

Routine in the laboratory was rigidly observed. Several mornings a week at 8 a.m. there was a *Literaturstunde*, at each of which a student, who had been given the task of scanning a particular series of journals, reported on papers that had significance, however remote, on our work. This was a somewhat painful experience for me. Apart from the inconveniently early hour, my knowledge of technical German was still insufficient and many of the terms were unknown to me, even in English. At the end of the hour we would repair to our several laboratories and attempt to get on with our experiments. At about 11 a.m. Professor Fajans would appear, accompanied by the servile Dr Wulff. I and the colleague who shared the room would spring to attention, click our heels and be interrogated about what we had done since his last visit. Perhaps a little encouragement would be meted out, or some mild admonition; but more often than not he would just depart silently to the next room and we would relax. While this is no caricature, supervision was not as draconian as the account might suggest. We were left undisturbed most of the time, and could slip out to do other things as long as a reasonable amount of experimental work was accomplished.

Outside the laboratory oasis of order and calm, Munich was turbulent and unsafe. Our institute in the Acisstrasse was just round the corner from the Braunhaus, Nazi Headquarters. On several occasions there were riots; streets were barricaded, across which mobs of Nazis and communists confronted each other, making it impossible to reach the laboratory. From our German colleagues we had discordant accounts of the situation; stories of great anger, suspicion and betrayal. It was difficult to be sure which side who was on. Herr Becker, the seldom-visible supervisor of the *Studentenheim*, appeared to us foreigners as a pleasant open man. We were unaware he was a Nazi, but much later I heard that he had been assassinated during the internecine Rohm putsch. This was the time of Marlene Dietrich and the Blaue Engel; an insidious decadence somehow intermingled with political and social malaise. The Munich atmosphere was the equivalent of that of Berlin a little later, as described by Isherwood.

Many small incidents in the streets and the countryside disturbed me. On one of my weekend excursions to the mountains, the train stopped at the market town of Murnau. From the window I looked with amusement at the antics of a motley crowd of peasants manoeuvring to be on the right spot as we came to a halt. The doors opened and people crushed together to mount the train. An old woman carrying a basket of produce slipped on the icy platform and fell to the ground. Nobody tried to help her; rather they pressed over her body to gain the entrance. I saw other instances of callous cruelty, but also acts of gentleness and generosity. I began to suspect that brutality and sentimentality were the two ends of the German stick.

In my cushy Scottish environment, I had never been aware of anti-Semitism; now I experienced it. With some of the German students, I attended a couple of Hitler's rallies, to be both excited and horrified by the atmosphere. His crude but effective rhetoric, the bursts of anger, the posturing, were much as I had expected. What astounded me was their impact on my German friends, whom I had learnt to appreciate as reasonable and balanced individuals. They were immediately infected by his charisma, joining in the shouting and cheering with the rest. In contrast, my non-German friends were completely unmoved by the mob hysteria he generated.

I also had an unexpected close encounter with the future Führer in the summer of 1930, strangely enough not in a beer hall, but a café on the edge of the *Englischer Garten*, where I had gone with some friends after the opera. We sat down at a table, ordered strawberries and cream, and were amazed to find Hitler with some of his cronies at the next table, drinking and chatting banalities. We half listened to their conversation, but excited by our opera experiences we were talking rather loudly. One of our group made a remark that could have been construed as tolerant to the Jews and it was overheard. Hitler rose, shouted abuse and banged the table. My immediate reaction was, 'this man is mad', but it was deeply sinister. Alarmed by the rumpus, we consumed our bowls of strawberries more hastily than they deserved and silently slipped away.

Munich then was still a big village rather than a city in the contemporary sense and an ideal place for student life. There were endless privileges, price-reductions and other advantages; students paid only half the normal price of a haircut. But the greatest boon of all was the low cost of tickets for operas, theatres and concerts. One could buy a standing-place ticket for the opera for next to nothing and lounge at a favourite location at the edge of the stalls. Five minutes before the performance was due to begin, a bell rang, doors were closed and late arrivals were excluded. Some seats were always unoccupied and a beeline was made for the best, not necessarily the nearest, and in nine cases out of ten proved an ideal location. Munich was, if not the Mecca of opera, at least its Medina, so the privilege of being a student there was enormous for any lover of that art form. With one or other friends, I attended the opera three or four times a week in Munich and by my return to England, I had seen more than 100 performances of 54 different operas, including three cycles of Wagner's *Ring of the Nieberlungs*, most of Mozart's operas and some interesting modern compositions.

Munich had three opera houses. The National in the city centre, next to the *Residenz*, performed an opera every day of the week and had a large general repertoire. The *Residenztheater*, within the palace, a perfect gem of the rococo period with its early revolving stage, was used for smaller Mozart pieces presented in authentic eighteenth-century setting, exactly as Mozart himself had seen at the first performance of *Cosi fan tutte* there. Sitting through that and also *Il Seraglio* in this perfect setting, and in the presence of old ghosts, was an ultimate cultural experience. The *Prinzregententheater*, a replica of the one at Bayreuth, was in the suburbs and used only for the Wagner festivals held each summer and for occasional plays such as *Peer*

Gynt, for which its large stage is particularly suitable. I attended the *Ring* there.

I appreciated greatly this peaceful but exciting world of opera, so different from the alien world of impending tragedy in the streets outside. I recall, 60 years later, the excitement of scrambling to find a good seat when the lights went down, the smell of Munich greasepaint, so different from London's theatres, and the lovely voice of Joseph Patzach, the *Heldentenor* that year.

Weekends found us seldom in the city. Bavaria had an infinity of interest: mountains, ancient cities, architectural beauties. Mostly in the company of Howard Robinson, we visited all the fairy castles built by mad King Ludwig II, took the train to Innsbruck, visited Berchtesgarten and the Konigsee, climbed the Watzmann and even went for a weekend to Venice, taking advantage of ridiculously low student air fares to make my first flight in a plane. As spring advanced there was a series of visits to the intriguing rococo churches sprinkled all over Upper Bavaria. I hate the word 'intriguing', but no other seems to express my curiosity as to why this over-elaborate and complicated mix of design and decoration suddenly appeared and proliferated in the eighteenth century in this peaceful pre-Alpine peasant community.

My first exposure to Bavarian rococo was the *Assamkirche* in the centre of Munich. I looked it up in my 1910 edition of Baedeker's *Southern Germany*, but it was mentioned only in a brief footnote that confirmed its existence but added that it was in such extremely bad taste that it should be avoided at all cost. By 1930, all had changed and it had become a major sightseer's venue, an eloquent comment on the transience of taste and fashion. It was a magnificent accomplishment, tasteful or not. The exuberance of the carving, the scrolls and shells, the cherubs, the piercing quality of the colour may have been the expression of a primitive artistic urge, but one that had genius, healthy or not. This church is a theatrical creation of a complicated, contradictory psyche; simple adoration juxtaposed with a statue of a self-satisfied, elaborately dressed pontiff, looking down from an elaborate 'box' with a sneer at Christ suffering on the cross. I found the rococo churches in the alpine fields much more touching as they diffused through my being not only their own tranquility, but that of the sun-steeped meadows supporting them. Somehow their over-ornate decoration managed to transmute into a special variety of simple piety.

Then there was Sally. A Scots colleague had discovered what he considered an exceptionally good course in conversational German and suggested I should try it to complement my technical vocabulary. I agreed and joined him in the lecture room, sitting beside a pretty young girl with russet-honey hair. This was Sally. We were introduced and the three of us chattered together lightly. She was an Anglo-Scot and lived in the wilds of Northumberland. I was uninterested in continuing the course and returned to the *Studentenheim* never giving Sally a thought. However, a few evenings later, I was invited to a dance by some friends. I usually avoided dances as I had always considered them pre-mating fairs and a waste of time. However, this time I went with them. Sally was there, looking rather forlorn and neglected. We danced only one dance and then sat down and started talking. We talked and talked for the rest of the evening and then continued the next day in a café.

She was a quiet but opinionated girl from a good family and well educated, with a recent degree in music. She had come to Munich ostensibly to learn German and listen to music, but also to find adventure and escape from Britain and life on the Northumbrian moors. She exuded elation at having discovered Europe and clearly regretted having to return home soon with little opportunity to see more. In the course of conversation, I told her about my cycling to Constantinople. She immediately showed interest, realising that here in Munich she was already half way to the Bosphorus. By now, the beginning of December, our little community of foreigners had consolidated. The *Studentenheim* was an all-male institution and having a girl in your room invited expulsion, but there were many young women students in rooms and *pensions* nearby who soon joined us in the activities of the *Studentenheim*. We sat about in cafés, arguing endlessly as students do. Our favourite was the *Pustastube* where, over glasses of sweet *Tokay*, sweet secrets were exchanged, doubts and aspirations aired and the characters of others torn to shreds.

Shortly after I met Sally, I went for a week's skiing above Kitzbuhl. On my return there was a message from her to say that she had decided to leave immediately on a quick trip to Istambul

Alexander King as a Bavarian Student

and Athens and would I help her with introductions and arrangements. I felt uneasily responsible. It seemed unusual for a young woman of 21 to want to go off to foreign parts alone in the middle of winter. However, she was quite determined and a couple of days later I saw her off on the train to Vienna and Budapest.

As for myself, I spent Christmas in London with my family and returned to Munich at New Year. As it happened, Bronwen, the girl I had met on the ship going to Helsinki, had decided to study German and had surprisingly (?) settled on Munich. I was uncomfortable about it, but when my mother saw me off at Victoria, Bronwen, was there. The *Studentenheim* was rather bleak and deserted and I felt lonely. I had an impulse to take the train to Venice where I knew Sally would be arriving a few day's later. A voice in me said, 'be careful; don't commit yourself'. Arriving in Venice on a very frosty morning, I scanned the maritime news for ships arriving from Athens via the Dalmatian Coast. There was one due at 8 a.m. the next morning, so I found a hotel and spent the day looking at churches. I was at the docks the next morning to find that the ship would be three hours late. It was a miserable wait prolonged by a blend of apprehension and anticipation, but eventually I began to see the passengers streaming out of the customs hall. Soon Sally emerged looking at me first in amazement and then with relief. She had not been expecting me but had a shadow of feeling that I might be there. Whatever her feelings towards me were, she could at last relax after weeks of solitary and sometimes dangerous travel. The stay in Venice was complete magic. There was deep snow on the Piazza of San Marco, not a tourist in sight. We seemed to have Venice to ourselves.

Returning to Munich, I had to plunge into laboratory work and producing *The Wicked Duke of Mantua* for public performance preoccupied my leisure time. I was fortunate in obtaining an introduction to the Director of the *Nationaltheater* and was allowed to borrow costumes and properties from the theatre's store. For instance, for the fatal duel between King Harry and the wicked Duke we had Sigmund's sword which, when a button was pressed, fell apart from a blow by the opponent. Howard took the part of the King, I, of the Wicked Duke and Bronwen who by now was firmly established in our group, was the Queen. My relations with her seemed friendly enough, but there was an underlying hostility and bitterness. There was another newcomer, a pretty English girl whom they called the Minx, a complete misnomer, for she was neither pert nor pushing. The performance went off well. We had some quite good reviews in the German press, but most of the foreigners in our audience seemed unsure whether to take the play seriously or as parody.

Sally returned to Northumberland at the beginning of February and I got down seriously to finishing the experimental laboratory work. It was soon time for Carnival or *Fasching* as they called it, a festival of noise, vulgarity and licence, healthy in its exuberance. The main merrymaking takes place at huge balls in beer houses. Ear-splitting bands blare out the night through and it is customary to end with breakfast at the railway station. My partner in all this jollity was the Minx, who I accompanied back to London as Easter approached. I knew I was madly in love with Sally.

Sally came south for the vacation and spent a couple of nights with my parents at Hatch End. I had hoped to entice her to descend the River Loire with me and Petroushka but, alas, the canoe's skin had shrunk and it was difficult to put her together. So she was abandoned and we went off instead for a week's walking in the Auvergne. At its end, we parted sadly in Paris, she from the Gare du Nord for England and I from the Gare de l'Est back to Munich.

The last months of my stay in Germany were rather bleak. All the girls had gone and most of the boys. I had to work intensely in the laboratory to be sure that I had all the data for the scientific paper which would justify my year of research. This was finally published a year later in the *Zeitschrift fur Kristallographie*. I also did a little teaching of chemistry in a private school, the headmaster of which was a leading person in some Bavarian royalist party. With all the trials of communism and Nazism, the Bavarians were finding it difficult to break into a new political era. Many of them retained a warm nostalgia for the Bavarian royal family, for its preposterous Ludwigs and Maximillians who had never done them much good. The headmaster apparently climbed to the summit of the *Zugspitze*, the highest mountain of Germany each year to light a fire in celebration of the birthday (or death) of Ludwig II. The head of the Wittlesbachs, Prinz Ruprecht von Bayern was said to be a credible expert in Persian art. I happened to be in the opera one evening when he attended. As he entered the royal box, the crowd rose to their feet and sang the Bavarian national anthem.

One final pleasure was the opportunity to attend the summer Wagner festival at the *Prinzregententheater*. On the last evening of the *Ring*, I was listening again to my fateful *Gotterdammerung*. During the last, long entr'acte, I strolled in the gardens where, by chance I ran into the Director of the *Nationaltheatre* who was chatting to a venerable gentleman whom he announced as Richard Strauss. The Director in turn introduced me as a young and upcoming British dramatist. I curled with shame but did not refuse an invitation from Strauss to sit with him in his box during the last act. I had no strong impression of Strauss. He said little and I ran off at the end as soon as I could.

The end of the university semester came. My parents joined me in Munich and I took them to Montenegro to share its sense of wildness and isolation. We went to the top of a hill above Cetinje to look down over Lake Scutari to the seaboard of Illyria. They even had the satisfaction of feeling frightened when a peasant shot his gun to warn us off his land. It was not a particularly happy summer, however. I was lovesick and hardly knew who I was. On leaving Munich I felt certain about one thing: War with Germany was inevitable.

Sitting by the waterside at Kotor, waiting for a ship to take us to Split, gave time to reflect on the year and assess its influence on me. In general it seemed highly successful. Munich had proved to be the ideal place to spend a student year abroad. Its cultural amenities, its location and the wonderful historic and dramatic surroundings are unsurpassed. On the whole I had made the most of the artistic possibilities in balance with the scientific facilities. My research topic was too far from subjects likely to attract my interest in the future, but useful as proof that I had done serious theoretical work and had a grasp of new techniques. Munich, that romantic 'big village' will always be close to my heart. Writing as I am now, over half a century after the events of this chapter, I would still, were I capable, sing a paean of praise to Munich, despite shameful acts that took place there under the Nazis. I would also seek out the ghost of the old *Studentenheim*, obliterated by an allied bomb during the war. As for my inner life, I had striven, not always successfully, to observe my own antics and thoughts with some objectivity. I continued to be aware of my inner contradictions and remained unfulfilled. I still did not know who I was.

Once back in London, I found my position in the family subtly changed from what it had been before I had departed. Harmony and affection remained as ever, but I began to feel slightly apart. The last strand of the umbilical cord had finally snapped and I had a new feeling of full responsibility for my acts and thoughts. This makes me feel I must express more explicitly my appreciation of my parents before they leave our story.

I had an enormous respect for my father, especially for the wise and determined way he insisted that I should have all the educational opportunities that were denied him. What a scenario there was for a frictional generation gap. He was a 'Wee Free' with all the inhibitions and dogmatic encumbrances that implies, teetotal, cautious and excessively thrifty, although undoubtedly generous towards others. I, on the other hand, immersed in the world of Nietzsche, Marx and Freud, nurtured by Ibsen, Strindberg and Pirandello, had never suffered poverty and with my scientific education was consumed with doubts as to the relevance of old myths and dogmas to explain the nature and aims of existence. But we had no difficulties of accommodation, he listened to my ideas, sympathised with my doubts and, over the years, his view of the world widened considerably. I was much closer to my mother. She was not very articulate but extremely sensitive. In all the years at Hatch End, my father left for work much earlier than me and this gave me the best part of an hour each weekday to talk to my mother, elaborate ideas with her and just think aloud. These hours were particularly precious to me. My mother died of cancer during the war. For many evenings I tried to soothe her pain by reading passages of the *Bhagavad Gita* to her on her deathbed. Much to the relief of the family, my father remarried with Mary Mackenzie, an old school friend of my mother and lived happy in retirement in Edinburgh, where he continued to be an elder of the Kirk.

7

A Dream Play

The Vanishing Virgin

In my dreams, there are many landscapes, beautiful, recurrent, persistent, that have become as real to me as well-loved scenes from the past when I was presumably awake. I recognise that visual recollections of incidents, and especially of important incidents from the past, are frequently distorted or blurred by emotions of love, distrust, guilt, insufficiency or satisfaction, to become less authentic than the landscapes of dreams. The episode I recount is of another type of psychic uncertainty that I record as indicating the multilayered nature of awareness – dreams, sleep throughout life while seemingly conscious, existence mainly through the flow of automatic associations, hallucinations, or finally the hope of being fully awake to one's self.

One Sunday in January, Sally and I had been skiing on the slopes of the *Alpspitze* above Kreuzeck, and returned in the evening to our little *Gasthof* in Partenkirchen. It was already dark when we reached the village, but the night, although young had a starry brilliance bringing the mountains so near they seemed about to collapse on us. Too beautiful to leave, we stood for a long time in the tiny portico of the miniature painted church, physically exhausted, but far too excited by the scene and with each other to feel anything beyond the first, creeping signs of tiredness. In the starlight, the painted facades of the houses and inns could be dimly discerned, while beyond them came sounds of peasant music and dancing, banal enough, but charged for us with foreign mystery. Yes, air and mountains were too precious to lose so early that night, so we started walking towards Partnachklamm, a stark, impressive gorge Sally had visited in the soft sunshine of an autumn afternoon, but I had never seen.

Just before reaching the gorge, we were attracted by a little inn and decided to rest there over supper before going further. We had a hearty meal of mushroom omelette and beer, followed by Apfelstrudel and coffee. There was nothing exceptional about the place; warm and cosy as only an alpine inn can be, its *Gemutlichkeit* was not of the sticky variety, but allowed us to retain our privacy.

Just beyond the inn, the hillside became steeper. On our right, the torrent issuing from the gorge roared along on its journey to the Black Sea. On the left, the path forked, one branch leading to the entrance to the gorge, the other mounting high over the gorge to join the main path through the gorge half a mile or so beyond. At the intersection stood a wayside shrine in peasant rococo, characteristic of this part of the Alps; behind a fixed pane of glass, stood an eight-inch high Virgin, dressed in the neatest of doll's clothes, topped by a gilded halo. Her background consisted of dried flowers and grasses, the ceiling of the shrine was of azure blue, with little sparkling stars to suggest the firmament. We paid our respects and continued by the lower path to the gorge. In a few moments we came to a rustic arch, the entrance to Partnachklam proper, where tourists paid their 50 pfennig fee in summer to view this commercialised wonder of German nature, but now closed of course. We passed through the open doorway and entered the gorge. Outside, the night continued its brilliant promise; the moon was rising and the snow sparkled, but in the gorge itself the cliffs rose steeply and little light penetrated. The pathway was frozen and slippery; it was cut out of the rock and below it the torrent rushed in darkness, while from the overhang of rock above our heads, great icicles hung menacingly, at times impeding our passage.

We realised it was foolish to continue, but the noise of the waters made communication impossible and somehow we just continued to advance cautiously, one false step could have had us in the chasm. I was very frightened and I knew that Sally must be feeling the same, it would have been inhuman not to feel afraid. For a time we clung to each other, but presently the path narrowed as the defile became the merest slit between high cliffs, forcing us to walk in single file. At one particularly slippery corner we lost contact. It seemed to me Sally had vanished into thin air. I halted and shouted her name at the top of my voice. The sound of the water dwarfed it to a whisper, even to me. There was no response. I peered into the cavern of darkness above the stream, realising there could be no thought of rescue. I groped my way step by step along the treacherous surface, past icicles pointing like ghastly fingers at me. At last, after what seemed an eternity, I emerged suddenly from the gorge into the broadening valley, moonlit on the sparkling snow. Sally was standing silently waiting for me, having come through the same chain of horror as me.

The tensions within our bodies suddenly released, a sense of relief swung us into an unreal hilarity, the falseness of this mood somehow illuminated by the mocking benevolence of the moon-drenched landscape. There was no question of returning through the gorge, so we continued the few yards until we reached the junction with the upper path into which we turned, to return by the easy path above the gorge. From below, only the distant, muffled roar of the torrent penetrated the peaceful night. Eventually, the path descended again to the original fork, near the inn where we had supped. By then our false jollity had evaporated in the cotton wool softness of the Christmas-card landscape, but there was an edge of fear in our hearts. We approached the little shrine at the crossroads and again advanced to pay our respects to the Virgin. A sharp uneasiness gripped us and we clung together instinctively – *the virgin had vanished.*

There was nothing to be said; it was definitely the same shrine with its background of dried flowers and grasses and the familiar starry ceiling. The glass front was unbroken, securely fixed as before. Neither of us spoke for a moment; we were assailed by that helpless shiver of fear provoked in most people by the very hint of the inexplicable, the supernatural. Moreover we both felt a compulsive urge to walk through that dreadful gorge once more. Hand in hand, therefore, we walked along the original path beside the running stream. The moon still lit up the winter scene splendidly. On we plodded, still dominated by a dream atmosphere, towards the commercial entrance to the gorge. But we never reached it. We persevered for what seemed to us ages, but there was no longer a gorge to be found. Instead, we were walking along a soft valley where a stream tumbled gently between low banks. Beyond, the outline of the mountains could just be discerned. The moon shone more brightly than ever on the barren beauty of this winter Alpine scene. There is something almost cosmic about such an atmosphere even though it may be the reflection of a landscape of a dream. Despite the mystery of these events, we were somehow no longer possessed by the unreasoning fear of the unknown. Within me there dawned a firm, though temporary, certainty that all things are one and governed by a single unity of purpose, normally concealed by everyday perception.

Eventually we had to accept the reality that the gorge no longer existed. At the time that seemed a perfectly ordinary, matter of fact conclusion to reach and it remained unquestioned. We retraced our steps back to the shrine that was still empty. Once beyond there, the spell was broken; we lost our unquestioning serenity and became possessed of a simple animal terror. We hurried back towards the village as quickly as our tired legs could carry us, past the inn again with its sounds of music. We now knew with unassailable certainty that there was no geographical mistake to explain the aberrations of that night. There was enormous satisfaction in observing that some at least of the familiar and man-made stood firm on their original foundations. Safely back in our hotel, we sipped comfort from glasses of Gluhwein.

It had been our intention to return early the next day to Munich and work, but in the grossly materialistic mood of the morning we decided that the disappearance of the Virgin and the trick transformation scene of Nature must have been exaggerated and cried for investigation. An elaborate research project was therefore planned, but after a pleasant daytime playing truant on the ski slopes until night approached.

It seemed proper to repeat the whole business again at the same time and with the same detail as on the previous day. So when the time came, we approached the inn near the gorge and ordered a supper identical with the last. Excitement hung in the air and we were happy to notice that the weather remained unchanged and the stars shone as brilliantly as before.

After supper we walked slowly along the path to the Virgin

who was safely back in her shrine amidst the flowers and grasses; she still smiled benevolently at us. Then we were through the arch and once more within the gorge. We saw little reason to revise our impression; it was hardly less terrifying than on the previous evening. The noise of the torrent deafened as before, the darkness was just as impenetrable, the icicles retained their menace. All was exactly as before. But there the detailed replication ceased because we could not force ourselves apart, but instinctively kept close so that we should be saved or lost together. To separate in the icy darkness was too much to ask, but we had invalidated the experiment by avoiding the shock that had presumably triggered the events of the previous evening.

Emerging into the moonlight once more, we did fancy we felt the same quality of relief and then the same jollity. The return journey by the upper path contributed nothing new. Here and there we recognised landmarks – a fallen tree, a curiously shaped rock at the side of the path – that reassured us that we were on the same route. Breathless with anticipation, we approached the shrine once again to find the Virgin still secure and serene, apparently immovable; we did fancy, however, that her smile had become somewhat equivocal. The experiment had been a failure. Nevertheless, for the sake of completion we did walk once more to the entrance of the gorge that we found as real as rock and torrent can be. The whole exercise had been futile and we should have known it. It may be possible to ensure that physical conditions remain constant, but not the exact states of mind and emotion of the human subjects. Gluhwein did not comfort us that night.

8

The Career Academic

September 1931. A new and hopeful phase of my life was opening that was to end abruptly in 1939. On my return from Munich I accepted an offer from Professor Baker of a post as demonstrator in the chemistry department of Imperial College, at the princely salary of £175 per annum. As I acquired experience in the years that followed, my teaching load and my responsibility in supervising the research of others gradually increased, and my post was elevated to assistant lecturer, lecturer, and finally senior lecturer, before my academic career was broken by the outbreak of war in 1939.

Before the term began, however, I determined to drive north to visit Sally in her natural habitat, the fastness of Northumberland. The Thompson home was in really gorgeous countryside, unexpectedly remote. From their house and across the small road alongside the North Tyne, it was about 35 miles to the next track, a wilderness unimaginable in overcrowded southern England. Their house was called 'The Sneep' a name I learned many years later is still current in Sweden to denote a spit of land just upstream from the confluence of two rivers, in this case the Tarset and Tarret Burns. As I found on my first night under their roof, the wind moans and twigs of trees tap on the windowpanes in true Wuthering Heights fashion. I understood immediately how the sparse melancholy and desolation of this rare country had enchanted Sally.

When my car drew up at The Sneep, I found that by mischance or mischief she was not at home. My embarrassment on plunging alone into this large family circle was acute, but her kindly father took me in hand and did his best to put me at ease.

I soon found I had landed in a formidable matriarchy, with the mother incessantly and unabashedly dominant, and the father a cheerful, quiet and beloved onlooker. There were seven daughters, most of whom were at home, and one son with his fiancée. By dinnertime Sally had arrived without apology and we all trooped into the dining room, a sombre place with heavy brown furniture and the walls covered by portraits of ancestors. I was placed between Sally and her father, who amused himself by drawing caricatures throughout the meal. I have no recollection of what we ate; conversation was what mattered. It was, of course, initiated and orchestrated by the mother. The main theme was a book, in vogue at the time but which I had not yet read, 'The Story of San Mikele'. They were all arguing about a chapter, 'the corpse conductor', which meant absolutely nothing to me. I felt small and ignorant and I sensed the mother was despising me. Sally was embarrassed and added little to the conversation. She was all too aware that they were assessing me as a potential son-in-law and I was making no showing.

Despite this ignominious introduction, the stay was thoroughly enjoyable. We walked the moors, climbed the hills, listened to the curlew's melancholy notes and I even began to like her mother. She was one of a clan of Stevensons, a Scottish family, many of whom had settled in Northumberland. She herself was a seventh child as was her youngest daughter who showed no sign of the legendary second sight supposed to be possessed by seventh daughters of seventh daughters. The founder of this dynasty was a certain James Stevenson of Paisley, a contemporary of my own west of Scotland forebears, some of whom he must have known. His grandson, James Cochrane Stevenson, Sally's grandfather, had been Liberal MP for South Shields for 30 years and founded the Jarrow Chemical Company which developed into the United Alkali Company, later one of the units that formed ICI. I know only one anecdote about him. He happened to be in Paris on some business during the Franco-Prussian War and was about to return home quickly as the Germans approached. Intending to reach the railway station, he hailed a cab and shouted 'a la Guerre'! During that visit he attended an auction of furniture from the Palais de Versailles and bought a Louis XV card-table which now sits in my home in London after having been nearly a century in

Northumberland and then back in Paris again with us for 35 years.

He had a large family, seven daughters all of whom were formidable women. One of them married a Moderator of the Church of Scotland, another the head of one of the biggest steamship lines, a third was one of the early women MPs and married to a senior Cabinet Minister, yet another became Professor of Economics at an English university as early as 1910, while the youngest, Sally's mother, after having raised a family of eight, became a magistrate, probably one of the most prejudiced in the land. Whether an accused was guilty or not often depended on whether he was a cherished Northumbrian or not. The influence of this 'monstrous regiment of women' was evident from my first entry to The Sneep.

Back at college I found that my modest role as demonstrator was to instruct a group of some 20 students in the art and science of inorganic chemical analysis in a huge vault of a laboratory. My students worked at benches in one corner, while three similar groups under their demonstrators filled up the rest. The atmosphere was congenial; the lengthy proximity to the students made individual instruction effective and I soon began to understand each one's personality, foibles and difficulties. There was ample opportunity for conversations on chemistry, and also on ambitions, passions, philosophy and all sorts of topics. The boys – for as yet there were few girls reading science – were but a few years younger than me, which made for an easy camaraderie.

This work took less than half my working hours, so there was ample time left for research. I started on a new line, based on an observation of the professor that in air at room temperature, active charcoal reacts slowly to produce some unidentified oxidized products. This led eventually to a whole series of papers on 'chemisorption'. After about a year of this easy existence, Baker called me to his office one morning and said, 'it is time you gave some lectures to the young men.' I was delighted and responded, 'What would you want me to talk to them about?' 'That's not for me to say', he replied. 'It is for you to decide what you can teach them.' Nevertheless he guided me very subtly to give a course on chemical calculations, a subject of little appeal to me but which was an excellent discipline and enhanced my own understanding. I discovered I had some talent as a lecturer. One of my colleagues, J. S. Anderson ran a parallel course and a year later we decided to write it up as a book. Together we worked out hundreds of exemplary sums to be solved by the students, with answers given at the end of the book. This was published in 1934 and became a minor bestseller. Many years later, just after the war, I was visiting the University of Melbourne on my first visit to Australia and had the gratification of seeing a white banner outside the university bookshop which read, 'King and Anderson are here again'.

Later it was suggested that I give a course on technical German. At this time, some knowledge of written French and German was required of degree candidates and questions in these languages (to be answered in English) were set in the final examination paper. Professor Philips, professor of physical chemistry, had previously given such a course and, as he was soon to retire, he asked me to take it on. He gave me his notes and I bought a small text on technical German, which gave little help, but there were some amusing phrases such as 'the professor has left his cast iron apparatus in the train' with which to amuse the students.

I very much enjoyed this course and I hoped that the students would be amused as well as instructed by it. It was a hybrid between serious academic lectures and a music-hall turn. I would do experiments, speaking only in German, demonstrating objects and ideas by actions, followed by the writing of their names in German on the blackboard. This course was given annually. Later in the 1930s it became a slight embarrassment as a considerable number of refugees fleeing from the Nazis entered Imperial College and had to submit to my inept attempts at teaching them their own language, I was never ridiculed, but I cravenly trimmed my histrionics.

This led to a further development. I was approached by a publisher to prepare an English–German Chemical Technology book. It was quite a challenge and planned as a short but comprehensive account of contemporary chemistry, in English on one page and German on the page opposite. Each new term had to be in context, printed in italics and indexed. As it happened, one of the Munich *Privatdozents*, Hans Fromherz had come to Oxford to do research on reaction kinetics. He agreed to collaborate in writing half the book in German while I wrote the other half in English. When this was completed, each of us translated the other's text. Fromherz was a large loud-spoken clumsy man, an excellent mountaineer. He had a huge, rather crude sense of humour, amused in a patronisingly appreciative way by everything English. He was an excellent partner. We spent many uproarious afternoons working on the book in the garden of my parents' house at Hatch End. The book was published in 1934 and sold surprisingly well. We brought out a revised and largely rewritten edition after the war. There were some 19,000 terms in the index, each entry identified not only by the page number, but also by a letter that indicated its place in the page. The rewriting was difficult because so much had happened in chemistry in the meantime with so many new terms, and I had become detached from day-to-day preoccupation with chemistry. As a consequence it was a difficult but useful re-education for me.

The need to write a revised edition brought Fromherz and I together after the separation of the war. A year before its outbreak, he had been drafted from academic life to work on chemical warfare projects at I. G. Farben. However, once hostilities had begun, his project had been abandoned and he found himself driving a truck in Poland. Apparently the Nazis had decided that the war was to be short and sharp so that the results of further research would come too late to be of use. The war, they assumed, would be won on existing technologies, a policy that finally favoured the allies who mobilised scientific effort fully and accelerated development as the war commenced, resulting in the atomic bomb and many other devices. By some means that I never understood, the Fromherz family managed to escape to Switzerland and when I met him after the war he was a professor in Switzerland.

The success of our terminology book illustrates a fundamental change that took place in science as a consequence of the war. In 1939 German was still the main language of the physical sciences and the book sold mainly to English speakers anxious to read the voluminous literature in German. After the war, sales were mainly in Germany where the new generation of scientists had to read English to have access to world literature. In these circumstances our book became the classic reference for translators and interpreters in the post-war international organisations.

I received useful royalties from the book until 1990. It is now quite obsolete.

My good, kindly, but largely out of date Professor Baker faded into retirement and was replaced by Professor Briscoe, a bright modern teacher who changed things entirely, introducing system into the research function, encouraging his lecturers to take on research candidates for Ph.D degrees. Under Baker, the Department had not encouraged the publication of research papers. He had seen the training of highly efficient research chemists as his primary function. With the new regime, publication of research findings became the main concern of scientists, as is now universally the case. Under Briscoe they appeared in quantity. Despite these reforms, I very much doubt if the human product of this important research school has improved.

Briscoe contributed greatly to bringing the chemistry department into the new age and was successful in attracting talent to his team. Indeed, the quality of the chemistry faculty was very high. Apart from Anderson, who later became a professor in Melbourne and then in Oxford, H. J. Emeleus had a distinguished part in the chemical side of nuclear development and became a professor in Cambridge after the war. Then there was Tommy Pearson, a pioneer of free radical research, a brilliant and attractive personality whose death a few years later was a great loss to chemistry. Another member of the group was Ronnie Purcell who became science adviser in the Ministry of Home Security during the war. We all realised that our education, although recent, was largely dépassé. The new concepts arising from physics had a revolutionary impact on chemistry, but were diffusing only slowly into British research schools. None of us had been taught the new chemistry, although we had picked up bits of it here and there. Our group realised that if we were to guide a new generation into the contemporary lifelines of science, we would have to start by re-educating ourselves. This was done, of course, by intensive reading. To deepen our understanding we organized a series of small seminars at each of which one or other of us would give an initial introduction to one of the elements of development relative to our needs: quantum theory, wave mechanics, elementary particles, chemical bonding and the like. We would then question, argue and debate the significance of each topic for the common benefit. These sessions resembled the *Literaturstunde* of the Munich Institute but were much more focussed.

My extramural activities during this period continued much as before, but there was much less time for theatre and concert-going. One performance that did impress me enormously was Ernest Milton in Pirandello's *Henry IV*; its playing around with reality and make-believe fitted in deliciously with my own pre-occupations. I was still pretty naive and uncommitted politically, although my interest in public and international events was growing. I learnt much from my students, many of whom were idealistically rather than ideologically inclined towards communism. The Spanish Civil war had broken out and a couple of my students went off to fight on the Republican side. This, as well as my foreboding about Germany, made me incline towards the Left, but never to espouse communism.

Sally had come to London. With much regret she had abandoned her musical ambitions having realised that she was just not good enough to become a concert pianist, while the prospect of teaching piano to little girls and reluctant boys for the rest of her life seemed bleak. She became a sort of apprentice in a strange little outfit that specialized in indexing, and had, as its bread and butter job, the indexing of Hansard. She accepted Reubke and Normanska tepidly, while they regarded her with a certain amount of trepidation and even awe. The vehemence of her views surprised them coming from that hitherto unfamiliar species – a woman. They named her the Valkyrie and I think that Normanska, who had seen her on horseback riding over the wild moors of The Sneep, had a veiled fear that she might attempt to carry him off to Valhalla.

At the beginning of 1933 Sally and I decided to marry.

9

The Valkyrie's Nest

Our marriage somehow divides the story of my academic years into two. Before it took place, I was relatively irresponsible, fancy free and childishly eager to grasp every new experience. As a married man I naturally and properly focused on family concerns, without abandoning earlier friendships or passions. I interrupt the narrative, therefore, to sketch the life conditions of young professionals at the time, introduced by a few externalised hints about our courtship.

Mating with a Valkyrie is not necessarily a smooth affair, but it certainly was an exciting one. In Munich and afterwards we had never discussed or thought about marriage. What mattered was to be together, ignoring the consequences, but somehow our common horror of the approaching war suggested the need for a permanent bond. I have to admit also that, on another level, I felt it would be delightful to have a beloved companion at hand, to make love with on any propitious occasion (how wrong I was!). Despite our foreboding about the future of the world, we felt that it would be marginally justifiable to create children who, we could hope, would be better humans than ourselves. Fundamentally, however, the reason for marriage was the shared conviction that we would always love each other, whatever obstacles might arise. I remember protesting at the time, that when we were old, in our nineties perhaps, we would still be in love. A happy fate allowed this to come true.

We were married in a simple ceremony in the Presbyterian church of Falstone on the Scottish border in March 1933, followed by a reception in The Sneep garden. It was a beautiful spring day with the daffodils in full bloom. Sally in an apricot velvet gown showed little sign of her Valkyrie nature. After a brief honeymoon at Yeavering Bell in the Cheviots, we departed for London and our new home, a fourth floor, no lift flat in Colville Square, just off Portobello Road. We did go canoeing to Russia that summer, about which more anon.

We had enough money for eating and living, but precious little for furniture. However, there were many generous wedding presents, the most welcome of which was a piano for Sally from my parents, a 1910 Blutner Grand. Reubke presented a smart chromium hall light. Years later after his death it had to be dismantled for repair, I found inside a little roll of paper on which there was a rhyme in Reubke's writing:

Who spilt the bean soup,
Knocks the plaster down?
Was once a Harlequin,
Is now a clown.
Mother, what makes the planet rumble so?
Child 'tis my own diapasons that blow.

This was obviously his attempt at an epithalamion, discovered 20 years after our wedding and ten years after its author's death. Professor Briscoe sent me a book on birth control and the physiology of sex. On receiving it I felt indignant. How could he imagine that I was so naive as to need such a work and, in any case, what could he know about the psychology or physiology of a Valkyrie? Later, however, I realised that it was a kind gesture of concern for my happiness.

The piano finally and with great difficulty, made it to the fourth floor where it occupied half the space of our living room. I walked across the park each day to my laboratory in South Kensington. Sally had no experience of housekeeping; I was much more domesticated than she. She resented having to get down on her knees to wash floors, but with moans and groans she struggled on and after a few weeks of working together, we were reasonably capable of running a small flat. Holidays were restricted by lack of funds, but we managed some lovely time at The Sneep, walking over the moors and climbing the hills; the wildness is a vivid memory. In the summer of 1935, a friend lent us her cottage in the Lake District, but we had no car now and although Sally was six months pregnant, we had long walks over this magnificent terrain. My parents were bereft of a driver and felt their movements restricted. They engaged a couple, a chauffeur-gardener and a cook-housekeeper, and enjoyed some years of pleasant comfort, until a new war disrupted their life once more.

Jane was born in September 1935. We were naturally anxious about the responsibility this pretty miniature human being's arrival had brought. On the practical level, the effort required in taking her up and down four flights of stairs was heavy, so we decided to park her in the fresh air, in a cage hung outside the window. Baby cages were in vogue with 'progressive' young parents at the time and one of Sally's sisters kept her baby in a cage attached to her flat in the Temple. The cage was securely fixed to the wall so as to be able to bear the full weight of a man as I courageously verified. Jane lay there happily when weather permitted, gurgling and screaming alternately. Neighbours and passers by on the streets looked up in horror and called the RSPCC to rescue her and punish us. The inspector knew all about baby cages and exonerated us, but to the frequent question, 'is it safe?' I soon learned to reply, 'more or less, but she's well insured'.

A year later, when we were visiting our White Russian friends, the Gubskys at their charming house just over Hammersmith Bridge, we learnt the house next door was vacant. We immediately contacted the owner who wanted to sell it. This was beyond our means and we were so clueless we had never even heard of mortgages. However, we managed to persuade him to rent it to us. I remember the evening we moved in, walking along the towpath in the direction of Barnes and Kew, watching the barges and pleasure craft moving with or against the tide of the river Thames. After the noise and bustle of Portabello Road, it seemed paradise, sights and smells of the country were within walking distance. Apart from a gap of some years during the war, we lived happily in Clavering Avenue until 1957. Catherine, our second daughter was born there in 1937 and Margaret, our third, just a week before war was declared in 1939.

I shall describe some of the practical details of life here, not simply to illuminate the trivia of my career, but as a sketch of the conditions facing a young professional family in the 1930s, so different from the situation today when the luxuries of yesterday have become the basic necessities for a decent life. Our house in Barnes was semi-detached, built in 1910 and sold at the then current price of £500. It had a small garden behind and a slip of forecourt in front, just big enough to support Petroushka. There was of course no central heating; walls were not insulated and the rooms were bitterly cold in winter. We had open coal fires in the sitting room and wretchedly insufficient gas fires in the dining room and two bedrooms, where they were essential for the drying of the daily wash of nappies. In the kitchen there was a coke stove that heated water and also a solid, black cast iron gas cooker. In the bathroom a spluttering geyser heated water. The kitchen stove required a great deal of attention, had to be raked out frequently and the clinker removed – one of my jobs. In the garden stood a concrete bunker for coal, another for coke. Coalmen brought fuel from time to time in black tarry sacks. They had to trample through the house and out the back door to discharge their sacks in the bunkers, leaving their muddy footprints on the carpets.

I recollect the winters being very cold. One year the Thames was frozen, but this was exceptional. Then there were the traditional London pea-soup fogs, a dreaded menace to everyday life. They came every year and often lasted for several days, casting a blight on all activity. It was difficult to go about one's business, to find one's way in the yellowish gloom with heavy heart, stinging eyes, with colds and spluttering coughs all round. I remember one evening, coming home from work, emerging from the Hammersmith tube station into one of these fogs, made even more sinister by yellow flares and the fog-muffled sounds of traffic moving at a snail's pace. Crossing, I found my way hesitantly to an island in the middle of the street, to be confronted by several large shapes emerging out of the fog. They were performing elephants from Bertram Mills circus at Olympia, being taken out for exercise. These tropical animals must have regarded the fog as yet another torture specially designed for them by the errant human species.

At that time domestic technology had begun to develop, but was still primitive or a matter of luxury. The carpet sweeper was there and the clothes wringer; the electric iron was beginning to be common and the vacuum cleaner was already proving a boon.

Off our scullery, there was a small pantry with a window to the garden to keep food fresh, refrigerators were yet unknown. Washing clothes was one of the hardest domestic chores. Garments were boiled in a large galvanized tub, pounded with a cone-shaped copper suction cup, scrubbed with a brush against a corrugated wooden board, and rinsed before being passed through a manual wringer. There were no aids for dishwashing, a constant chore for a family of five, but detergents were beginning to appear.

Being a housewife was a full-time job then; much of the work demanded hard physical energy. Sally, as an avant garde mother, took child raising seriously and there were frequent journeys to the Babies' Club in Chelsea to monitor growth and seek advice on the current fads. As the children grew they had to be taken to and from home and school on foot or by bus. There was little time for external pursuits or dalliances. In the evenings Sally had the constant task of maintenance. Synthetic fibres, led first by nylon, appeared only after the war; little toes poked little holes in little socks, so that a couple of mending and darning sessions were essential every week.

We could not afford a telephone; that came a few years later. Television was in the future, but we had radio. I had earlier possessed a crystal set and enjoyed poking around with the cats whisker to find stations. Now bulky but effective valve sets were common, but we did not have one. Theatres and restaurants were surprisingly inexpensive. One could have a good, six-course dinner at the Regent Palace Hotel for half a crown (10.5 pence) per head. Excursions to town had to be very infrequent.

All the drudgery was accepted cheerfully as the norm in what we perceived as an advanced society. Indeed, we felt privileged in comparison to the harder lives of earlier generations. It made for a happy community, all engaged in common family welfare. There was quite a lot of entertaining of friends and colleagues. Childish illnesses came and went, with few complications.

But this cosy bourgeois bliss was only part of the story. Petroushka was still an important element of our lives. In the summer of 1933, the year of our marriage, she took us to Russia, on an exciting voyage that I shall report later. In 1935 we were able to leave the baby at The Sneep with her grandmother, and set off for a walk in Lapland. At that time, before the Finno-Russian war had stripped Finland of much of her territory, that country stretched northward as far as the Arctic seas, with the tiny port of Petsamo just east of the Norwegian frontier. We went by boat up the scenic Norwegian coast all the way to the North Cape, then disembarked to walk overland to Petsamo where we were able to load up with food for many days, including dried reindeer meat. Heading east into the wilderness, there were no tracks, inhabitants or even reindeer flocks and without maps of the interior, the country was difficult, with no obvious system of valleys and watercourses. Large lakes had to be skirted, rapid rivers waded and high, bare mountains traversed; often made up of steep screes that trembled with every footfall. We were there during the period of the midnight sun, but it soon becomes monotonous. Camping was painful because of clouds of voracious mosquitoes while large numbers of lemmings tried to raid our food reserves during the so-called nights. We proceeded as near to the coast as feasible, coming to the only partially demarcated frontier of the Soviet Union. We advanced further into Russian territory until we fancied we could see the smokestacks of Murmansk on the horizon. This was exceedingly foolhardy as the border region was probably seeded with land mines. We were lucky and, starting the return journey, diverted to the 'Fishers' Peninsula' and returned to Petsamo by boat from the tiny village of Virolahti

We had another voyage in Petroushka in 1934. Fromherz brought his canoe from Germany and together we descended the Rhone river, from just below Geneva, to the Mediterranean, through beautiful landscapes and past historic cities. However, Petroushka's main tasks were thoroughly domestic. She lived in front of our house and was wheeled the few hundred yards to the towpath of the Thames from which she was launched, just beside Hammersmith Bridge. We would paddle up with the tide to Kew Gardens or Richmond or downstream through the centre of London, as far as Greenwich. There we would stop at a cafe which offered 'dainty teas' and wait for the tide to turn to help us in the paddle home. One warm summer day we had a picnic supper on the river steps of the House of Lords. As the children grew up, they participated gleefully in these short expeditions. Eventually we needed more space than the aging Petroushka could provide so we bought a second one so the whole family could be waterborne.

This was perhaps the happiest, hardest-working phase of our domestic life. We now had three happy little girls and would have liked a son also, but with the outbreak of war, it seemed wise to stop breeding. The girls grew up, went to school, then university, married and had their own families in due course. Details of their future careers and fortunes will have to wait until they publish their own memoirs. Alas this does not apply to Margaret, the youngest, who died in 1963. As a little girl of 10, she decided that her life's work would be to play the oboe. She consulted a friend of ours, Geoffry Gilbert the flautist, who advised her that her lungs were not yet strong enough to sustain that instrument and that for a couple of years she should content herself with the recorder. Fortunately her school, St Paul's, was strong in music and supported her, providing a good teacher of the oboe. Eventually she took her LRAM and then followed us to Paris where she gained a place in the Conservatoire. There she gained the premier prix, premier nommé for her oboe. She fell in love with a boy, one of the Japanese musical whiz kids, Hirohiko Kato. He was a superb flautist, also premier prix at the Conservatoire and a pupil of Rampall. Margaret played briefly in the BBC orchestra and, now married, moved with Hirohiko to Bausoleil on the Mediterranean when he was appointed first flute of the Monte Carlo orchestra. Unfortunately he was an enthusiastic mountaineer who had been through the guide's school at Chamonix. One bright day that summer he took Margaret for a walk in the high alps. Without warning the weather changed and they were trapped on a ridge on Mont Blanc and were never seen again. Ten days passed before a helicopter was able to search for them. Ten days of agonised waiting for us. Hiro's body was found a couple of years later. Margaret still lies buried in the snows. Life was never the same again for Sally and me.

10

Academic Continuum

We sensed the black clouds of war rolling closer as my academic years flew by. The apparent stupidity of the politicians appalled me and I began to feel indignant concerning both national and international scenes. In 1937 I met an intelligent MP, Dr Hayden Guest, secretary of the Leverhulme Trust to which I had applied for support for a scientific expedition to the Arctic that I was then trying to organise. We became friendly and he fed me with socialist ideas some of which seemed sensible, others fanatical. I joined the New Fabian Society and avidly read their material. Slowly I began to understand a little about this charade we dignify with the name politics. I was even approached to be a Labour candidate at a by-election, but rejected the invitation. A political career was not for me.

My activities at the College continued to expand. I gave a couple of new courses including one on colloids. Also, a number of postgraduate students elected to do research with me and in due course there was a crop of Ph.D. theses. I never bothered to write one myself, conditions for doing research were still very poor. While chemistry had gone well beyond test tube and retort stage, the level of instrumental sophistication was very low compared with that of research laboratories today. We had little specialised equipment and much depended on our skill as glass blowers. I managed to obtain some instruments with various research funds, but my equipment was still inadequate. Infrastructural support simply did not exist, and there was no secretarial help. I had to walk 200 yards to make telephone calls at a payphone at the entrance to our institute and had no access to incoming calls.

I had continued my interest in adsorption and published quite a few papers in this field. A Canadian postgraduate research fellow brought us a technique very relevant to my research, namely a method of spinning very thin silica fibres and, using a hydrogen flame, winding them round carbon rods to produce very sensitive spring balances. These registered quite small weight changes to substances suspended from them in enclosed spaces such as glass tubes and exposed to different gases at different temperatures, yielding hitherto inaccessible results. With another student we examined the magnetic properties of some metal oxides, indicating an unexpected variability of composition, thus apparently contravening the law of constant proportions. This was particularly interesting then at the dawn of solid state physics. Through casual observation of a colleague, I began working on an emulsion system that, according to the method of shaking, would produce an oil in water or a water in oil emulsion. It was amusing that shaking a tube of the ingredients with the right hand always produced one of these and, with the left hand, the other. This led to an interest in emulsification in general. Emulsions are used in a wide variety of industries and I was invited to advise on industrial problems, becoming a regular consultant to one company that had approached the College authorities for help. This led to a small improvement in my financial situation. Then there was a common project in which most of my senior colleagues took part. Heavy water (Deuterium oxide, D_2O) had been discovered recently, and we devised a high-precision density method of determining the D_2O content of waters from various sources. We did a vast survey of all kinds of waters from the oceans, the Antarctic icecap, lakes high in Tibet, animals and vegetables, cancerous tumours, electrolyte baths etc. As part of this programme I found there was a minute difference in retention of the two isotopes by active charcoal and silica gel. On the basis of this I took out a patent with ICI for the preparation of heavy water by a differential adsorption process. It proved less efficient than the electrolytic separation process adopted for atomic energy purposes.

During these long years, my travel lust remained latent, but our experiences in Lapland with its wonderful emptiness, oblique lighting and midnight sun had given me the urge to experience an arctic summer again. I was also convinced there was a place for shorter, research-oriented expeditions as little was known in detail about remote places. One day I happened to read a classic, Lord Duffryn's *Letters from High Latitudes*. In the previous century, this intrepid diplomat had taken his yacht far into the North and there was a drawing that immediately caught my imagination. It was of his ship in a mist-bound sea, with a break in the clouds displaying a snow-covered volcanic peak. This was the Beerenberg on the small island of Jan Mayen in the Barents Sea. Immediately I knew that I had to go there to experience the bleak enchantment the picture suggested.

I contacted James Wordie of St John's College, Cambridge, who I found had explored this island. He was a friendly, engaging personality whose description of Jan Mayen confirmed my intention. Moreover he convinced me that a scientific expedition would be quite feasible, there was much work to be done there. So in early 1936 I began to make contacts and attempted to raise funds for an expedition to Jan Mayen the next year. I also identified a number of senior research students from various College departments anxious to take part, and together we sketched out an attractive programme of investigations. Obviously I had no credentials to justify research grants for such an expedition, nor were my research interests appropriate to the investigations proposed. Not surprisingly, I found no response. I did not lose heart and decided to try again the next year. I managed in the meantime to elicit the interest of Sir Henry Tizard, Rector of the College. Tizard felt that the expedition might provide valuable field experience for a few senior students. My persistence had already made my name familiar to some of the grant giving bodies, and doors now opened a little more easily. Especially helpful was Dr Hayden Guest, already mentioned, and with his help I was awarded a Leverhulme fellowship to organise the expedition, at which point others felt able to respond. The Royal Society gave me a research grant as did the University of London. The Royal Geographical Society lent us some useful equipment, including tents. All this contributed to make possible the launching of the expedition in 1938. Had the cat turned round again?

The island of Jan Mayen is situated in the Greenland Sea 250 miles east of the Liverpool Coast of Greenland and 300 miles north of Iceland. It was known to the Vikings and visited by them on their voyages to and from Greenland. The Beerenberg being visible more than 100 miles off at sea means it would have been difficult for them to have missed it. They named the island Svalbard (cold coast), a name now generally applied to Spitzbergen.

With the decline of Norway in the fourteenth century, the existence of Jan Mayen was forgotten and it was not until the beginning of the seventeenth century, when the Dutch and English were rivals in the Arctic whaling grounds, that it was rediscovered. It was sighted independently by several ships, the most famous being that of Hudson. Credit is now given to the Dutch captain Jan Jacobszoon May van Schellinkhout who landed there in 1614, after whom the island is named. In 1615 Robert Fotherby went ashore and reported seeing tracks of foxes and bears; he named it Sir Thomas Smith's Island, but in 1618 King James I, in granting a whaling monopoly to a Hull company refers to it as Trinity Island. As late as 1630 Richelieu gave a similar licence in the name of the King of France to a company to whale off the Isle de Richelieu. We were to see the remains of huts and the skeletons of many whales half buried in the sand.

Normally the island was evacuated by 28 August each year, but in 1632 Basque pirates plundered it, so it was decided to leave a wintering party during 1632–34, but all perished of scurvy before the spring. With the decline of whaling in the Arctic in 1642, Jan Mayen was deserted, visited only by the curious who reported volcanic eruptions or brought back fantastic accounts of the size and shape of the Beerenberg. Later it had minor economic value to hunters of its only mammal, the arctic fox, and the occasional bear stranded there by receding ice. During the first two decades of the twentieth century there was still a take of arctic fox and the occasional bear. The Norwegian government now controls the hunting of foxes on Jan Mayen and their population has probably increased greatly in recent years, thanks to an inexhaustible supply of fulmer petrels, easily taked from their nests on the cliffs.

In 1921 the Norwegian Government established a meteorological and wireless station, filling a meteorologically important gap on the world's weather maps. Norwegian sovereignty was proclaimed in 1929. Jan Mayen is of considerable scientific interest on account of its isolated position, recent geological origin and the Beerenberg, which has suffered a major eruption since our visit. A number of scientific and mapping expeditions have worked on it since 1882. In 1921 James Wordie's party spent a month there and climbed the Beerenberg.

With the help of the Norwegian Arctic authorities we arranged a passage to Jan Mayen from Tromsoe in the early summer of 1938, on an aging sealing ship, the *Fortuna*, passing near to it on its way to East Greenland. A return passage was guaranteed by another ship that would be returning from Greenland in the autumn. I had assembled a team of scientists from various disciplines. I was fortunate to have Scott Russell, a biologist and experienced mountaineer; with two botanists to assist him. There was also a marine biologist, a glaciologist, a surveyor and a geologist. A doctor from Middlesex hospital came for medical emergencies, which happily never arose, but he was also an experienced amateur ornithologist and brought with him another colleague in that field. I was to be responsible for soil analyses. By taking such a large party we hoped to be able to make a comprehensive survey of the main features of Jan Mayen during its short Arctic summer. I went up to Tromsoe a week in advance to buy hydrogen cylinders and other equipment, smoked salmon and other arctic food stores. Joined by the other members of the team we set sail in the *Fortuna* which, by favourable winds, brought us to our island in four and a half days. The expedition and its work were described in detail in the *Geographical Journal* of August 1939. Here I shall give only a few personal impressions.

My initial impression of Jan Mayen was of utter desolation, relieved by the elevated and elevating vision of the Beerenberg whenever the weather allowed visibilty, which was not often. The terrain around us was entirely volcanic; sand and dirty grey ash with glistening crystals of olivine. It was treacherous too, all too easy to fall through the surface into hidden hollows. The island is about 35 miles long and has three distinct parts. The central region where the meteorological station is situated is as narrow as two miles, it is low lying and consists of many large lava floes and volcanic sand deserts. There is practically no vegetation as there is no soil and no nutrients. There are many long beaches of volcanic sand, strewn with driftwood from the great rivers of Siberia and a surprising number of glass balls, floats from fishing nets. There are also three lagoons, freshwater lakes separated from the sea by narrow sandbars. In summer these are a favourite resort of waders. Less than a mile west of the radio station, Egg Bluff rises 500 feet sheer from the sea; on its top are a series of steam blows also observed in other parts of the island.

The southern part, to which we made several expeditions, is seldom free from mists for more than a few hours at a time and is a dark forbidding country, devoid of all colour. It consists of a mass of irregularly placed peaks and volcanic craters of rugged outline, rising to as high as 3,000 feet, many of them falling steeply to the sea. Throughout the island there are frequent cliff falls, the volcanic rock being everywhere unsafe. With no regular valley systems, progress is made across innumerable screes that look like huge heaps of uncemented ashes.

The northern part is entirely occupied by the Beerenberg, its ice fields and glaciers, most of which descend to the sea. This splendid mountain which rises 7,680 feet, is the dominant feature with an ice-filled crater, found by our surveyor to be over a mile in diameter, forming the nevé of a glacier, named the Weyprecht. On the Northern side of the crater rim there is a huge gash, walled by sheer rock faces from which an almost continuous series of icefalls descend to the sea. The highest peak of the mountain lies immediately to the west and we proposed to name this summit Haakon Peak after the then King of Norway, and this has been accepted by the authorities. Immediately across the glacier is the second highest peak, which we named Hakluyt Peak, thus bringing back to the map the name given by Fotherby to the whole mountain. The crater rim, a serrated horseshoe ridge rising sheer from the ice-filled basin and two and a half miles long, connects the two peaks. Climbing parties from our expedition, led by Russell ascended the high peak twice and the hitherto unclimbed Hakluyt Peak once.

The climate of Jan Mayen can only be described as atrocious. A spur of the Gulf Stream reflected off Spitzbergen meets the Polar Drift near the island and produces turbulent conditions. The radio operators told us that the summer of 1938 had been exceptionally good according to island standards which seems to be borne out by the fact that we were able to make two ascents of the mountain. Nevertheless, conditions during our visit made life at camp difficult enough. During July there was more or less continuous mist at base camp, broken only by fierce gales and heavy rain. The beginning of August saw some improvement in conditions, but a glimpse of the sun was always the exception. One night we went to sleep in a thunderstorm of almost tropical violence and were wakened in the morning by an earthquake. Towards the end of that month we experienced a hurricane of 48 hours duration, during which our tents collapsed, their poles snapped like matchsticks. Storms of volcanic sand are common and ships miles off at sea have collected inches of such sand on their decks. But weather conditions on Jan Mayen are surprisingly local. Thus, during the hurricane at base camp that prevented work for two days, our surveyor was busy at the South Glacier, three miles away, in relatively calm weather. The south coast is generally mist-bound, while at a short distance across the spine of the island, the north-west coast has a much higher percentage of calm, clear weather.

It was important we should attempt to climb the Beerenberg as soon as possible after our arrival as we had apparatus for the registration of cosmic rays that had to be exposed as long as possible. The conjunction of high altitude and high latitude is propitious for such work, providing the thinnest layer of atmosphere for the rays to pass unabsorbed. For a successful ascent, calm weather is essential and there were very few days in which conditions were favourable, with the exception of those on which our climbs were actually made. We left base camp on the grey evening of 12 July in reasonable visibility, ascended the volcanic hillside and crossed the lateral moraine to meet the South Glacier at a height of about 800 feet. The mountain was in full view, the glacier leading gently up to a height of about 5,000 feet, from which steep steps led to the crater rim. After a tedious march of several hours up the glacier, we reached a rocky outcrop just below the steeper slopes and halted for a meal. Conditions on the upper slopes were worsening; a thick mist made route finding difficult and snow conditions were bad. I was terrified as I had never experienced such terrain. Great care had to be taken on the broken slopes to avoid crevasses, dangerously hidden by thick layers of soft snow. We moved with long rope lengths between us, but even then there were many falls into crevasses. After an exhausting climb during which we frequently sank above the knees in soft snow, we finally reached the crater rim.

Jan Mayan, the Beerenberg from South Glacier

Thick clouds obscured the view, but, through gaps, brief glimpses could be seen of the opposite side of the crater rim. In the other direction, the peaks of south Jan Mayen could be seen sticking out of a sea of cloud. We then climbed along a narrow ridge to the highest point, a rounded mound covered with coarse crystals of ice. By this time snow was falling and, although it was a moment of triumph, I was sad to realise that it did not fulfil my vision of Lord Duffryn's sun-lit peak that had brought me there. It was obvious that no improvement of visibility could be expected, so we deposited the cosmic ray apparatus and descended in worsening snow conditions to the top of the South Glacier. This was the most unpleasant part of the whole expedition. We had a series of adventures inside the crevasses, during one of which I found myself perilously perched on a ledge of ice, looking down at the white hell beneath and up at the small hole through which my companions were trying to communicate with me. Verbal contact was impossible as the snow and ice muffled voices. We reached base camp in heavy rain after 18 hours on the mountain.

The second ascent of Haakon Peak, in which I did not take part, was pleasanter than the first. It was rewarded with sunshine at the summit, with magnificent views of the great ice-filled crater which made it possible to photograph the complete rim. Climbing Haakon Peak was part of one of our most ambitious expeditions, an attempt to explore the largely unknown north-east corner of the island, a small triangular piece of land cut off from the rest by the flanks of the Beerenberg. We might hope then to complete the circuit of the mountain at a low level and finally, circumstances permitting, climb Hakluyt Peak. Accordingly, a party of five set out along the south coast that consists mainly of cliffs, some of them rising sheer to 2,000 feet and nearly all unscaleable. However, progress was easy above the cliffs on gentle slopes that are the most richly vegetated part of the island. No such route was possible on the steeper east coast, but a gully of loose rock was found down which it was possible to scramble to the shore. We then marched on the narrow beach to the north-easterly point of the island, reached after 14 hours of hard going. The next day was spent exploring the region in lamentable weather. There seemed little prospect of improvement and a delay of more than two days would have caused a food shortage and abandonment of our plans. We decided therefore to gamble on an improvement in weather later in the week and limit the circuit of the mountain to only two, Russell and Jennings the glaciologist. The rest would return to base camp, leaving our food supply to the two climbers.

This return journey was quite our least pleasant journey and quite frightening. We had to wade in the surf just under the snouts of glaciers about to calve, were seriously menaced by rock falls and, after climbing up the gully to the top of the cliffs, were blanketed in thick mist through which we had to strike by compass over difficult country. Happily the weather eventually favoured the climbers who accomplished their objectives including the ascent of Hakluyt Peak.

Huge numbers of fulmer petrels, guillemots, puffins, glaucous and kittiwake gulls and little auks nest on the cliffs, and their behaviour was a constant delight. We also identified a number of species of migrant birds, blown off-course and landed on the island. The place of the bird population in the overall ecology of the island became clear to us. One of my most exhilarating moments on Jan Mayen was to come across a magnificent rockery of flowers at the top of a cliff, as bright as a seedsman's packet, their colours contrasting brilliantly with the uniform grey of the desert. They were mainly *Ranunculus glacialis*, the arctic buttercup, with its pale translucence, and the vivid azure of *Martensia maritima*. The soil here, in contrast to the surrounding terrain, had a significant content of nitrogenous nutrient and was a dramatic instance of the biological cycles. The sea beneath was rich in plankton, that fed the fish which supported the birds. Arctic foxes pawed birds from the nests and their droppings provided sufficient nitrogen to produce this unexpected floral display.

The end of the season came all too quickly and conditions began to deteriorate. When our ship, the *Polarbjorn* arrived from East Greenland, a large swell made it impossible to embark near the campsite, so we had to carry two tons of equipment five miles to the sheltered side of the island. When we were many miles away on the *Polarbjorn*, I had a last glimpse of the Beerenberg, still bright above the mist. It was good-bye to Jan Mayen and, had I known it, to rugged travel in general. I did see Jan Mayen once again, 30 years later when my plane on the polar route to Japan happened to pass near it. The pilot kindly allowed me a nostalgic good view. This was soon after a major eruption of the Beerenberg. I fancied I could detect changes in its outline. I had found the expedition so fulfilling that I was determined to mount a scientific expedition to some interesting corner of the planet every few years as a means of giving groups of research students real field experience as well as character building. But it was not to be. Soon I was engulfed in the war and later, in its consequences.

Having reached the European mainland, we all began to return to so called normality, to rest, clean up and celebrate. I was heart broken to shave off my exuberant beard, a fiery red that made a mockery of my black hair. Before doing so, I had it photographed. It appeared later in a cosmetic trade journal, to accompany a paper I had written on cosmetic emulsions. Back warmly in Clavering Avenue, after a spate of stories about our adventures I lapsed into the old life; it was as if I had never been away. Shortly after my return to London I was asked to give a BBC television interview for the programme *In town tonight*. Television in 1938 was quite new; it was still the only regular TV service in the world, with very few viewers. I was driven out to the studios in Alexandra Palace, while Sally was invited to see the transmission in Broadcasting House in the West End. I regretted my beardlessness for the occasion, but was subjected to very heavy make-up. The interviewer indicated a large square outlined in white on the floor, and warned me to stay firmly within it during the interview because of the inflexibility of the cameras. The live interview then commenced. I felt at ease and it went well. As I warmed to my subject I could not stand still, but began to walk about. The interviewer signalled to me to keep still, but I had forgotten about the square and ignored him. Suddenly I felt something grip my ankle and pull me back to the centre. I looked down with alarm to see a grappling arm doing its job. When I reached home, Sally's first comment was, 'That went well. But why did you suddenly take fright and look down as if a crab had caught your toe.' How different this primitive TV studio was from those of today with their sophisticated backgrounds and equipment.

The threat of war became ever closer as Hitler swallowed up Czechoslovakia. I happened to be shopping in King Street,

Hammersmith as Neville Chamberlain passed back from the airport, returning from the Munich talks with 'peace in our time'. The crowds in the street were mildly enthusiastic and applauded him, only to revile him some months later. When I was in the TV studio I watched a news bulletin, a report of the failure of Lord Runciman, Sally's uncle, to solve the Czechoslovakian problem with the Sudeten Germans. Chamberlain had asked him on two occasions to undertake a mission to Prague, but he had refused, knowing it was doomed to failure. Finally the Prime Minister persuaded King George V to insist, and he felt that he had to go through with it, despite his conviction of its futility. It was an abject and humiliating experience that damaged his reputation at the end of a long and honourable career and finally hastened his death.

There were a number of secondary consequences of the Jan Mayen expedition; I was invited to lecture about it to the Royal Geographical Society. This went well and was published in their journal. Then they awarded me the Gill memorial prize for the effort and elected me to the Council of the Society to represent young professionals. I attended meetings of the Council assiduously until wartime obligations forced me to leave England. I enjoyed my acquaintance with the Geographers of the Council. I venerated Freya Stark who was charming and so knowledgeable about all things geographical. Then there was Sir Dudley Stamp, a great geographer and St John Philby the Arabian explorer and father of the infamous Kim Philby. In the course of the next few years, quite a few scientific papers appeared on aspects of the Jan Mayen work. Thus ended my career as a geographer.

Returning to the laboratory after these months of freedom was quite difficult, especially as I, in common with the other members of the Jan Mayen party, developed bouts of flu with nasty coughs as a result of losses of immunity during even this short stay in the Arctic. My researches continued well, several of my students being awarded Ph.Ds, and I was well settled into the academic community. To my great surprise the Chemical Society awarded me the Harrison Prize, an annual research award made to someone under the age of 30. I had not applied for it and had no idea who put me up for it. It gave a welcome boost to my self-confidence. This award again brought me to the attention of

Tizard who had been so supportive of my organisation of the Jan Mayen expedition. This double exposure was to have vital consequences for me in the not-too-distant future. At the dinner of the Chemical Society at which the Harrison Prize was presented, I sat next to Sir Ian Heilbron who had recently been appointed Professor of Organic Chemistry at Imperial College. He was receiving the senior research award of the Society. This was the initiation of another friendship that again influenced my future.

It looked as if 1938 was going to be my most successful year yet. I had pulled off the expedition, won a couple of prizes and been promoted to a senior lectureship. I could begin to apply for a professorial chair. Nevertheless, I was unlikely to become a high-flying research scientist. I was competent and hard working, but I accepted that somehow I lacked that elusive something, that intuitive creative imagination that distinguishes the work of the great scientists. Nevertheless, with my undoubted research skills and my ability as a lecturer, I felt I could be a successful professor with a considerable research reputation. Such a career, I thought, would satisfy me and leave plenty of time for public duties and personal delights. How little did I know myself!

In the summer of 1939, just before war broke out, I did a second descent of the Rhone with Dan Piggott. I had seen little of him these last years because after taking his science tripos at Cambridge, he joined the Colonial Service and became an assistant district commissioner in Tanganyka where he no doubt did an excellent job. But I felt sad that his scientific gifts – considerably greater than mine – were largely wasted. We saw him occasionally when he was home on leave. He had never regarded Sally as a Valkyrie and now he called on her to vet the suitability of a number of girls whom he might marry to partner him, the lonely man in Africa. Dan saw, as I did, the inevitability of war, and realised we might not meet again for many years, so, before returning to Africa from leave, he planned to canoe the Rhone with me and board his ship at Marseilles. We had a delightful ten days together and off he went. It was his farewell to Petroushka.

Two months later war was declared and my whole complacent world shattered.

A Flash from a Parallel Scenario

Through my Russian friend, Nicolai Gubsky, I heard about an interesting if bizarre individual, G. I. Gurdjieff, who seemed to be a mystic, philosopher, magician or charlatan; possibly a mixture of all four. I found some of his ideas as communicated by Gubsky, interesting and challenging. Apparently he believed that Man did not possess a soul, but only the potentiality of one, towards the actualising of which Gurdjieff had developed a series of inner exercises. Born of an Armenian mother and Greek father, he was living in Paris at the time and I determined to meet him if possible. On Dan's departure from Marseilles, I therefore packed up Petroushka and took her by train to Paris *en route* for London,

I had an appointment to meet Gurdjieff at his 'office' which was, in fact, a table adjacent to the busy Boulevard des Capucines in the Cafe de la Paix. I reached it at 10 a.m. and began looking for him. I did not expect any difficulty identifying him since I had seen a photograph of him with a high, bald cranium and heavy, drooping Middle-East moustaches. Indeed, there he was, just as I had imagined him. He was in deep conversation with a young woman who spoke with a shrill American voice. He conversed in a mixture of English and French together with occasional words foreign to me. He was obviously trying to get her to part with a large sum of money and she was vaguely resisting or bargaining

Obviously I could not interrupt this intimate situation, so I took a seat at an adjacent table, ordered a coffee and observed. Eventually the young American yielded, opened her handbag and produced a sizable wad of banknotes that she had no doubt prepared for this eventuality. I was too far away to read their denominations, but I heard him several times using the word 'zeros' as he gathered up the notes. He seemed satisfied, but did not thank her. He said goodbye and asked her to sit at a nearby table. He placed the pile of notes ostentatiously on the table and waited. This seemed to be an established practice and in no time at all, people dropped by; beggars, prostitutes and idle people who were each given a few pieces of her money until it had all disappeared. Then he made a gesture and she walked away. I was informed later that this was not a unique situation. She was being given a lesson of self-liberation. The money was certainly not for his personal benefit.

In due course I approached, introduced myself and was invited to take the seat vacated by the young woman. He was very talkative, gushing about all sorts of trivialities, boasting about the quick time of his drive up from Vichy the previous evening. He asked had I a woman and children, while every now and then he distracted me, shuffling over the table a special chocolate, wrapped in cellophane through which one could glimpse a miniature photograph of himself. This performance and the torrent of superficial remarks, I found disturbing, as he no doubt intended. I had prepared some questions to which I wanted answers, but how to get them in I quite despaired. At last I burst out, 'very interesting, but I did not come to you to discuss little things'. He smiled and changed his approach, talked seriously and answered some of my questions before I could ask them, as if reading my thoughts. The distribution of chocolates ceased. I was so saturated with impressions that I could absorb no more; I was unable to ask other questions.

Soon I took my leave. He said, 'if your wish is to be Man, come and see me again in Paris. Just ask for the little dancing master.' My mind surged with images of Diaghilev. I had a solitary lunch at a restaurant in the Gare du Nord, before boarding my train for Calais and London. It occurred to me that I had been hypnotised, and indeed a sense of unreality permeated me. This state remained with me for a couple of weeks. I gave the chocolates to Sally and the girls. Yes, I mused, I shall certainly go and see him after the war.

George Ivanovich Gurdjieff – seer or charlatan, both or neither

11

The Blast of War: From Laboratory to Bureaucracy

When war was finally declared, a sort of orderly panic erupted. People dreaded an immediate strike from the air. The reaction was to get the children out of the cities as quickly as possible. The railway stations were thronged by thousands of bewildered children being seen off by their parents to evacuation havens in the countryside, while in rural areas volunteers scurried to find billets in which to lodge them. Air raid wardens, already in place, were busy enforcing their local tyranny while awaiting the expected bombing. But no bombs fell; we were at the beginning of a long phoney war while Hitler was busy with the Poles and then the Russians.

Our two elder girls were already at The Sneep, so I drove Sally and two-weeks-old Margaret up immediately. I then spent a few days driving a truck on some local emergency, the nature of which I no longer recollect. Then I returned to London to see what was happening in the Capital. Everything was fairly normal, with more bustle than usual and no child to be seen. As required, I registered immediately for national service. I surmised that, as a scientist, I would be put into uniform, given a rifle and sent to guard a depot of chemicals. Fortunately the authorities were not that stupid. Unlike the Germans, they expected a long war and reserved people with my qualifications for appropriate tasks as they appeared. In fact, I never heard from them again. I was, however, summoned with some of my colleagues to listen to the Minister of Home Security outline the strategy to combat chemical warfare. I resumed my normal activities.

At the College uncertainty reigned. The rumour was that we were to be evacuated to the University of Edinburgh. However a single German bomb, the first to hit Britain, happened to drop on that city and evacuation plans were dropped. The College functioned, if marginally, throughout the war, and Edinburgh never experienced another bomb. So I continued with my research and waited. By that time I had gathered together a small research school around my own work and I had a team of keen young people ready to be used as required.

The spring and summer of 1940 were as beautiful as I could remember and everything was quiet. A false sense of security descended on London and the children came trickling back from the country. My own family returned to Clavering Avenue, but we knew that the calm could not last. Hitler's Russian campaign was in full swing and doing well, so he now turned his attention to the West. The so-called Maginot line had been turned and the Allies' Western Front collapsed. We began to look for some more or less stable refuge from the city, preferably easily accessible from London so that I could join the evacuated family at weekends. After a few failed attempts to find somewhere, we found a suitable small house in Newbury where Sally and the children were to spend a couple of years, with frequent visits by me.

Just about that time Lord Rothschild, who was in charge of a unit named Military Intelligence Research (MIR), approached Sir Ian Heilbron for advice. Heilbron invited me to help him. Here began a short interruption in my career pattern, utterly unprofessional, probably futile, but great fun. We were supposed to research sabotage and counter-sabotage. Our first task was, in fact, one of self-sabotage. The evacuation from Dunkirk was approaching and we feared a German invasion. Should this occur, it was important that our petrol supplies be put out of use, as the enemy would undoubtedly take them over. The traditional means of stopping an engine was to put sugar in the petrol tank, so in south-eastern England, sugar, already rationed and in short supply, was requisitioned and placed in sacks beside petrol pumps. England's wet climate would soon dissolve the sugar. It was important therefore, to discover a chemical that, added to petrol in small quantities, would bring vehicles to a standstill. This was our task.

Victor Rothschild was a real cloak-and-dagger type. Our group held its meetings at his headquarters in Wormwood Scrubs Gaol, in its macabre condemned cell with the signatures of notorious murderers scrawled on the walls. I was given a car for our experiments and drove off boldly from the prison, down the slope towards Acton High Street. This was fine, but as I approached the first red traffic light, I discovered with horror that the car had no brakes. I ran straight into a passing Rolls Royce, which proved to be the property of the Marquis of Bath. MIR had to clear up the mess, material and legal. After this initial skirmish I was given another car which did have brakes. We were now able to commence our 'field testing' of various additives to the car's petrol.

The bombing of London had started. The family was already safely in the country and I was sleeping in a protected cupboard under the stairs during air raids. It was very frightening at first, but the body very quickly accustomed itself to the bangs and blasts. I remember well being alone and apprehensive the night the first bombs fell and the eerie quiet next morning. I had to deliver a paper to the War Office and crossing St James's Park to reach it I was amazed to find all the birds, ducks and pelicans fast asleep, having been kept awake all night by the noise of exploding bombs.

Working from South Kensington, the obvious 'field' for our test runs was the relatively calm and traffic-light-free system of roads within Hyde Park and Kensington Gardens. There was an excellent circuit entering from Exhibition Road, over the Serpentine Bridge, round to Marble Arch and Hyde Park Corner and westwards back to the starting point. We began quite well on the first test run with a tankful of doped fuel; the motor was running well and after a few circuits the difficulty of bringing a car to a halt when one wanted to began to sink in. Then a shrill whistle and we were stopped by the police who had been observing us and concluded that we were cruising around looking for girls to pick up. Our dignity and high sense of public duty were punctured and only partially restored when the police accepted our explanation. We continued our testing on subsequent days, with little success. The car seemed to enjoy every foul mixture we fed it. The Dunkirk evacuation was now taking place and we joined those who drove to the coast to retrieve the escaped service men. We filled up the car with petrol diluted with rubber solution, set off to Dover with the engine rejoicing and returned in the triumphant vehicle with a load of soldiers and a sense of humiliation.

Our approach was obviously wrong so we went back to the drawing board and I had what I thought was a bright idea. Nickel carbonyl, a liquid intermediate used in the recovery of nickel from its ore, if dissolved in small quantities in petrol should on combustion deposit a fine film of nickel on the spark plug and bring the engine to a halt. The only trouble about nickel carbonyl is that it is spontaneously flammable in air and oxidises to very poisonous carbon monoxide. I was aquainted with the research director of the Mond Nickel Company whose factory was in Swansea. I telephoned him and explained my idea. He encouraged me to try it and suggested that they had a technique by which the chemical could be sealed into small glass bubbles that would easily break when dropped into a petrol tank. I visited him at Swansea and was given a Winchester quart of nickel carbonyl to experiment with. The sample was in a wooden box, packed with active charcoal to adsorb the gas in case the bottle broke. I proceeded to the station to await the delayed London train. When it arrived, I found a compartment and sat with the precious sample on the floor between my knees – so far, so good. The train moved slowly towards Cardiff. There was an air raid in progress and we came to a stop between two gasometers. I could see and hear planes dropping bombs in the dock area a couple of miles away. Then a bomb fell somewhat nearer and the blast shattered a window in the train. I panicked. What should I do? If I was to sit still and the train was to be damaged, my bottle would probably explode and the other passengers be poisoned by carbon monoxide. Were I to throw it out of the window it would catch fire and I would certainly be lynched as a saboteur by my fellow travellers. Being a coward, I just sat still and hoped for the best. Soon the bombing stopped and the train continued. We finally reached Paddington not at our scheduled time of 10 p.m., but at 1.45 a.m. Another air raid was in progress, so I descended with my burden to a shelter and then, there being no underground trains or taxis, I had no alternative but to walk the few miles home, carrying my quite heavy 'bomb'. I decided the current raid was on a distant part of the city so set off through the unlit streets and wearily reached Hammersmith Bridge about 4.30 a.m. Clavering Avenue was almost in sight, but I had forgotten there would be an armed guard on the bridge. Of course a soldier stopped me and asked me to open my burden. There was no point in explaining that if I opened it, he would be either burnt or gassed; so I just refused. I spent the remainder of the night in custody until telephone calls released me in the morning. How I longed for the privacy of my cupboard bedroom.

We had plenty of other ploys. We were like a bunch of kids after their first chemistry lessons doing dramatic, if macabre, experiments. I became adept at making products which looked like ordinary aspirin tablets or cosmetic creams which, if introduced into chemical vats, would eat holes through them, and delayed action incendiaries of all kinds. More interesting was our work on decoy fires to be lit after a German pathfinder plane had dropped a flare near a suitable target to guide the bombers following. Our fires consisted mainly of piles of wood treated so that they would catch fire easily. But that was all too simple; our fires could easily be seen as the bonfires they were, something closer to the irregularities of a building set on fire by a bomb was required. I rigged up a pulley system that allowed an occasional milk carton filled with liquid oxygen to slide down a wire and empty on to the blaze, causing a sudden and intense burst of light. My proudest moment was when I extended this idea by throwing little packets of metal salts into the fire. One night in the Midlands copper salts were used. I was told that the next day Lord Haw Haw had boasted over the German radio, that a non-ferrous metal plant had been destroyed in the region, identified, no doubt by the green flashes of our illusion. Of course the Germans tumbled quickly to the decoy purpose of the fires and were certainly doing the same to deceive us. We were therefore soon busy devising simple spectroscopes to enable our pilots to distinguish between the real and the artificial.

This nonsensical phase passed for me quite quickly as the consequence of a telephone call from Sir Henry Tizard. In the critical years leading up to the war the Rector of Imperial College had been chairman of the Aeronautical Research Council, the paramount body responsible for ensuring the technical readiness of the RAF. Its work had ensured, for instance, that the Force possessed smaller numbers of Spitfires and Hurricanes rather than larger numbers of older, slower planes. It had also supported the development of Radar or Radio Direction Finding (RDF) and placed particular emphasis on microwave radar. The enemy was far behind on this and was unable to intercept it. When Chamberlain resigned, to be succeeded by Churchill, Tizard had to relinquish his post because of the new Prime Minister's determination to rely for scientific advice on Professor Lindeman of Oxford, later Lord Charwell. The relationship between Churchill and Lindeman was very close and his influence on wartime science was great; we referred to Downing Street as Unter den Lindeman! Tizard retired as Master of Magdalen College, Oxford. But he was involved in many matters on the side.

Tizard and Lindeman were bitter rivals and each had his coterie of supporters amongst the war scientists. In earlier days, both had been research students at Gottingen and apparently good friends who frequently played tennis together. Some unknown conflict had arisen between them and thereafter their mutual antipathy was intense. Tizard's enthusiasm for microwave radar was apparently not shared by Lindeman, but it was to prove critically important in permitting a numerically inferior RAF to win the Battle of Britain. Indeed after the war, C. P. Snow in his Godkin Lectures at Harvard, attempted to prove that had Churchill come to power six months earlier, Britain would have lost that battle.

In 1940 Churchill asked Tizard to undertake a special mission to the United States. Before Pearl Harbour, the Americans had very little advanced military research in progress and the British range of new products was well in advance of American weaponry. Churchill, believing America would be dragged into the war sooner or later, felt it was important to get them updating their military hardware, and Tizard was despatched to Washington with a letter of introduction to President Roosevelt and a 'black box' of British war technology. In fact, it was a black lacquered tin trunk containing British war secrets, including samples of the improved explosive RDX (Research Department eXplosive), radar circuits and the first microwave tubes, the new RAF bomb-sight, a proximity fuse that would explode a missile without it actually touching its target, and a carefully argued case that a successful development of the atom bomb was possible. Much later the black trunk with Tizard's name inscribed on it came into my possession. I intended to give it to the Imperial War Museum, but it was stolen from the attic of my house.

The mission was a success. Tizard found the President warm and sympathetic. He was introduced to the Chiefs of Staff of the three Services and to the heads of the great technology-based corporations including GM and GE. Arrangements were made for an immediate visit of a high level American team to Britain to see our war technology facilities at first hand and establish a system of day-to-day contacts. In fact the mission made it possible for US government and industry to institute a war technology development programme significantly before Pearl Harbour when they were forced into the war.

Tizard's telephone call reached me one evening at a moment when I was considering if the nightly raid was sufficiently near to force me under the stairs. He said, 'Do you know about my recent mission to the United States?' I answered, 'Yes, I've heard rumours about it.' He added that an American team, headed by James Conant, the President of Harvard would be landing at Southampton later that evening. 'I want you to take them round all the defence research and development installations in the country and see they meet all the right people. You should go to the Ministry of Supply at 10 tomorrow morning and get them to sign you up before the Americans arrive'. I was stunned. I protested my incompetence. 'I know nothing about the national war effort in science and technology. I have never entered the doors of a government department in my life and I don't even know where the Ministry of Supply is located.' He replied, 'you are a fool. Do you not see that if you take on this challenge to bear-lead them through all the experimental establishments, you will soon know more than anyone about the scope and totality of our national programmes?'

Reluctantly I was cajoled into agreeing and next morning duly turned up at the Ministry. I was greeted briefly by the Director General of Scientific Research, signed the Official Secrets Act, and was appointed Assistant Director of Scientific Research 7 (ADSR 7). I was quite proud to have a title. The arrangement was that I should devote 50 per cent of my time to this new task, leaving the rest for my normal work at College. I was allocated a bare room with a trestle table, a few chairs, an ominous looking filing cabinet, and a dour looking executive assistant, although I had no idea what that meant, and a secretary. Apparently these were the automatic entitlements of my new rank. I sat down disconsolately not knowing how to address the assistant and with no work for the secretary. However, at 10.30 my frustration was broken; a knock at the door and an American face peered round at me. This was Caroll Wilson, the secretary of the mission, who came to make preliminary arrangements for their visit. As fate decreed, I was to know him professionally and in friendship for the next 40 years until he died.

A few days later, once all the necessary arrangements had been made, I met the full mission. Jim Conant, with his drawn New England face, sharp eyes and quick reactions impressed me enormously; Warren Weaver, Director of the Rockefeller Foundation, was a quiet, friendly man who emanated academic concerns; and finally Larry Havsted who appeared just as I had expected, an agreeable entrepreneur with a lively innovative mind. It was a good team and they missed little. We started our round of visits which took us all over the country and the whole system was opened without reservations to them and me. The job was easy; it required a good background scientific knowledge, common sense and much physical stamina.

I began to have some idea of the scope of the Ministry's scientific research department and to meet its personalities. I took an instant dislike to the Director General, Dr Gough. He seemed a dictatorial, suspicious, narrow bureaucrat, uncommunicative because of his own insecurities. Quite often I had to dictate letters for his signature, in reply to correspondence from senior officials abroad. Years later, C. J. Mackenzie, the President of Canada's National Research Council, related to me that at the beginning of the war, he had a lengthy correspondence with Gough about establishing co-operation between the scientific efforts of Britain and Canada. Gough's replies were always terse, perfunctory and with little substance. All of this suddenly changed; Mackenzie began to receive friendly, 'come-hither' letters with suggestions for common activities. This change of tone dated from my arrival at the Ministry of Supply. Gough must have signed my letters without bothering to read them.

There were two Deputy Directors General; Dr Paris for physics and Dr Roffey for chemistry. Paris was a genial, constructive man, but my contacts with him were superficial. Dr Roffey was open, intelligent and warm and we quickly learnt to work together. The others were a mixed bag. On the physics side was

Sir John Cockroft, Nobel Prize-winning nuclear scientist from Cambridge, shy, taciturn and effective. Over the years our professional relationship gradually softened to friendship. I greatly respected his integrity, loyalty and intellect. Then there was the bomb disposal unit under the Earl of Suffolk. It was exciting as well as a little unnerving to mix with these intrepid men whose heroic work was so critical during the long months of the blitz.

Scientific missions from Canada, Australia and New Zealand arrived in London and I became their point of liaison. Then Basil Schonland, a great South African geophysicist, arrived in Brigadier's uniform and was added to my charges. By this time also scientists from European countries overrun by the Nazis began to appear in England, attached to governments in exile, and my contacts extended to the free French, Dutch, Poles and Norwegians. Scientists from the Dutch army set up a research unit in the same corridor as my own laboratory at Imperial College. One incident, which I had no inkling of at the time, was the smuggling to London of Niels Bohr masquerading as a Mr Baker.

The initial policy mission of the Americans was soon replaced by a larger, more permanent group of specialists who showered us with questions and requests for introductions daily. I took on as an assistant, Dr Wilfrid Mann, a physics colleague from the College, of whom more later. Apart from simple liaison work, we were able to help in many discrete ways. For instance, we forged a relationship between scientists of the American mission who represented the new research structures of Washington with their compatriots of the American Army Chemical Corps, a relationship which barely existed at home.

Then came another turn of fate. There was considerable concern in Parliament that co-operation between the army, navy and air force was insufficient, each tending to go its own way. A story current at the time illustrates public perception: a senior naval officer, with urgent need to deliver an important message to the Admiralty, but ignorant of London's geography, rushed uncertainly down Whitehall. Encountering a red-tabbed army staff officer, he stopped and asked, 'Can you tell me which side the Admiralty is on?' The latter replied laconically, 'Well in the last war I think it was on our side. This time I'm not so sure.' The army's materiel was provided through the Ministry of

Supply, the air force by the Ministry of Aircraft Production and of the navy by the Admiralty. This seemed to many a recipe for duplication and waste. The problem was particularly evident with the scientific and technical components of service requirements; radar was being developed by all three services simultaneously. While the requirements of all three were necessarily different, there was obviously much in common. Professor Blacket, navy chief scientist, made a strong plea for a more integrated approach, but Churchill favoured continuation of the *status quo* and resisted all attempts to introduce an integrated defence procurement system. However, pressure to do so became very strong and the situation was politically sensitive so finally a Ministry of Production was established, if not to integrate, at least to harmonise the needs of the three services,

In the new system the needs of science and technology were met by the appointment of advisers, Sir Thomas Merton for physics, Sir William Stanier for engineering and my friend Sir Ian Heilbron for chemistry. Each was to have a younger deputy and to form a team of six, able to cover a wide field of requirements. A naval commander was added to keep them in administrative order. Heilbron invited me to be his deputy. In practice I took with me nearly all my previous functions from the Ministry of Supply that continued to occupy a considerable part of my time. Officially I was still on a 50 per cent basis, but my work at the College became less and less. The seven of us met daily and worked happily together, but each tended to have individual projects and interests.

The new system was doomed to failure. The vested interests of the three services and their procurement agencies were so powerful and rooted in history that they simply ignored us, combining to resist any interference from a small naive group implanted by Parliament against their wishes. The senior advisers soon recognised this and referred to themselves as the three blind mice. We had long discussions about our dilemma and decided that we should carve out a role for ourselves, namely to look at *new* problems with scientific content as they arose. These might be of secondary importance and certainly not what we were appointed for, but it could be a useful service.

Initially we had meetings with the Ministers each Monday morning in the presence of the Permanent Secretary (head civil servant), John Henry Woods. These were very formal, but I learnt much about the ignorance of Ministers and the futility of scientific advisers in the absence of a common language. The Minister would outline a problem to us, thinking we might proffer advice. Heilbron tended to respond spontaneously without reflection, at times putting forward some bright, if simplistic ideas. The Minister would then order Woods to follow up the suggestions and report to the next meeting, by which time Heilbron, who had probably forgotten his advice of the previous week, would often say that he had a much better idea this time and new enquiries would be started. The Minister understandably grew tired of these charades and soon lost interest in us. I quickly came to question the value of outstanding academics as scientific advisers with their perfectionist instincts, but my affection for Sir Ian grew and I spent much time with him and his charming wife. But much more about scientific advisers later.

We did some useful work in the reorganisation of the camouflage experimental station at Leamington Spa where we found that while artistic designs were exceptionally good, insufficient attention had been paid to the ease of detection by infrared photography. Meanwhile my international activities were taking up most of my time and made my new position more effective, amongst other tasks I was responsible for staffing the British scientific mission in Washington set up following the Tizard mission. I appointed as its head Tommy Taylor of Brazenose College Oxford.

Then a new and serious problem came to my attention. The Japanese had just entered the war and there would obviously be fighting in tropical environments where mosquitoes and other insects would be a major menace to soldiers on both sides. The Surgeon-General's Office of the army discovered that the only insecticide available to us was pyrethrum, imported mainly from Japan. Here was a problem waiting for us to tackle that was not in anybody else's *chasse gardé*.

My War Against Insects

From my earliest days insects have been my enemies. In the 1920s I could not enter a railway carriage without being bitten by fleas. During the summer of 1921 after we moved to London, I was taken for a weekend in the country. As soon as I entered the train, fleas jumped on to me and, by the time we reached our destination, I was massively infested with my body covered with itching sores. Once settled in our lodgings, my parents took me into a field, stripped me naked and removed the fleas from my garments, freeing me until the train journey home. In later years, fleas seemed to abandon train journeys and mosquitoes replaced them as my special enemy, particularly when canoeing holidays entailed camping on the banks of rivers. This caused many wretched nights, striving to avoid scratching the bites.

The mosquito menace reached its zenith during Sally's and my Lapland visit in 1935. The insects were large, vicious and innumerable. They flew in clouds so dense as to shade the sun and they were desperate, lusting for blood in order to reproduce – with little animal blood around apart from lemmings and us. We knew of this danger in advance and wore puttees to protect our legs and gauze veils our faces, but they still harried us. The only insect repellent at the time was citronella and we doused ourselves with it, but the mosquitoes did not understand they were supposed to avoid it. The worst moment of each day was pitching our tent, designed to be impenetrable to them. However, while erecting it, we inevitably trapped a few hundred. I remember one evening slaughtering the memorable number of 365 before we could hope to settle for the sun-lit night. I can understand how these flying fiends could drive an unprotected person mad. Fortunately the arctic ones are not disease carriers like those that gave me dengue in Greece. I tended to suffer much more than Sally, whose blood I expected to be sweeter than mine. Towards the end of this trip I must have developed some small immunity against the bites, although nothing like that of the inhabitants of the Lapp villages we reached as we left the wilderness who seemed not even to notice them. That year in my morbid interest concerning insect foes, I read *Rats, Lice and History* by H. Zinnser, which attempted to show that, throughout history, insect-borne diseases in wars had always killed more people than the enemy, mainly through typhus transmitted by the lice on rats, including in World War I. When war broke out in 1939 then, I was aware that insect control was likely to be important, especially if it spread to tropical areas where the anopheline mosquito would become an ally of the louse.

When Japan suddenly declared war against us, we were faced with a sorry situation. The main insecticide to which we could safely expose men was the pyrethrum mainly imported from Japan; not very effective against mosquitoes, yet even before Japan's entry the mosquito had already begun to strike. A convoy of ships coming out of Lagos had many crew members stricken with malaria, causing the enfeebled convoy to be an easy target for German submarines. While intensive trench warfare seemed unlikely, a land war in Europe, with large masses of men was still vulnerable to typhus. There were, in addition, several other needs for insecticides during the war, including home agriculture to reduce reliance on imported foods.

We therefore decided, immediately we heard about the lack of insecticides, to set up a research programme into the mosquito and louse, in the hope of finding some improved insecticide and to develop repellents. My hostility against insects greatly fanned my enthusiasm. Professor Buxton of the London School of Hygiene and Tropical Medicine took responsibility for the breeding of lice and testing the effect on them of various compounds we would supply. For the mosquito component we unearthed Sir Rickard Christopher, retired to Cambridge after a lifetime of malaria control in India. He was an amiable eccentric, a man in his mid-eighties who apparently bred anopheline mosquitoes as a hobby, feeding them daily with his own blood. He was glad to sacrifice some of that blood for this patriotic task. We also formed a programme committee of entomologists and officials, under Sir Edward Appleton, to oversee the work.

Some new candidate insecticides were already under development by Shell and ICI, but they lacked the persistence of action we required. Some progress was made by the use of adjuvants to pyrethrum, but this was still insufficient. We tested many chemicals, amongst them a series of chlorinated hydrocarbons synthesised for us in Heilbron's laboratory. Several of these proved excellent insect killers, but were equally lethal against humans. Then one day a letter arrived on my desk, intercepted by the censor in Liverpool. The Geigy Company of Basel had sent it to its subsidiary in Manchester, announcing a new patent for a compound that had proved excellent for the mothproofing of furs. The letter noted that copies had been sent to Geigy New York and Frankfurt. The chemical had been given the trade name Neocid, but its generic name was also mentioned; dichlor-diethyl-trichlorethane. Here then was a chlorinated hydrocarbon, apparently safe enough to use on furs. It was potentially what we were looking for. We had a few grams synthesised in Heilbron's laboratory and sent to the two testing centres. It was immediately clear that it was an exceedingly effective insecticide against both the mosquito and louse and ad great persistence; in fact, just those properties we were seeking. The generic name was far too long for common communication so I coined the abbreviation DDT, the adoption of these three letters into common usage worldwide is, no doubt, my greatest claim to fame as well as infamy. I have often been grateful to the woman in the censor's office who diverted this letter so relevantly to my desk, wondering if she realised the importance of her action.

A sample of DDT was sent to Professor Cameron at University College Hospital for toxicological testing. To our delighted surprise it passed as fit for use by humans. Government representatives were sent to Manchester, to announce to the Geigy people news of their own discovery, and direct them to set up a pilot plant to deliver a ton of DDT in one month. This was duly accomplished and DDT found its first operational use within three months of the censored letter reaching my desk. Emulsions of a DDT solution in oil were used in washing army shirts in field laundries in Europe. The British, American and Canadian troops in the field were practically lice-free, while the German forces who missed DDT were heavily infested.

We undertook other developments aimed at protecting troops on tropical fronts often forced into 'foxholes'. For this we made DDT sprays, operated by compressed carbon dioxide 'sparklets' then commonly used for the gasification of soda water. At the end of the war we were informed by Eastern Command that DDT had saved them a huge number of casualties. There was also considerable progress in the identification of new insect repellents. I felt that I had my revenge for the insect scourge of my youth. Strangely enough I have not suffered from insects since then, although frequently exposed to them.

Having consolidated this work, it was decided I should go to America to describe it to them and discuss how we could co-operate in malaria protection in the Eastern Theatres. I travelled on the Queen Mary and reached New York by a circuitous route to dodge submarines. The excitement of passing the Statue of Liberty and finally witnessing the famous skyline was much greater than I had anticipated; landing on American soil, I surprised myself by converting it into a symbolic ceremonial, though resisting the impulse to kneel down to kiss the ground. Tommy Taylor was at the dockside to meet me and took me to lunch at a restaurant overlooking the ice rink at Rockefeller Center. The peaceful scene of the elegantly costumed skaters cavorting at our feet, the leisurely diners and the more than ample portions of well-presented food was too much for me arriving from the drabness and the rationed meals of London. My stomach, long adapted to austerity could not take the richness of the food and, for the first time in my life, I fainted.

The next morning we went by train to Washington and I checked in at the British scientific mission, formally known as the British Central Scientific Office (BCSO). I soon discovered that the Americans already knew about DDT, having intercepted the Geigy letter and had worked on lines quite similar to our own.

After a series of fruitful discussions with the experts of the U.S. Bureau of Entomology, I flew to Orlando, Florida, to visit the experimental station carrying out their DDT experiments. I participated in measuring the mosquito larva populations of various streams and pools, and the wading in alligator infested swamps. I also witnessed their method of feeding their experimental lice, which lived burrowed in felt mats in a kind of large Petri dish covered with a bell jar. Their food, in this case human blood, was presented in the form of robust, but poor, young African-American men who lay naked, outstretched on beds, face down. The glass containers were brought in and the squares of felt placed on the men's backs. A few moments later the mats were removed to uncover the swarms of puce coloured lice clinging to the human flesh, some five minutes later the colour of the lice had changed to bright red. The felt was reapplied and after a few moments, the lice were back in the felt and returned to their glass domes. The men seemed sleepy but were clearly not uncomfortable. Their backs were swabbed with an analgesic, and they were given a couple of dollar bills.

Returning to Washington, another unexpected turn of fate intervened. Taylor had just been informed of his transfer as science adviser to High Command in South East Asia. I was instructed to take charge of BCSO until they found a successor. They never did. My mission to Washington, expected to last a couple of weeks, in reality extended to several years.

While in Washington, I was busy with other things and had little to do with insects. However, as a postscript to the DDT story I should relate an unexpected incident towards the end of my stay in America that led to the first planned use of DDT in post-war society. The Colonial Office was considering the future development of the British Caribbean Colonies and decided to send a mission to investigate the resource potential and scientific problems of the West Indian territories. The members of this mission, Sir Robert Robinson, the President of the Royal Society, and Professor Simonsen, chairman of the Colonial Products Research Council, duly checked in at BCSO on their way South and I was co-opted as a member of their team and invited to travel with them. In the course of this mission we found ourselves unexpectedly stranded in Port au Prince in Haiti at the beginning of a weekend. We wondered how we could usefully use the next two days. I had a primitive interest in British Guyana, mainly, I think because of a postage stamp I had collected as a small boy depicting the Kaieteur Falls, the highest waterfall in the world, which had appealed to my imagination. Guyana was not in our itinerary, but I suggested to my colleagues that it might be an interesting place to visit for our enquiry, and they agreed. We found the means to reach it and telegraphed the Governor of the Colony to ask if he would welcome a visit from us. An affirmative reply came immediately and we set off by plane via Trinidad to Georgetown. On our way we flew down the coast of Venezuela, passing the mouth of the great Orinoco River. There I observed a curious phenomenon. Suspended silt was precipitated from the river where it met the salt sea. This was to be expected; what was surprising was that it did not appear as a single whitish band at the interface, but rather as a series of white lines at logarithmically spaced distances from each other. I realised, with some excitement that I was witnessing a demonstration of the Liesigang phenomenon that I had been investigating not so long ago with test tubes. I assumed it was a rare observation, made possible by unusually calm sea conditions on that particular day. Later I had a letter published in the journal *Nature* describing the phenomenon.

At dinner in Government House later, conversation turned to the malarial problems of the colony, and I mentioned the role of DDT in the war in the East. The Governor pricked up his ears when I mentioned how the chemical retained its insecticidal properties so persistently. He explained that in Guyana, which was suffering much from malaria, the infected anophelenes were domesticated; they lived in the homes of people, while the wild mosquitoes of the forest were free of the parasite. If this were really true, treatment of doors and walls with DDT-treated whitewash should keep the houses permanently free of malaria. He arranged, there and then, for me to meet the malariologists and medical officers of the Colony at 6.30 the next morning to put the case before them. This duly took place and he then cabled the Colonial Office in London asking that Guyana be accepted for the first post-war tests of DDT for civilian malaria control. Later that morning we left Georgetown to continue our mission and, on return to the United States, I arranged for a team from Columbia University to undertake a pilot experiment in Guyana. I never got within 100 miles of the Kaieteur Falls!

Several years later the telephone rang as I was sitting at my desk in London. The caller asked if I still remembered my weekend in Guyana. He was Dr Gigioli, the malariologist I had talked to at that early-morning meeting years before. I invited him to lunch, where he described to me how the follow-up had been successful in virtually eliminating malaria from the villages, saving lives and greatly increasing the health and vigour of the inhabitants. As a consequence there had been a dramatic increase in the birth rate and hence in population. I related this to Sally that evening. She looked quizzically at me and said, ‘not bad for one weekend’.

On reflection I felt rather disturbed by the Guyana incident. By this time, *Silent Spring* had appeared and DDT was beginning to be seen as an environmental danger, finding its way into the food chain, yet I had inadvertently helped to bring into the world many people likely to have lives of poverty and misery. DDT had proved powerful for both good and evil, as is true for so many scientific discoveries. Was the arrogance of the scientist encouraging me, as so many others, to play God? In any case it would have little significance cosmically; the insect was beginning to change genetically to become immune to DDT and malaria would rise again and be more difficult to fight; nature would regain its balance.

12

The Science Councillor

So here I was, sitting at a desk in Washington, in an 'acting' position, with no idea what fate awaited me, and Sally and the children in Northumberland 3,000 miles away. It was relatively easy for me to take up the challenge of the British Central Scientific Office (BCSO); I had been associated with it from its beginning, been at the receiving end in London and appointed its present staff. With the chaos in London, I knew the authorities would not find it easy to recruit a successor more appropriate than me and hence, unless I made a mess of things, it was unlikely they would even try. I decided that I would probably be in Washington for some time and should take the job seriously, and try to increase the usefulness and prestige of the office. In reality the question of my tenure was never raised again and the adjective 'acting' was dropped.

Our function was to collect as much information as possible relevant to the war effort, provide early signals to London of new US developments and policy changes and channel them to appropriate individuals or government departments at home. We had also to transmit British scientific data and information to the Americans, forward their enquiries and act as a Washington base for the continuous flow through the Office of British experts. Then there was an occasional special task of identifying appropriate US universities or firms, to undertake research or development projects for which our bombed facilities at home were inadequate.

The organisation of the small office was quite complicated. I had a small nucleus staff with me; Wilfrid Mann, my old friend from Imperial College days, as deputy, two chemists and a back-up group of American and Canadian women responsible for the library, travel arrangements and the like. Wilfrid was particularly concerned with American research on proximity fuses. Then there were a number of independent individuals using our common services, sent out by ministries in London. One of these was a metallurgist, another an expert on chemical warfare and a third on construction techniques. Finally, there was an atomic energy unit, kept apart for security reasons, under Sir John Chadwick, the discoverer of the neutron who acted as liaison with the Manhattan Project. A stream of British nuclear scientists passed through this unit on their way to join the Los Alamos centre or the Canadian Montreal Group.

My first job was to establish relations with the large number of US bodies engaged in war research. While we were permitted to approach US services, agencies and laboratories directly, our main contact was with the new central research structure built up for the war emergency, the Office of Scientific Research and Development (OSRD). This body attempted to bring together and utilise the scientific manpower of the nation, whether in universities, independent institutes or industry. One of my first calls was on its director, Dr Vannevar Bush. Immediately after Anglo-American co-operation had been established by Tizard, a Canadian, W. H. Webster had set up BCSO very competently, but the British authorities had assumed it would be appropriate to appoint some eminent scientific figure to head the Office, Consequently, Sir Charles Darwin had been sent out and later Sir Alfred Egerton, Foreign Secretary of the Royal Society. They did not relish the somewhat pedestrian work of running the office, dealing with the flow of highly specific requests from London. They therefore tended to indulge in high-level talks with their US peers. When Taylor later took charge, Bush and his colleague were relieved that the level was now about right and their time not wasted with courtesy meetings. He therefore welcomed me warmly, knowing I was unlikely to waste his time. Bush was a warm, lively, pawky person with rugged New England features. He had a delightful, sharp wit and was very direct and uncompromising. On our first meeting, I sat opposite him at his large desk. He asked me to turn round and look at the wall behind me, where there was a black-framed mock heraldic coat of arms with a retort superimposed on a benzene ring, with a scroll beneath, with the words *illigitimus non carborundum*. He explained that he had to spend too much time talking to senators, congressmen and other pretentious people, and the emblem comforted him. It reminded him, 'don't let the bastards wear you down.' I had continuing good relations with him and worked on a day-to-day basis with his two assistants, Carrol Wilson and Caryl Haskins, later to become head of the Carnegie Foundation. After the war I had continued dealings with Bush, concerning his hobbyhorses, the alga *chlorella* as a possible foodstuff, and the Bush Rapid Selector, first of the new technologies for mechanical selection of information.

Fresh from my London experience of scientific co-operation between countries of the Commonwealth, I began to enquire about the activities of the scientific representatives of these countries here. The situation of Canada as an adjacent country was obviously one of close and intimate contacts with Washington. The British had located a former colleague of mine from Imperial College at the National Reseach Council (NRC) in Ottawa, the Nobel laureate Professor G. P. Thomson, the nuclear physicist, an appropriate appointment in view of both British participation in the Montreal project and our interest in the Canadian atomic energy programme. Thomson's presence in Ottawa became less necessary after the arrival of Chadwick in Washington so he returned to England. Soon after my arrival I went to Ottawa to meet Dean Mackenzie at the NRC with whom I had earlier had prolonged correspondence (unwittingly signed by Dr Gough). It was arranged I should visit Ottawa for a few days every six weeks and also have a formal link with the British High Commissioner's Office. Mackenzie and I became firm friends, a relationship that lasted until his death in the 1960s.

In Washington I discovered that Australia and New Zealand had established small scientific liaison offices, while two first-class scientists had recently arrived at the South African Embassy. All these scientists, in addition to us at BCSO were asking the Americans questions – often the same questions. I came to the conclusion that it would benefit everyone if we could all be located together and our name BCSO change from British *Central* to British *Commonwealth* Scientific Office. Not only would this simplify life for the Americans, but also provide a wider range of scientific skills and disciplines than the current system. There were some difficulties, especially on security, since, apart from Canada, the Commonwealth countries were not privy to the atomic programme, but this could be overcome by internal security within BCSO, isolating Chadwick's unit.

I discussed this plan with all the ambassadors concerned and with Caryl Haskins in Bush's office and found easy consent. I sent London a blow-by-blow account of all these negotiations, but never had any reaction to my cables. I even found offices to accommodate the new organisation within the OSRD building at Dupont Circle – but still no response. We were all set to go, but still no reaction from London. After weeks of frustration, during which the Commonwealth scientists were gradually losing faith in my ability to implement the plan, I finally sent an ultimatum to London informing them that I had taken accommodation and that the Commonwealth Office would start work on the first day of the next month unless they prohibited it. This was a bombshell. Clearly, my London recipients had not even bothered to read any of my many cables. Their reaction now was (and I paraphrase), 'What the Hell do you think you are doing? Come back immediately to explain.'

This was extremely awkward. I had just made arrangements for Sally and the girls to join me in Washington. In fact, I only received the London rebuff as I was on the way to New York to meet their ship's arrival. I quickly arranged to fly back to London a few days later, and arrived in New York just in time to welcome them to the New World. Washington was grossly overcrowded. I had searched for days for accommodation for the family, but in vain. Finally, with the help of Wilfrid Mann, we found a small floor space in a wretched lodging house where we could doss down for a few nights. Fortunately a family friend who was living in Washington offered to take in Sally and one of the girls and persuaded two of her neighbours to take one each of the others while I was away. I then flew back to London in great trepidation, both professionally and personally. Actually it proved to be quite easy; Appleton, Heilbron and even the Cabinet Secretariat saw sense in my plan when it was fully understood. It would even save money! So I returned in triumph. The cat had turned round again.

The British Commonwealth Scientific Office came into operation as planned. It was extremely successful. The presence of the

Commonwealth scientists in no way jeopardised the scope or security of my UK activities, nor did the UK presence inhibit the independence of the others.

I must record some details about our difficulties in that first year of family life in Washington. On my return from England I found the children very happy, each had been easily incorporated into the lives of their various host families and remained with them for several weeks until, with the help of the embassy, we found a suitable house to rent. But at this point the financial realities struck. When in the Ministry in London I had still been on half-time pay from Imperial College and that, of course, ceased when I left. In the excitement of my new work and new environment I had not bothered to question the adequacy of my part-time government salary. Now I was left with only £1,000 per annum. That was, alas, only part of the difficulty. Some local decision by the treasury decreed that British salaries in Washington of less than £1,000 were tax-free while those above that level were normally taxed. This was fair for the locally recruited office assistants, but for me it meant that my take-home pay was less than that of my secretary and had to cover entertainment, which, although modest, was a necessary part of my job. Not yet having diplomatic status, I could not even claim duty-free liquor so necessary for Washington hospitality. I protested immediately my need for a salary appropriate to my full-time work and asked that in the meantime they reduce my pay to £999 to provide tax relief until we agreed a permanent settlement of my difficulties. This they denounced as 'cheating' and we began negotiating on the salary as such. However, with the difficulties in London in the buzz-bomb period, this took many months; we were quite desperate. The only solution was to take in lodgers. Two 'singles' from the British missions came to stay with us. Sally was very supportive even to the extent of preparing breakfast for them each morning. It was nearly a year before I received a salary reasonable for my status plus diplomatic privileges, being by then appointed Scientific Councillor at the Embassy.

Meanwhile my professional life became ever more hectic as contacts grew, as did demands for information from England. Our domestic life flourished and we quickly became part of the local community with many American friends. One local contact proved to be professionally useful in later years. Paul Sifton, the father of the family that had taken in Jane when they first arrived, was the Washington representative of the Union of Automobile Workers. Through him I met many of the American labour leaders. I became a member of the Cosmos Club where I met many American personalities and took part in the meetings of a small 'brains trust', a group of a dozen people, including my friend Carrol Wilson and also Buckminster Fuller. We lunched together weekly, excitedly discussing all sorts of current and future issues.

We spent many weekends with our American scientific colleagues and enjoyed excursions in the nearby Virginia countryside. One Sunday we took part in a reading of *The Tempest* on the summit of Old Ragg, one of the peaks of the Blue Ridge Mountains. We roamed the Virginia countryside and even within the District there was much to explore. On our first Christmas in Washington, we went for a long walk to prepare ourselves for dinner. This was in the snow-covered Rock Creek Park, an extensive wild tract stretching out from the centre of the city. It was a beautiful day with sunlight on the snow, but the park was completely empty; we saw no other walkers in the course of a couple of miles. We were commenting on the American reluctance to walk when we spied a couple of people approaching from the distance. 'They must be British', said Sally, and sure enough it was the British Ambassador, Lord Halifax and his wife.

The children quickly became assimilated as Little Americans. They adopted the accents of their friends when at school, but remained bilingual, speaking British English at home. When Roosevelt died, Catherine was chosen to give the school obituary speech, eulogising 'our President'. I asked the headmistress why they had selected a little English girl; was it not inappropriate. She replied, 'but she has such a charming accent'. Jane attended Junior High School, soon lost the diffidence of the small English child, but rapidly fell back academically to the level of the American school. She spent summers at a camp in Maine. As the war in Europe came to an end and that in Japan continued, there was considerable anti-British feeling. Sally was once turned out of a taxi when the driver heard her accent, mainly because the American public felt we were celebrating victory too exuberantly, while they were still fighting the Japanese and suffering many casualties. This sentiment reached a head with American criticism of Churchill's actions in Greece. They felt he was being soft to the communists.

Throughout my stay in Washington I frequently travelled all over North America, visiting government laboratories, universities and industrial firms. The flow of UK and Commonwealth visitors using BCSO facilities never ceased; indeed, as the war neared its end they increased and diversified as attention veered to post-war concerns. Alexander Fleming was sent to talk on penicillin, and J. M. Keynes turned up with his ballerina wife. Even Professor Lindeman arrived for tripartite nuclear talks chaired by General Marshall that I attended. Robert Robinson, President of the Royal Society, spent several agreeable days with us and also Appleton, seeking US experience in relation to his planning the future of the Department of Scientific and Industrial Research – plus many others with complicated needs for contacts and travel. Our work in the office changed noticeably with the end of war in Europe. There was useful exchange of information on the tropicalisation of instruments and equipment and on miniaturisation and other subjects of post-war significance. We also did our best to assess US intentions with regard to the atom bomb.

When finally Hiroshima took place, we had to report on the attitudes of specialists and the American public, and later I was sent to meetings of the Atomic Energy Committee of the United Nations at Lake Success, a very boring chore, since it was mainly to listen to interminable speeches on procedural matters by Gromyko, the Soviet Foreign Minister. Simultaneous interpretation had not yet arrived, so consecutive reiteration of sterile argument in several languages made listening intolerable. The war over, I was anxious about what I should be doing next and keen to get back to London, but had to agree to stay on in Washington for some time for post-war planning which was going on nationally and internationally. This proved an interesting phase with a great variety of tasks, and during this period I joined Sir Robert Robinson's mission to the Caribbean and did a worldwide tour of consultation and preparation for a Commonwealth Science conference to be held the next year, all discussed in other chapters.

The government had invited a group of leading scientists from India to visit British research centres and government laboratories in preparation for India's independence. They also wanted to give them the opportunity to discuss the organisation of science and technology in North America and I was asked to receive them and make arrangements for their visits there. The leader of the Indians was Sir Shanti Bhatnagar, with whom I had published a paper on magnetochemistry in my Imperial College days. The group had been feted in London and dined with the King. In Washington their reception was very low key and they felt a bit deflated. Nevertheless they visited many research centres and were given information about US post-war plans for science, then far from complete. In their discussion with American scientists, the Indians felt that something of great importance was imminent, but being withheld from them. This was just before the bomb was unleashed on Hiroshima, a time when atomic plans were particularly closely guarded. However, from shreds of information, hints and silences, the Indian scientists put two and two together and before they left they were well aware of the situation. Returning to Washington, the group felt unwanted; they sensed the colour prejudice and, as it was now autumn, they began to shiver. On their last day in Washington, Bhatnagar said to me, 'I feel the cold, but I'm damned if I will buy an overcoat here. I will wait till tomorrow to do so when I am back in the Empire', and he departed for Canada. We had a warm relationship with the Indians, most of whom I met again later.

Then there was a visit from Joseph Needham, the endocrinologist and student of the history of Chinese science. He had been stationed in Nanking for the British Council and was now working for the preparatory commission trying to set up the future UNESCO. At this stage the organisation was to be for Education and Culture and the Americans suggested provisionally that it be called UNECO. Needham was convinced, as was I, that S for science had to be included in its title. We visited a number of American agencies with influence on the commission, to try to persuade them. We were successful, but perhaps not for the right reason. UNECO was apparently the name of a well-known biscuit so an undignified UNECO gave way to the anodyne UNESCO.

In Washington the Americans were preoccupied with the need

for a reorganisation of their scientific research structures. OSRD had been a great success in spearheading the technology that had won the war and an equivalent system might be important for civilian needs. They realised that a high proportion of the scientific innovations leading to critically important hardware and strategic developments had originated in Europe. American efforts in basic research were relatively feeble in contrast to their speed and success in technological development. Vannevar Bush wrote a report that is now a classic, *Science the Endless Frontier*, foreseeing the impact new discoveries would have on both economy and society and pleading for adequate finances and infrastructure. This led to a proposal to create a National Science Foundation that immediately ran into a maze of difficulties due to the near impossibility of reconciling the interests of the large numbers of departments, agencies and institutes concerned with various elements of the total scientific effort, with deep disagreement about how it should fit into government structure. A bill creating the Foundation was passed by Congress, but vetoed by President Truman.

As the British were also in the middle of discussions on post-war structures for science and were anxious to understand the American attitude, I was instructed to follow the debate and, indeed, took part in it to some extent. Argument concerning the National Science Foundation seemed interminable so interim measures were taken, on the initiative of a very enterprising Naval Captain Conrad with the support of Admiral Fuhrer, whereby research projects were temporarily financed by the Bureau of Naval Research.

BCSO activities were increasingly of the type normally dealt with by embassies and my day-to-day relationship with members of the staff of the British Embassy grew immensely, but I began to consider future employment: back to the laboratory or to some quite new activity? I was by now well known in government circles and the scientific community. There were two possible Chairs of Chemistry in Britain for which I was supported. I was approached as a possible Vice-Chancellor of the University of New Zealand, and I was invited to be Secretary of the Royal Society, an honour I declined, suggesting an alternative who was accepted. There would be plenty of other opportunities, I thought. Then Sir Henry Tizard appeared on the scene once more and resolved my uncertainty. He had been appointed Chief Science Adviser to the new Labour Government under Attlee, responsible for both defence and civil science. He asked me to be his deputy on the civil side. I accepted immediately out of respect and loyalty to a man I so much admired.

13

The End of Petroushka and Other Sinister Matters

This chapter relates a series of events that took place in different periods of my life and is conveniently placed here at the point of return to London with no Petroushka in the front garden of Clavering Avenue or elsewhere. Petroushka had aged rapidly becoming very frail after a long and adventurous life. She had paddled the lakes and rivers of Europe, shot rapids, sailed across the Baltic, capsized ignominiously in the Thames off Wapping Old Stairs, visited Jan Mayen and crossed Lake Ladoga to visit the holy island of Valamo. Until the outbreak of war, we had left Petroushka lying peacefully in the front garden with her double stowed snugly in the loft.

One day, fairly early in the war the Admiralty telephoned me to say that Naval Intelligence had informed them I had a kayak they wished to requisition. Passers by would have seen a canoe in the garden so the call did not surprise me. I had to comply and it was with great sorrow that I sacrificed Petroushka to a higher cause. However, hardly a week passed before they telephoned me again, politely thanking me and adding that they understood I had a second canoe in better shape hidden in my loft and they wanted that too. I was indignant. This was too much; my privacy was being violated. It was no use. They took away Petroushka II in triumph. There had been a hint of this invasion of my private life before. When we returned from our exploration of the Arctic Coast of Finland in 1935, I had been telephoned and invited to give as much detail as possible of that terrain. How did 'they' know that I, a private citizen had been there? Then again, in 1938 when we were on Jan Mayen, a German tourist liner had approached the coast. This was part of Hitler's *Kraft durch*

Freude scheme. A group of men dressed as civilians, but of marked military bearing, came ashore. They were surprised to find the remote island in the occupation of a number of young Britons. They asked many very silly questions about the island before returning to the ship, but we blessed them because they left us a crate of fresh lettuce, a welcome addition to our monotonous diet. On my return home, Naval Intelligence invited me to visit them and tell them what the Germans had done and asked us that day. I can only conclude that we had an agent on board ship. Of course I responded willingly, but years later I was irritated to think that this train of events may have led to the seizure of a canoe from my loft.

What the eventual fate of my two Petroushkas was I shall never know. However, an event I heard about on the radio may give a clue. A number of German ships had been sunk in the harbour of Bordeaux by limpet mines stuck to their hulls in the dark by Allied saboteurs in kayaks. This may be irrelevant, but I fervently hope that the Petroushkas had made the supreme sacrifice helping to win the war against Hitler.

After the war I again suffered an uncomfortable feeling I was being watched, but why and by whom I could not fathom. One evening, on returning to Clavering Avenue from the theatre, we found a large photograph of a wedding party, staring at us from the centre of the mantelpiece. We had never seen it before and recognized no one in it. The children were away from home and we had left the house locked. Some stranger had broken in, placed the mysterious photograph in position and left, leaving no trace. Then, not long after, there was a still more baffling incident. I was about to go to the United States on a mission that was to last ten days. Before departure I deposited my keyring and driving licence in a drawer of my desk upstairs. When I returned, Sally mentioned that our bank manager had telephoned, asking me to call in to retrieve my keys and driving licence which had been handed in at the bank. I said, 'that's impossible. I know exactly where I left them in a drawer upstairs.' I dashed up and confirmed they had disappeared. I understood why the bank was involved, as a matter of of convenience I had given it as my address in applying for driving licence renewal, but how had it got to the bank? The next day I asked the manager how he came to have them. He explained they had been handed in by a gentleman, speaking with a foreign accent, wearing a black overcoat with an astrakhan collar who said he had found them in his coat pocket in Claridges's cloakroom. What did all this mean? How did 'they' enter the house, go upstairs and leave without Sally knowing? Why did they purloin only two articles of no value? And who were 'they'? I had absolutely no information of security importance in the house to attract foreign agents.

I came to the conclusion the break in must have been done by British intelligence, intended as a warning that I was closely observed. But why? I had been aware of the existence of the British and American atomic energy programmes, but only in quite general terms. I had some understanding of atomic politics, but no knowledge of technical detail. In any case all my 'secrets' were already obsolete. There had never been any question of my loyalty or reliability. Nevertheless, in a casual conversation between the British Ambassador in Moscow and a friend of mine, my name had come up and apparently the Ambassador had said that I had communist leanings, which I certainly never had. Years later, I was able to confront the Ambassador with this story but he denied it. I obviously had a few enemies, but never quite figured out who they were.

I continue from these slightly sinister events to another disagreeable area, namely my contacts with some of the 'atomic traitors'. As I became integrated into the embassy staff in Washington, I naturally became acquainted with the Head of Chancellery, Donald Maclean. As a colleague, I participated with him in numerous meetings with the Ambassador. On one occasion, at a meeting with General Marshall on the tripartite atomic strategy, I sat behind the Ambassador as his science adviser, alongside Donald as political adviser. I had absolutely no hint he had any loyalties other than those of his job, nor did I sense anything of his troubled domestic life nor his problem with alcohol. We even had a slight social relationship with Donald and his wife and dined with them a couple of times. They both seemed normal, cheerful people. On one occasion he invited us home to meet his mother who was in Washington on a visit. Her husband had been a close colleague in the British cabinet with Sally's uncle Walter Runciman, so we felt we had some background in common with the Macleans. I was utterly shocked when I read of his defection to Moscow in 1951. I had run into him by chance on the steps of the Athenaeum a few weeks before his disappearance and he seemed to me as urbane and unruffled as ever. In later years, when I was frequently in Moscow, I hoped I might run into Donald on the street, but never did. I feel sure that his motivation was idealism transmuted into ideology. One can sympathise with those who decry the corruption and inequity of our society, while doubting their sanity in choosing the Soviet model as superior. The story that, when faced with the realities of communist society, Maclean became disillusioned and finally a dissident, seems to bear this out. I never met Burgess.

A few years after my return from America, my colleague Wilfrid Mann returned to Washington, attached to the Embassy on scientific aspects of nuclear intelligence, and was much involved with Burgess and Philby, whose loyalty had not then been questioned. In 1979, some three decades later, an English journalist, Andrew Boyle published a book entitled *The Climate of Treason: Five who spied for Russia*. This story, serialised in the *Observer*, centres around the already well known 'atomic traitors', Burgess, Maclean and Philby as well as two more, yet unmasked, code-named 'Maurice' and 'Basil'. An article in the *New Statesman* appeared immediately, identifying Anthony Blunt as 'Maurice' and Dr Wilfrid Basil Mann, 'a modestly obscure physicist' as 'Basil'. It continued, 'Both men, as far as anybody can tell, are wholly innocent.' In fact, the public exposure of Blunt's treachery came immediately, although it emerged that the secret service had been aware of it 25 years earlier but kept quiet for counter-intelligence advantages. In his book, Boyle never actually identified Dr Mann as 'Basil', but the insinuation was obvious, although the facts he quoted contradicted his case. For example, 'Basil' is supposed to have supplied Maclean with technical data to pass on to the Russians. In fact, Mann had never met Donald in his BCSO period, during which he had, in any case, no contact with nuclear matters; Maclean had already left America before Mann arrived there to take up his intelligence work.

The insinuations of Boyle's book were naturally nectar for the media and the matter reached the House of Commons where an MP asked the Attorney General 'if he will refer to the Director of Public Prosecutions with a view to prosecuting Dr Wilfrid Mann for breaches of the Official Secrets Act, in connection with the defection of Burgess and Maclean.' The Attorney General replied that he had 'no information relating to the conduct of Dr Wilfrid Mann that would justify my making any reference to the Director of Public Prosecutions'. This cautious defence did little to help Mann, and the pressure from the media mounted. He did, however, have lively and unsolicited support from scientific colleagues in England and America. I happened to be in Washington at this time and was staying with the Manns. The scene outside their house was one of chaos with a large crowd of jostling TV camera crews, a real siege. Wilfrid was able to escape to go to his laboratory only by jumping over the fence behind his house, into the garden of a neighbour, to be picked up by a friend waiting in a car in the next street. After the dust had settled, it was accepted that Wilfrid was not implicated and that the 'fifth man' had never existed.

I had a further brush with an 'atom traitor' in the person of Bruno Pontecorvo, the contact once more Dr Mann. In July 1946, Wilfrid longed to get back to his old career as an experimental physicist and, through our good friend Sir John Cockroft, had accepted a post at the Atomic Energy Project Laboratories of the National Research Council of Canada. These had recently moved from Montreal to Chalk River 120 miles up the Ottawa River from the Capital. Sally and two of the girls had spent their summer holiday in Nova Scotia and drove over to Ottawa to join me after one of my routine visits to Canada. We went up to Chalk River to spend a few days with the Manns. They were living at Deep River the 'company town' of scientists, situated a few miles from the labs at Chalk River. Each morning I accompanied Wilfrid to his workplace in the pool car he shared with two colleagues, one of whom was Bruno Pontecorvo. I found him lively and intelligent, very Italian, flippant and flirtatious. He and his slightly astringent Swedish wife had settled down happily in this community of scientists and were very popular. That he might be a Russian agent was unthinkable. The work of Chalk River was concerned with nuclear power generation and had nothing to do with the military. Also this isolated point in the Canadian backwoods would have been about the worst possible place for contact with other agents or transmission of information to Moscow.

Having left Chalk River and after a period at Harwell,

Pontecorvo took up a professorship at the University of Liverpool. He seemed settled there and had been seeking advice concerning his children's education. They decided to go for a family holiday in Italy and left Liverpool, apparently fully expecting to return, even to the extent of informing the milkman to leave the required number of pints on the day they were due back. But they did not return; they all went to Moscow instead. What happened during that fateful holiday in Italy to make him defect remains a mystery, as does the use he might have been to the Soviets. It is difficult to see this pleasant, debonair man within the dark introspection of Moscow.

With all these superficial contacts with defectors, I began to feel a creeping guilt by association, but in my next suspicious encounter I felt a real guilt of omission. I had given a contract to Professor Raymond Boyer of McGill University, Montreal, to undertake research on an aspect of the DDT programme and had an appointment to visit him to discuss its progress. I spent the day at the NRC in Ottawa before returning to Montreal to visit Boyer, and mentioned my plan to dine with Boyer at his house that evening to Jack Mackenzie, the NRC President. Mackenzie reacted strangely; stood up, locked the door and took the telephone off its hook. 'I must warn you', he said, 'that Boyer is to be arrested for treason tomorrow morning. Be careful what you say over dinner, because his apartment is bugged.' My immediate reaction was one of shock; what could this apparently simple academic have done? 'I will cancel the appointment and go back to Washington by the night train', I said. 'That you must not do', said Mackenzie, 'he would smell a rat and probably do a bunk.'

I duly dined with Boyer, feeling like the rat he did not smell. My conversation was, to say the least, stilted. There was another guest at dinner, a scientist from London, very left wing, whose inflammatory jargon, common at the time in red fanatic circles, must have charmed the security people listening. I learned from the TV next morning that Boyer had been arrested. It was all so futile. The information he transmitted to the Russians was utterly unimportant and had probably been passed to them through official channels already. I have felt guilty about this event ever since. How frail is friendship: I wanted to help him but did not know how, or else lacked the energy or indignation to do so. I puzzled about his motivation for passing this trivial information to an allied power that was regarded as an enemy. I knew him well enough to be sure that his ego was insufficient to make him believe that his action was significant. Some years later I was told that for some personal reason connected with his marriage, Boyer, born into Quebec high society had been rejected by that society that he saw as false and corrupt. Perhaps his giving so-called secret information to the Russians was a mistaken gesture of good will towards those whom he felt were trying to forge a better world.

About this time two Canadian physicists working in a government laboratory had likewise been found guilty of treason for passing classified information to the Soviets. It appeared the papers they transmitted were of quite minor significance; they had access to much more important data but had not divulged it. Obviously, their intention had been to make a gesture of friendship towards the Russians who were suffering from the German invasion, but not to endanger their own nation as was within their power. This contrasts with the attitude of arch-traitors such as Klaus Fuchs who was apparently an acute megalomaniac, feeling himself above and beyond the constraints of loyalty. At the time of his arrest he was working at Harwell. His reaction on arrest was, 'what will Harwell do without me?'

14

Commonwealth Connections

One of the most striking features of our century has been the dissolution of the British Empire into which I was ushered at my beginning. I was conditioned less by my parents than by the general social environment to accept its vastness and significance as the natural state of things. As soon as I could read, I had to admire on the atlas the considerable proportion of the earth's surface in splendid red, while another large expanse was shown in a dreary brownish purple, belonging to the rival French Empire. It was long before doubts began to arise in me about colonialism and domination. We had swarmed all over the planet not to subdue other peoples, but to bring them health, teach them to wear clothes with the morality these confer and, above all, convert them to the uniquely true religion with its passport to paradise. Had I not been uplifted by innumerable missionary sermons in church? Had I not had contact with the sincerity of this vision through my mother's cousins, devoted medical missionaries on leave from the colonies?

Close contact had been kept with my Canadian relatives throughout my childhood, with one of my uncles visiting us every year and the family of a second Canadian uncle once spent a summer with us in Highgate. Then a couple of Australian cousins turned up in uniform during the war. So I had the impression of the Empire as exclusively Anglo-Saxon. It seemed regrettable that America had broken away due to some misunderstanding on the part of George III.

By the time World War II broke out, this initial conditioning had been wrenched away. The pendulum had swung and I was probably as inflamed by the guilt of colonialism as I had been previously by its glory. It was clear the Empire could not survive in its earlier form. I began to see some looser form of co-operation emerging between the diverse fragments of Empire, a premonition of the Commonwealth to come. This was strengthened by my experiences in the Ministry of Supply as liaison between Commonwealth countries, then called Dominions, as well as American scientists with the British war effort. This exercise was exclusively Anglo-Saxon, since only the old Dominions had their own relevant efforts in military research and application.

Canada had been associated with defence science since the early war days, when C. J. Mackenzie, Acting President of the NRC, had joined Tizard in Washington so that the talks with the Americans had been essentially tripartite. Many Canadian scientists, such as Omand Solandt and Tuzo Wilson, had already been attached to British scientific units and there was a constant flow of visiting Canadian specialists. To cope with the increasing volume of contacts, Mackenzie decided to open a scientific office in London. The first appointment of a Canadian scientific representative was not too successful. He was a young chemist, probably brilliant and certainly ingratiating. His entertaining at the expense of the NRC was excessively lavish and he spent his leisure time assisting his midwife girl-friend delivering babies. He was soon replaced by Dr Hewlett, a senior officer of the Council who did an excellent job. Soon thereafter, an Australian scientific delegation, led by Sir John Madson arrived in London. Between games of billiards in the basement of the Athenaeum, Madson established friendly, effective relationships with the leaders of British war science and, on his departure, a continuing Australian scientific mission remained. New Zealand also joined in the co-operation. Dr Marsden, a pupil of Rutherford, head of their Department of Scientific and Industrial Research, set up his scientific liaison office under a young, eager Neville Wright. Dr (Brigadier) Basil Schonland of South Africa, an outstanding geophysicist in London on military duties, also took part in the liaison arrangements.

The various Commonwealth scientific missions soon established effective contacts throughout the London establishment and the various governmental experimental stations, as well as a harmonious relationship between themselves. It became clear that personal contacts achieved a more sensitive and rapid flow of information than would have been possible by correspondence through normal diplomatic channels. This enabled detailed replies to be obtained quickly to specific requests, and facilitated the choice and effectiveness of visits by their experts. The scientific representatives of these countries frequently met socially, often with me present, to share experiences and discuss the significance of their work. They were all convinced that the kind of relationships established by the exigencies of war and the methods devised for obtaining information from one another had potentially great importance for a peacetime world.

Although these were the darkest days of the war, with bombs falling on London every night and the war's successful outcome uncertain, we remained optimistic and thought much about the future when peace returned. We began serious debate on how post-war arrangements might be shaped. Although each member of our circle was a government representative, we felt that the academic community should be alerted to our ideas, so we determined to approach the Royal Society. I was in a suitable position to take the initiative because, although I was in effect working for the Ministry full time, I was still on half-time employment at Imperial College. I knew Dr Gough would not approve of my approach, but he could hardly object to my talking to scientists in the private part of my assignments. Accordingly, I paid a series of visits to Professor A. V. Hill at the Royal Society, accompanied by Neville Wright. Hill was interested in our ideas and soon involved the President of the Society, Sir Henry Dale, as well as its second honorary secretary Sir Alfred Egerton. They agreed to create a British Commonwealth Science Committee to discuss how to launch our ideas into practice. This was not a formal committee of the Society, but an independent group meeting under its auspices with me acting as its convenor.

We met on a number of occasions in the apartments of the Royal Society in Burlington House, generally in the evening and mostly in the relative safety of its basement. Eventually we issued a report putting forward a series of proposals and suggesting that a British Commonwealth Science Conference should be called within a year of cessation of hostilities. The report was widely disseminated in Britain and Commonwealth countries and was implemented. Unfortunately my copies were lost with other papers in the chaos of war, but copies exist no doubt in the archives of the Royal Society.

When I moved to Washington, my Commonwealth interests intensified. The creation of a British Commonwealth Scientific Office in that city, already described, was a logical extension of my London efforts. This body operated successfully for many years in war and peace. With the end of hostilities in Europe, its UK element became involved in Commonwealth issues much wider than those seen by the old Dominions as their direct concern. With the end of war in Europe, Commonwealth matters avalanched into my interests. In 1945 Sir Edward Appleton paid a visit to BCSO to sound American opinion on peacetime structures for research. He was delighted to see the Office in operation and told me that the decision to hold the Commonwealth Science Conference was now about to be taken formally. I gave a large cocktail party for him, which I remember vividly because Lady Appleton arrived two hours early when the hostess, my wife, was still in her bath. Appleton was an exceedingly warm and friendly person, basking in his fame, immortalised in the naming of the Appleton Layer of the stratosphere.

Later on a new series of contacts arose with the Commonwealth Science Council and, still more actively, CPTM being established under the Commonwealth Secretariat.

Shortly after, the Indian mission, already described, arrived. Just after their departure to Canada, Sir Robert Robinson and Professor Simonsen came to Washington to discuss their mission with the Caribbean Commission and were housed in BCSO. As it happened, the Science Director of the Ministry of Aircraft Production, Sir Ben Lockspeiser, whom I knew well, was in town. Here was an opportunity too good to miss; an exceptional congregation of important people concerned with Commonwealth science. So I decided it would be useful to hold a preliminary, informal meeting on post-war Commonwealth science co-operation. I telephoned my friend Jack Mackenzie who was preparing to receive the Indian mission in Ottawa. He readily agreed to host the conference at the National Research Council and it took place with scientists from the United Kingdom, Australia, Canada, India, New Zealand and South Africa. It was a constructive and friendly occasion, a rehearsal for the official conference that took place in Britain the next year. The London authorities were, of course, not too pleased with this event which was a maverick act outside their control, but as the meeting had no power to take decisions, or intention to do so, and as so many politically respectable senior officials were present, there were no ripples.

A few days after returning from Ottawa, we set off for the Caribbean. The mission consisted of Robinson, Simonsen, me and a young Trinidadian secretary, Dr Eric Williams, who later, when his country became independent, became its first Prime Minister. Our group was friendly and cohesive. Robinson, one of the greatest chemists of his time, was quite naive about world affairs, but I found him relaxed and full of humour as we laughed our way from island to island. Simonsen, nice and sensible, was hardly more worldly wise. He radiated a respect for Robinson, akin to awe. Williams was constantly amused by our antics.

Our task was to visit some of the British colonies in the region, survey their natural resources and suggest how scientific research might assist in the process of economic development. At the outset I was quite sceptical about our competence for this task. We were three basic chemists to whom development meant the process of moving from scientific discovery to technology; none of us had any economic background and little worldly wisdom. The Caribbean Commission had suggested that we should begin by looking at the strategy and implementation of a novel development plan in the American territory of Puerto Rico derived from the innovative activity of its former Governor, Rex Tugwell, well known as a chief architect of Roosevelt's New Deal. Puerto Rico was seen as a case model of economic development, so we set off to San Juan. Tugwell's successor as Governer, a Puerto Rican, explained the development methodologies to us and we visited many projects in the Island as well as in the nearby American Virgin Islands. This was an exceedingly useful education as an introduction to our mission. For me, lacking any economic education, it was a real initiation to the concept of economic development and aroused a curiosity about the relations between science, economy and society that has coloured my thinking ever since.

We next went to Trinidad, where we spent the best part of a week in discussions with officials, scientists and agriculturists.

This was the central and most useful part of our task. Then, after the Guyana diversion, we returned to Trinidad. As time was running short we decided to break ranks. I was to go on to Jamaica for talks while the others went to Barbados. To reach Jamaica I had to spend a night in Baranquilla in Colombia, before catching a flight to Kingston. I found Baranquilla a depressing city set in a flat plain, horribly hot and humid, with vultures perched attentively on all the telegraph lines. The hotel guests appeared to be mainly Germans and it was disturbing in 1945 to listen through the flimsy walls of my bedroom, to angry conversations in that language. I presumed they were Nazis who had escaped the fatherland just before its fall. My days in Jamaica were especially positive. The engineers and administrators with whom I talked were serious and innovative, resource possibilities were more significant than elsewhere on our journey and one could sense an energy in the people. We identified some areas where applied research held promise of real development.

Returning to Washington, I had to deal with the problems and enquires that had accumulated during my absence as well as write my part of the report to the Colonial Office. But my routine life in BCSO was soon interrupted again. Early in 1946 I received a cable from Sir Edward Appleton, which delighted me by asking me to undertake a tour of some of the Commonwealth countries in preparation for the scientific conference which was now being planned for 1947. I had discussions with Dr Mackenkie and his staff, which were easy since we were already fully in agreement on the aims of the 1947 conference. Then I set off for Australia. At San Francisco I boarded a US Air Force bomber in the company of a number of Australian army officers returning home on leave. Our first stop was Hawaii where we spent the night. It was a moist cloudy day and Waikiki beach was deserted, desolate and litter-strewn. The next morning we set off in sunshine from a perfect tropical landscape. My companions talked endlessly in anticipation of their first meal of their national dish, steak and eggs. We landed on Johnson Island, an atoll that was little more than a runway with some huts, to spend the night and in the morning, as the desalination still had broken down, we were served cans of beer with which to shave. With brief stops at Fiji and New Caledonia, we reached Brisbane.

I had my first discussions with the Queensland office of the

Commonwealth Scientific and Industrial Research Organisation (CSIRO), then off to its headquarters in Melbourne, where Sir David Rivett its director gave me a warm welcome. We held a series of talks about topics they hoped would feature in the forthcoming London conference and I visited some of the CSIRO laboratories in Victoria and New South Wales for discussions. I was particularly privileged to be taken by Sir David to a meeting in Sydney of representatives of all the Australian Universities to discuss the national situation on research. CSIRO was very strong and well funded by the federal government, undertaking research on agriculture, fisheries, earth science and industry, and also fundamental science subjects like radio-astronomy and chemical physics. The Australian universities, all their representatives testified, were seen as underfunded by the State Governments responsible for them, and their research facilities were miserable. This was the first of a series of meetings that eventually led to the creation of the Australian National University at Canberra, a post-graduate institution which proved highly successful in reviving academic research.

From Sydney I flew to New Zealand where I found Dr Marsden and his colleagues equally enthusiastic about Commonwealth co-operation and the conference. I then returned to Australia and, after final talks in Melbourne, visited CSIRO installations in Tasmania and Adelaide. Then came the long flight over the desert to Perth, where I gave a lecture in the university. The next stop was Ceylon, at that time not possible to reach in a single hop. The flight was interesting; first along the barren coast of Western Australia and then over the Indian Ocean to the Cocos Islands, still owned by the Clunes Ross family. I spent the night there in a tent, where I could see from my bed creatures I imagined to be tree crabs, climbing up and down the palm trees outside.

Arriving at Colombo, I was greeted by Dr Karanuratna, an elderly medical man, then the doyen of the colony's scientific community. He was warm and intelligent, famous for his discovery of a rare tropical disease. We developed a close friendship and he later visited me several times at home in Clavering Avenue, bearing offerings of Ceylon tea in ornamental boxes. I was introduced to other scientists, having pleasant discussions with them on the shape and objectives of the conference we were planning, but there was little of scientific interest to see. I did visit the University of Ceylon, situated in beautiful tea grove scenery on the way to Kandy with its venerated Temple of the Tooth. The university was residential and based on the Oxbridge system of intense personal tutelage.

The scientific interest may have been slim, but Karunaratna's social hospitality was terrific. I was taken on a visit to a home for the dying, where I tried to give comfort to an elderly and terminally ill Scot. My next call, at a house for the rehabilitation of what was then termed 'fallen women', was certainly less gloomy. The social highpoint, was a dinner of the old girls of his wife's school. This was apparently an elite institution, based on the English public school. We sat at tables outdoors on a hot and humid tropical evening, the husbands all sweltering in dinner jackets with black ties. The Mistress of Ceremonies was Mrs Bandaranaike, later Prime Minister of liberated Sri Lanka. The speeches were charmingly formal, the main one being by the President of the alumni association, the beautiful wife of my host. I found the occasion quite touching and my sentiments were far from patronising. I was quite sad to leave Ceylon with its warmth and simplicity that was so different from the friendly efficiency in Australia.

I then took off for New Delhi where Sir Shanti, with habitual exuberance, greeted me as an old friend and guest in the Bhatnagar household. I had a useful and interesting time in India, visiting a number of research institutes and excited by the depth and passion of intellectual discussions. We were already in full agreement on the London conference, having discussed it in detail in Washington and Ottawa, so the concern moved to the design of a research structure for India after independence. Shanti was to be in charge of this and had strong ideas. His mind was set on the creation of a number of large National Laboratories, modelled on England's National Physical Laboratory. I argued against spending scarce resources on bricks and mortar too early in the game that might determine the pattern of Indian science for generations, and advocated a much looser structure, capable of evolving as priorities emerged. A couple of years later, I was invited by the Indian Government to be an advisor on the same subject. Bhatnagar and I continued the debate in friendly, but opposing terms. My advice was not accepted and the construction of the large national laboratories soon commenced. I remain convinced my advice was sound.

Reporting back to London, I found that prospects for the conference and for the construction of continuing machinery for co-operation appeared bright. I was happy to learn it was planned to hold the meeting in two stages; first, a non-governmental conference organised by the Royal Society aimed at surveying the whole spectrum of research on which co-operation between the countries would be useful, and second, an intergovernmental meeting to ensure that co-operation between national laboratories and programmes would be effective. The conference duly took place, attended by leading scientists from national institutions and academia from both the older Dominions and a large number of territories about to become sovereign nations within the Commonwealth. There was even a delegation from the mandated territory of Palestine that gave me useful introductions to what would soon become the scientific community of Israel. The Conference was a considerable success and led to a sound co-operative structure. At the end of the final session, Professor A. V. Hill recounted his initial talk with Neville Wright and me that had provided the small spark to ignite a substantial fire. Learning that it had been decided to create a BCSO in London similar to my Washington model was one of my happiest moments.

On my return to London I became preoccupied with UK domestic problems, but still retained responsibility for international scientific co-operation and attended BCSO meetings as UK representative. It is surprising that this liaison function had come with me from my installation in the Ministry of Supply, thanks to Tizard, then to the Ministry of Production, the Lord President's Office and finally as we shall see later to the Department of Scientific and Industrial Research (DSIR). In later years, my job caused me to concentrate on European problems and I had little opportunity to follow Commonwealth scientific affairs in detail. I did, however, participate in the second large Commonwealth Science Conference held in Australia, enabling me to renew acquaintance with Sir David River, Ian Clunes Ross and other Australian friends. Then, much later again, after I had retired from OECD, I led a team to review for it the scientific and technological policies of Australia, now a member of OECD.

When I reached my eighties I had a brief 'Indian summer' of Commonwealth activity thanks mainly to my friendship with that remarkable entrepreneurial woman Dr Mihaela Smith, who was working with the Commonwealth Science Council and later on the Club of Rome, before eventually setting up CPTM. I attended meetings arranged by the Commonwealth Science Council in Sri Lanka and twice in Kuala Lumpur. Also, I acted as an adviser to Mihaela's group advising the smaller Commonwealth Countries and helping them to devise a 'smart partnership' between industrial and government leaders for their economic development. In 1996 I and other CPTM Fellows, including Brian Locke (another member of the Club of Rome) advised the President of Botswana, Sir Ketumile Masire on his planning for 'Vision 20–20'. The Commonwealth Science Council was closed in 2003, but as a Companion of CPTM, I am still called on for my views by leaders from the around the Commonwealth, even as recently as 2006.

I have been disappointed at the Commonwealth Organisation's failure to pursue scientific and technological development with the vigour that our 1946 efforts suggested. Nevertheless, there is still great utility in the Commonwealth approach despite its bureaucracy ensuring the dominance of conventional thinking. One potential strength of the Commonwealth Organisation is that, with its heterogeneous membership, it is politically weak. This makes it better placed to examine with minimum prejudice, longer-term problems on their intrinsic merits.

15

Science in the Ascendant

This chapter attempts to sketch some features of the advancement of science in relation to society, as a background to much of my career. At the outset it is necessary to talk about the relationship between science and technology and between science-based technology and economic development – relationships that are still far from fully appreciated by the public and by politicians. Simply stated, science is seen as knowledge through rational experimentation. Technology is the mechanism by which new scientific discovery or empirical invention is applied for economic, social or medical purposes.

The real situation is much more complicated. Pure science, mainly pursued in universities and some independent research institutes, has as its sole objective the extension of human knowledge. It is assumed that those who practise it give no thought to the potential applications of their discoveries, although this is frequently not the case. Applied science is cultivated in industrial, governmental and other research institutes and attempts to exploit the possible application of particular discoveries or of the corpus of accumulated scientific knowledge for practical (e.g., economic) objectives. Then again, there is oriented fundamental research, use-directed and supported or commissioned by user interests, that endeavours to identify elements of new knowledge needed to include in the mosaic of existing science in order to make planned applications possible. The final stage on the path to economically viable production is technological development, the design of a pilot chemical plant or engineering prototype, a job for the engineer rather than the scientist.

While in the past technology was largely a matter of

invention, today it is hardly possible without scientific input. In important developments the individual inventive genius has become a rarity; most major innovations are based on the work of well-funded teams of scientists and engineers. It is somewhat of a paradox, however, that while technology is so often based on fundamental research, such research, especially in 'big Science' fields like high energy physics or astronomy, demands its own novel technology for the construction of its sophisticated and at times very expensive experimental equipment. The system of science is very different from that of technology, although the two are intimately interactive. The findings of fundamental research are freely published and distributed internationally; each contribution is an element, however small, of the extending fabric of human knowledge and as such part of the common heritage. In contrast, advances in technology, whether hard- or software, are economic products, protected by patents and fiercely guarded.

Homo Sapiens appears to have possessed a capacity for innovation from its earliest days, fuelled presumably by the universal struggle for existence. From the shaping of the first stone implements, technological development, trial and error, has been man's main means of advancement. Each important innovation brought about a modification in societies. It is probable that a clever woman somewhere stumbled on how planting of seeds could yield reliable crops of food. This discovery spread over the millennia at a painfully slow rate throughout the world, leading to the replacement of hunter/gatherer society with one of settled agriculture and animal breeding. Technological development still drives social change as is obvious in the information revolution. It must be remembered also that there have been two main paths through which technological development has generated social change. The first, as in agriculture, has raised living standards, the other, through the improvement of weaponry, has created and destroyed civilisations by war.

It is not known where, how or by whom scientific thinking entered the human consciousness. It probably arose as a product of imagination and curiosity as people observed the movement of the celestial bodies and the cycle of the seasons. It may also have come from contemplation of man's early technologies; building structures for which measurement was essential, and concepts of dimensions and their relationship leading to geometry. Very likely a sense of the rational, mingling with the aesthetic, led to proto-scientific speculation in Sumerian times and in the great civilisation of the Indus valley. In classical Greece it flowered magnificently. However profound and seminal it was, Hellenic philosophy had little to do with science as we consider it today. It was dominantly speculative and, although brilliant in exploring the nature of being and of things, little attempt was made to test its probings in practice. 2000 years had to pass before the reality of the Hellenic concept of the atom could be demonstrated and then it was very different from the model Democritus had imagined. It can be argued that the isolation of the sublime heights of Greek speculation from everyday reality and rejection of experimentation retarded the advancement of science 2000 years. The social stratification of classical Greece, with the elite minority regarding physical work as unworthy and the disfranchised majority fettered by despised physical activities, made it inevitable that experimental science would be ignored as an intellectual pursuit of truth. Despite the high period of Arab science and the work of a few medieval alchemists who stumbled into genuine scientific thinking, it was not until the Renaissance that experimental science began to flourish.

The two great and contrasting figures in the science of the sixteenth and seventeenth centuries were Galileo and Francis Bacon who stand as role models for the two trends in scientific activity that have persisted since: Galileo for the purity of scientific thinking, searching for truth for its own sake despite disapproval; Bacon, prime advocate of applied research leading to technology, advocated experimentation for the benefits, material and otherwise, that new knowledge would generate for the human condition.

Associations of scholars began to consolidate. The first Academy was the Academia dei Lincei, of which Galileo was a founding member. Then in 1648, Charles II – a dissolute monarch with intellectual pretensions who enjoyed the company of savants – created The Royal Society. He had also an ulterior motive; he sought their advice in fighting the Dutch war, which turned out to be one of the most ignominious in British history. In one sense, therefore, the Society was the first example of a national Science Advisory Council. The early academies recruited from a wide range of disciplines; philosophy, history, classical learning and theology as well as natural philosophy, or the natural sciences. Indeed the word science – *La science*, *Wissenschaft*, *Nauk*, was regarded as a synonym for knowledge, an equivalency still assumed by most academies of science today which include the physical, biological and social sciences as well as philosophy and the humanities. Only the Anglo-Saxon countries have limited the term science to the experimental disciplines, excluding the social sciences and humanities. The Royal Society consolidated this situation by honouring achievement in fundamental physical and biological research and ignoring outstanding contributions in applied research. This emphasis goes far in explaining the magnificent achievement of Britain in fundamental research and the lower prestige offered to practical accomplishments.

The Industrial Revolution erupted towards the end of the eighteenth century mainly through empirical invention that had already absorbed Newtonian mechanics. While in Britain the elite scientific bodies distanced themselves from engineering, scientists and industrial leaders came together in less prestigious bodies. One such was the Lunar Society of Birmingham whose members, coming from different parts of the Midlands, dined together for many years, when the moon was full. Amongst its members were the industrial pioneers Boulton, Wedgewood and Watt, and scientists such as Erasmus Darwin and Joseph Priestly, the discoverer of oxygen. They discussed topics of science, technology (although they did not call it that), the nature of society and the meaning and aim of human existence. These men saw in Watt's improvement of the steam engine, the arising of a new type of industrial society capable of producing wealth sufficient to abolish poverty. They were, indeed, the Fathers of the Industrial Revolution. History shows that these sages were wildly optimistic. Their revolution was highly successful economically, but merely transmuted rural poverty into urban misery, as the actions of the Luddites and machine wreckers testified. Many decades were to pass before science accumulated sufficient substance to give reality to their technocratic vision. I shall return to the Lunar Society later.

Discovery and change were rapid throughout the eighteenth century, following the abandonment of the phlogiston theory and understanding of the nature of combustion. Chemistry became a science and established new and important industries; articulation of the laws of thermodynamics gave a sound theoretical basis in engineering. Electricity was a discovery of experimental science and even Gladstone questioned its uses. Fundamental research tended to remain perfectionist within its ivory tower. An indication of this was the reaction of the famous physicist, Clark Maxwell on learning of Graham Bell's invention of the telephone, who remarked with disdain that it was a mere gadget embodying no novel principle. As the twentieth century approached, governments began to see that science, both pure and applied, had some place in national life, and for the first time began to organise to use its benefits. Fundamental research, obviously related to education, was encouraged mainly within the universities. The Geological and National Physical Laboratories, with directly utilitarian functions, were established as independent bodies.

The so-called Great War of 1914 was still a classical conflict between standing armies, dominated by the horrors of the trenches. But with a heavy element of technology in the use of explosives, tanks, poison gas, zeppelins and the airplane, all of which, developed by research, depended on massive industrial infrastructure. The British government had created a structure of research organisations including Medical and Agricultural Research Councils as well as the DSIR, which operated a series of national laboratories and Research Associations for many sectors in co-operation with industry. The defence services possessed their own laboratories. The three Research Councils all reported to the Lord President of the Council who thus became *de facto* Minister for Science. During the interwar period all these bodies prospered. Science teaching became widespread in schools, university research attracted modest support and industry expanded, consolidated and developed more sophisticated technologies. When war was declared again in 1939, Britain had much of the infrastructure needed for a science-supported conflict as was demonstrated by the contents of the 'black box' that Tizard took to America. However, these achievements and the organisations that produced them were quite insufficient for the science-dominated war which was to develop. Many new research and development facilities were created quickly for radar, anti-submarine warfare, guided missiles and, of course, nuclear energy.

The impressive use of science and technological development in helping to win World War II raised hopes that they might be equally successful in the period of reconstruction, and thereafter by building prosperity and social progress in a peaceful world. Particular wartime products such as radar, the jet engine, the computer and nuclear energy, as well as discoveries like penicillin and DDT, had immediate and obvious uses in the civil economy, while new research seemed to have potential for many other industrial developments. There was also much interest in the use of so called Operational Research in peacetime development. The name, Operational Research (OR), was applied during the war to the use of scientific logic and simple statistical methods for the analysis and optimisation of defence operations. It had done so much in upgrading the effectiveness of the small RAF in the critical Battle of Britain, in clearing the Bay of Biscay of German submarines and in many other operations, as to suggest its use in the operations of industry and society.

Before 1939 science had little impact on the public mind or appeal for politicians, but all was soon to change. Science in the war had been 'a good thing', its structures and policies had now to be modified to ensure that it met the needs of civil society. In Washington a huge debate had taken place (Chapter 12) on how to rectify the balance of effort between science and technology by organising a massive increase in facilities for fundamental research. In Britain, the need was just the opposite. The British had noticed with surprise and admiration the speed and efficiency of the mammoth US war production in comparison with the slower, more painful UK development of original ideas into mass-produced hardware. While this was partly due to the difficulty of maintaining production in a continuously bombed island, there was an obvious need to stimulate technological development and production skills. A study made at the end of the war showed that for every research worker in America, there were two and a half development and design engineers. In Britain the ratio was reversed.

Even in the darkest days of the war a committee was set up to discuss what structures and policy machinery would be appropriate for the peace that would follow. This was chaired by Lord

Hankey the famous war cabinet secretary of World War I. I had the opportunity of giving evidence to it on one of my debriefing visits from Washington. Many issues demanded exploration and clarification. There was much confusion concerning the place and influence of scientific and technological systems and the role of government in each. How relevant were the government's own laboratories? How could relations between industries and universities be developed? Should priorities be established for funding research and what should be the criteria for determining the distribution of resources amongst very diverse claimants. Now, 60 years later, these issues remain unresolved. With new importance given to research, the scientists were all too aware they were in for a good time. Very soon their demands for new and sophisticated research tools far exceeded the increased finances then available.

Very little thought had been given to the role of science in national development or to how its discoveries could be converted to economic development. Science was seen as a good thing, but that was about the extent of the understanding. The economists had paid scant attention to the place of science in economic growth; Keynes, for example, hardly mentions it. Their general view, seldom stated explicitly, was that scientific discovery followed by technological development flows from the interaction of economic forces. Scientists, with their own clan arrogance, saw things otherwise. Their image was of the professor, working in a university laboratory making a significant discovery in his search for new knowledge and without interest in its economic exploitation. Some greedy entrepreneur, possibly in a distant country, might be appraised of a potentially lucrative application of said discovery, employ a team of applied researchers to explore its practicality, see it through a pilot stage to produce a patented, marketable product and make a fortune, without the professor even being aware of it.

There is a grain of reality in both these models: new technology rather than scientific discovery is only viable if economic conditions are propitious. For instance, the forklift truck was ready to enter the market as early as 1912, but it was a couple of decades later before labour costs had risen sufficiently to make it economically viable. On the other hand it is often dangerous to rely on the 'technological fix'. Most issues have technical components and, in some cases, technological solutions, but frequently the fundamental science on which to base these solutions does not yet exist. The lead time from discovery, through applied research and development to the technology is very long; it may take as much as 30 or 40 years, which will be much too late. If there is a strong political will to provide a 'technological fix', the lead time can be drastically shortened by the use of exceptionally high resources. This was the case with the Manhattan Project that delivered the atom bomb. Another example was the American 'man on the moon' project, highly prioritised for national prestige as well as defence reasons. But for smaller yet still important economic issues, the 'technological fix' cannot be relied on. Economists also conveniently ignore the fact that great industries have arisen directly from breakthrough discoveries in the fundamental research laboratories, such as those uncovering the secrets of solid state physics leading to microelectronics, or the discovery of DNA giving rise to new biological industries.

Things are very different today. Scientists soon discarded their innocence, accepting with little questioning support from industry that permits them to pursue fundamental knowledge, well aware of its orientation towards economic goals. There are also examples of scientists developing entrepreneurial qualities in exploiting applications of their own discoveries. Thus, in the post-war years, science and technology were gradually seen in terms of investment by industry, government and the public. In government, awareness of this concept was general and ill defined. The idea of a national science budget is difficult because of the extreme diversity of scientific effort towards targets of very different priority. Initially very few administrations attempted to integrate science and technology spending in defence, agriculture or the public services into a comprehensive total of investment. Rather, allotments for the various sectors were agreed by negotiations in Cabinet, favouring powerful ministers and dominated by the finance minister.

The investment concept was quite obvious to large corporations in science-based industries such as chemicals, pharmaceuticals or electronics, but for many firms science came through routine chemical analysis, testing, quality control and trouble-shooting. The more adventurous had research departments, but research was often a euphemism for routine tasks. For such firms research was revered as a 'good thing', to have a research department was a matter of prestige and was seldom considered in terms of innovation. Nevertheless, the recruitment of even a miniscule scientific staff upgraded the technical competence of firms and kept them in touch with developments throughout their industry.

As late as the 1940s, studies in America indicated that the reasons for undertaking research by a majority of companies were not laid down in terms of corporate policy. For many it was simply prestige, of 'keeping up with the Jones's of their industry. There was rarely a close relationship between research departments and top management. Many research managers confessed ignorance of corporate objectives, differentiating such firms from those with a real research policy integrated into corporate policy and awareness of market research findings. The isolation of research departments tended to create defensive attitudes and NIH, 'not invented here' syndrome, resistant towards innovation from outside. Such circumstances also pointed up the need to develop research management as distinct from financial and general management, notably how to adapt research to moving targets and the need to bring to an end projects that were promising on scientific grounds but had dubious economic prospects.

The culture of industry, with its management styles in research as in other areas, differs greatly from country to country. They are influenced, and in some cases determined, by their national education system and social traditions; for instance, a particularly favourable environment seems to have been provided for US technological development. A young and vigorous country, it retains much of its earlier pioneering spirit. With an ostensibly classless society, the 'log cabin to White House' ambition persists, the dollar is worshipped and a wide range of educational facilities exist to cater for every type of intelligence and skill. It is easy to see how this dynamic and pragmatic nation should have stressed the technological, rather than the discovery elements of development.

The French environment is strikingly different. France, with its deeply rooted sense of cultural superiority, the *Code Napoleon*, the Cartesian heritage and the intellectual quality of the elitist *Hautes Ecoles*, is instinctively conservative despite the French Revolution and frequent expressions of extreme political ideology. Intellectual achievement is regarded as an important aspect of national prestige. This has led to much successful innovation and quality, although top management in industry has, until recently, been somewhat over-theoretical and rigid. The legalistic basis of French institutions is somewhat heavy.

In Germany, both the scientist and the engineer have long been held in high esteem. From the mid-nineteenth century German scientific research led the world and the quality of its universities was unsurpassed. But they also developed a series of *Technische Hochschule*, technical universities of high quality to train science-based engineers. Thus Germany was well served by pools of skills both for scientific discovery and production technology, aided by a national enthusiastic devotion to the work ethic.

Britain, on the other hand, still suffered from having pioneered the Industrial Revolution, profiting from its fruits and building its enormous, if ephemeral, Empire. Years of high prosperity and imperial grandeur seem to have nurtured a deep-set national complacency. Industry, hedged by investment from the past, began to lose its competitive edge, particularly against American and German vigour. While leaders in French industry were mainly educated in law and those of Germany were engineers, British firms were dominated by accountants. In Britain the engineer had a relatively low prestige. This was partly due to terminological inexactitude; we use the term engineer to include individuals with post-graduate or graduate qualifications, well-trained technicians and quite low-skilled craftsmen. The distinction between these is not easily appreciated by the public.

The industrial, like the political scene, is greatly influenced by our notorious class system. The fee-paying private schools that the British, to the bewilderment of foreigners, refer to as ‘public schools’, educate and perpetuate an elite of money and caste which sharply differentiates its members from the majority of citizens who attend the free state schools. Teaching quality in the public schools is undoubtedly high, but it has been based on the classical tradition into which science teaching has penetrated but slowly. As a consequence, the leaders of society tend to be classicists, lawyers, historians and financiers, often scientifically illiterate and uninterested. The lower standing of scientists and engineers has generally resulted in inferior salaries compared with those of the generalist elite. The sharp cultural differentiation of those with a classical or literary education from the scientists and the difficulties of interaction between the two was highlighted by C. P. Snow in his book, *The Two Cultures*. Margaret Mead, the American anthropologist, pointed out that this is really a problem of a small segment of Anglo-Saxon society. It is not significant elsewhere. This is obviously a consequence of the narrow connotation we give to the word science. Whilst in Britain science is seldom seen by most people in cultural terms, in other countries it is regarded as an element of *the culture*. The cultural divide in Britain has had a profound influence on our national development; an issue that is no doubt being resolved by the passage of time.

Public attitudes towards science have fluctuated wildly in the period under discussion. Its contributions to the winning of World War II were regarded with pride, some surprise, and little understanding of what was involved if it was to be the key to recovery and world status. In the following decades, spectacular developments such as radio telescopes and the huge atom smashers of the nuclear physicists were seen as elements of a mystique inaccessible to the general public. Still later, during the Cold War, fear of the Damocles sword of nuclear annihilation unveiled to the public the evil and dangerous aspects of scientific development. Also many began to fear the dominance of science-based technology as a threat to the environment and as bleeding away the quality of life. Nevertheless, although the alienation of a minority was vocally and at times violently expressed, the majority of citizens as well as the body politic seemed hell bent on the creation of an utterly materialist consumer society.

This then was my sense of the background to my immersion in the politics of science, technology and society as I slipped back to London from Washington in 1946.

16

A Maverick in Their Midst

Chapter 12 mentioned efforts made in Washington to design policies and structures for US science in the post-war era. Vannevar Bush's *Science: The Endless Frontier* animated the debate. While there was general agreement that a new institution was required, negotiations were lengthy due to conflicting departmental and power interests. A similar exercise had been tried several times in the past; the need for more vigorous scientific advice became evident in every war. The resolution was always the same, 'create a new, more vital and responsive science advisory body'. While the Smithsonian Institution was the first of these, its advisory function was taken over by a newly created National Academy of Science. During World War I a National Science Council was created that yielded place in World War II to the temporary and effective OSRD. Yet all the older institutions were retained, even though their influence was diminished or their functions changed – a solution that only America is sufficiently large and rich to afford. In the end, a Presidential Science Adviser was appointed and the National Science Foundation (NSF) created to meet the Bush concerns, while an invigorated National Academy performed a useful and responsible role forwarding the views of the independent science community.

In Britain the situation was more complicated. By 1939 there appeared to be a settled division of responsibility between the various national bodies concerned with science. The DSIR had grown to become a mini Ministry of Science, responsible to the Lord President for virtually all aspects of civil science, except agriculture and medicine for which separate Councils existed. The Royal Society retained its historical advisory role to the government. There was also a detached network of defence research establishments. Nearly all the Whitehall departments had some scientific activities about which they sought advice on an independent *ad hoc* basis. The Departments thus gave little encouragement to those who felt that some central co-ordination of science policy was necessary. This was also the instinctive view of the non-partisan Civil Service.

A Scientific Advisory Committee of the Cabinet was established in 1940, but its mandate was limited to matters that did not fall within the jurisdiction of the three Research Councils; consequently topics discussed by it were peripheral. While it was clear that Whitehall saw little need for central control of science policy, individual scientists and the (non-official) Parliamentary and Scientific Committee began to press for just that. However, for two years immediately after the war this issue was quite theoretical. Churchill, who proclaimed 'science should be on tap, but never on top', saw no need to augment the power of the SAC and relied on scientific advice from Professor Lindeman (later Lord Cherwell).

The general election of 1945 was a dramatic watershed event. Churchill, who was at the zenith of his popularity as the successful war leader, was quietly voted out in favour of Attlee's Labour Party. I was in London during the election on one of my reporting visits from Washington. The day after the results were declared I was in Parliament Square which was thronged by excited, surprised people. A thanksgiving service had just ended in St Margarets and the congregation was spilling into the street. Out came Attlee with his colleagues about to become ministers of the new government. The public, who had just voted them in, gave them warm applause. Then followed Winston Churchill who had been rejected, the crowd enthusiastically acclaimed him. I was touched and proud of this sign of political maturity. The people were enthusiastic in their praise of a man who had done so much to win the war, but recognised that a new leadership style and new policies were needed for peacetime.

Science policy considerations also changed abruptly. Lindeman was out and Tizard, his old enemy, reappeared on the scene. The New Lord President of the Council was Herbert Morrison, a man known to respect intellectuals, who in lectures and constituency meetings expressed great optimism about the potential contributions of science and technology to human well-being. He reacted quickly by setting up a committee on Future Science Policy. Conflicting suggestions poured in from the Treasury, the Research Councils, Tizard, the Royal Society and the scientists' lobby. These demanded much scrutiny and compromise. Two years of difficult negotiations followed before an agreement was reached in January 1947, establishing an Advisory Council on Scientific Policy (ACSP), as a cabinet committee serving all ministers. Its secretariat, the Central Scientific Secretariat, was, however, not to be in the Cabinet Office but with the Lord President, who felt he needed scientific staff in his own office. A proposal to appoint a single chief scientist to the government was rejected, perhaps to dissipate shadows of the advice monopoly of the Lindeman years. Nevertheless, Tizard was appointed full time as chairman of the ACSP and of the newly established Defence Science Committee (DSC) so becoming *de facto* the government's chief science adviser. This was somewhat ironic; although he was a voting Conservative, he could only hold office with the Socialists because of the rift with Lindeman.

Another event had much relevance to my future work. A committee chaired by Sir Alan Barlow, Second Secretary of the Treasury, had just issued its report on scientific manpower. Its recommendations, immediately accepted, created the Scientific Civil Service. This aimed to provide conditions and prospects conducive to creativity for scientists in government employment that the normal establishment was unable to furnish. Furthermore it hoped to facilitate mobility of employment of scientists between government laboratories, Research Associations and universities. The reasons for my own appointment as head of the Central Scientific Secretariat are obvious in retrospect. In this tricky situation of compromise, I was one of the few people in good relationship with all the parties whose views and aspirations had to be reconciled. I was well known to the civil service from my Ministry of Production and Washington days: Appleton had already supported and worked with me; I was friendly with Dale and the secretaries of the Royal Society; and I was a protégé of Tizard. I must admit it would have been difficult at that time to find another academic scientist fool enough to take on this sensitive job. In fact my good relations with these diverse interest groups survived the strains of compromise, with one exception. The terms of office of the President and Secretaries of the Royal Society had come to an end. Their successors hardly knew me and were very suspicious of my functions and the new organisation as diminishing their advisory role. Treated with quiet reticence, I sadly never recovered their confidence.

My new job, as Head of the Central Scientific Secretariat, Secretary of the ACSP and personal scientific adviser to the Lord President, was a non-political appointment so I had to become a member of the civil service. I joined with some distaste and prejudice; it was painful to become one of the guardians of the *status quo*. I settled down in a room in the Cabinet Offices, welcomed by the head of the Lord President's office, Max Nicholson, a friendly, wise man who, happily as I soon discovered, was no conformist. We worked together harmoniously over the years. He was extremely useful because his realistic appreciation of people and situations curbed some of my wilder intentions. I discovered that my staff, mysteriously selected and already installed, consisted of a physicist, Wardley Smith, Mr Murray, an academic philosopher, an administrative assistant secretary to keep us on the rails and my personal secretary, Denise Franklin, a loyal and intelligent friend who was to work with me for years. My initial reaction was one of frustration. How could this small team cope with its mandate? The Central Scientific Secretariat had been envisaged as 'the sole body in a position to maintain a continuous survey of the whole field of Government civil science. Unless it functioned well the ACSP might perish for lack of roots. Faced with this *fait accompli*, I should perhaps have protested, but I feared that doing so I would be branded a difficult colleague, an unfortunate image that would activate opposition to every idea I put forward.

Soon I was introduced to the Minister, Herbert Morrison, a cheerful overworked man. He welcomed me warmly and I took an immediate liking to him. As our working contacts grew I learned to appreciate him and respect his honesty. Then there was Tizard, who I was delighted to be with again. He was certainly one of the few people throughout my career for whom I have unstinted admiration. He was not an outstanding scientist, indeed one knew next to nothing about his personal research achievements, but he had a profound feeling for the scientific method, and was a clever administrator with enormous common sense and wisdom. In particular, he had a special gift for assessing the abilities of people. He told me on one occasion about a discussion he had with the headmaster of Winchester about how to guide pupils into lines of study appropriate to their qualities: 'I would ask each young man', he said, 'to describe the construction and functioning of (a) a pair of scissors and (b) an umbrella. From the answers it would be easy to foresee whether he should pursue a scientific or humanities education.' Yet, although our relationship remained warm and co-operative, it was strangely impersonal. We were never on first name terms, while with Morrison, he was Herbert and I Alex.

With Solly (later Lord) Zuckerman, the Vice-Chairman of ACSP, things were quite different. I had previously met him only superficially, now face-to-face I took an instant dislike to him which I am sure was mutual. We worked well enough together and I admired greatly his intelligence and capacity for sustained hard work. Nevertheless, I felt that somehow he was an enemy. I have had few overt enemies in my life, but I am sure he was one.

The Advisory Council, a heterogeneous mix of academics, industrialists, officials and trade unionists proved to be surprisingly harmonious. The senior official member was Sir Alan Barlow. I was pleasantly surprised to find him constructive and knowledgeable. The Council began by considering the inner workings of the government's own civil science activities. Our first annual report outlined coherent institutional proposals, attempting to convince ministers of the need to take the long-term view on research – always difficult for national administrations with impending elections and the demand for quick success. It stressed also the vital importance of the technological development phase leading from research to production. There was an awareness of the danger that research undertaken by government for its own needs can be forced to respond too frequently to immediate issues and political events, and thus degenerate to pedestrian research and trouble-shooting activity. This we saw as a trap, especially in Britain where there existed many mission-orientated government laboratories.

The ACSP recommended that the main thrust of government

research should be under the three research councils – the Agricultural and Medical Research Councils and the DSIR – the DSIR having by then developed into a conglomerate of research councils via a series of research boards for its various laboratories, whose members consisted mainly of academics and leaders from the private sector. These three bodies, as the ACSP itself, reported directly to the Lord President and hence interaction between them could be expected to be relatively easy. The objective of this stress on the work of the Councils was to ensure that long-range investigations would happen without undue interruption or diversion as a result of short-term technical or political preoccupations of the Departments with their changing policies and demands for services. In this scheme road research came under The Road Research Laboratory of DSIR and not the Ministry of Transport; building research came under the corresponding DSIR Board with its own laboratory and not the Ministry of Works; medical research under the Medical Research Council and not the Ministry of Health; agricultural research under the Agricultural Research Council and not the Ministry of Agriculture. It was accepted that ministries such as Agriculture would continue to need their own facilities for applied research on more immediate problems. To complement this system and ensure the closest of relations between research bodies and users, it was proposed that each ministry should appoint a Chief Scientist to be responsible both for identification and formulation of any foreseeable problems in their sector, to which research might contribute and keeping their ministry and its sector informed of new research. They would take part as an assessor on the appropriate research council or board. This ACSP advice was accepted in principle. Several departments did appoint chief scientists: Agriculture, Works, Fuel, Power, the Home Office and even, for a time, the Scottish Office. It is significant that the Board of Trade and the Foreign Office ignored the advice.

These general arrangements functioned for many years and represented perhaps the nearest the country has ever approached to a coherent science policy. The main criticism (and weakness) of the system, levelled against it later by the Rothschild Report, was that in placing research too far from the user of its results, programmes could develop that were too theoretical, too long term in their significance and not close enough to practical needs. This is indeed one of the central dilemmas of science policy. Both short- and long-term approaches are necessary and means have to be found to develop both in close interaction. In all political affairs, electoral pressures force the administration to face immediate issues visible to the public so that longer-term, and possibly more important, projects may be sacrificed. Given the inherent long-term nature of the research and development processes, this can all too easily lead to ineffectual use of a country's scientific resources.

Following the Rothschild report, a customer-contractor principle was applied to the Research Councils, while later that of the Trend Committee led to the dissolution of the DSIR and the placing of responsibility mainly with the executive departments. The mechanisms for the provision of central advice changed several times subsequently. The main issues of science policy today remain unresolved, much the same as they were 50 years ago. Ministers responsible for science activities are immediately confronted with the problem of allocating resources amongst the wide spectrum of activities demanding support. The rapid growth of scientific activity and its institutions after the war, flourished under a mystique. It was assumed in quite general terms that more science would yield more prosperity, better health, more security and enhanced national prestige. This induced Parliament to vote large sums to programmes and projects about which they knew next to nothing. The scientists themselves were not slow to exploit this mystique to attract the resources that their unquenchable enthusiasms demanded. To them it was obvious that funds should be distributed entirely on the basis of the promise of particular projects and on the established reputations of their eminent leaders. Resources were growing rapidly, but in most countries it was soon obvious they were considerably less than demands for research support. Choices would have to be made, priorities established and policies evolved. It was unthinkable all this could be left to the scientists themselves. Science may well be a good thing but to be a national investment, it has to be planned and managed like other investments. Ministers, very few of whom had any scientific background, would have to educate themselves and take advice; hence councils for this purpose sprang up in country after country. How to devise resource-allocation machinery became their initial task.

Initially scientists were very suspicious of the concept of science policy; it connoted a bureaucratisation of science and a dirigism that would inhibit creativity. The very idea of research priorities was initially horrifying to them, but it soon became clear, for instance in radioastronomy as in other 'big sciences', that each professor of astronomy could not be given funds to build and maintain the very expensive installations which their ambitions required. Indeed, many smaller countries could not afford a single example of such equipment. National co-operative schemes as well as international cost-sharing solutions began to appear. In most cases, governments knew better than to interfere in the substance and detail of research, and restricted their role to allocation between broad fields and negotiations on particularly expensive projects. Decisions on smaller projects and applications for support by individual researchers were wisely left to the scientists themselves through committees exerting the 'judgement of peers'.

The early work of the ACSP was by no means restricted to these problems of policy and structure. Different ministers took an interest in our work. For example, Nye Bevin, the Minister of Health, asked Morrison if I could help him with some difficult interviews. This emotional Welshman was easily attracted by way-out ideas and schemes and responded to appeals from a range of charlatans, seers and crazy inventors. However, he was wise enough to protect himself from commitments that he might regret, so I had to sit in on some of his talks and warn him if their ideas were too far out to be taken seriously. I also supported Harold Wilson in answering questions at his press conference on the creation of the National Research Development Corporation (NRDC), designed to facilitate promising ideas from government and university laboratories through the development phase to commercial production.

We gradually formed an impression of ministers' characters and idiosyncrasies from our quite superficial contacts with them. Churchill's wickedly witty vignettes of his wartime colleagues – Attlee, 'a sheep in sheep's clothing; Ernie Bevan, 'a good working-class John Bull'; Cripps 'there but for the grace of God goes God' – certainly illuminate a fractional truth, but are cruelly diverting. Cripps, who I had known in the Ministry of Production, was certainly an ascetic, somewhat austere and forbidding in appearance and personality, but essentially humane and understanding, his apparent coldness redeemed by a vivid sense of humour. He was the arch debater and would argue convincingly on any subject whether he believed in it or its opposite. Morrison told me one morning returning from a cabinet meeting, how Cripps had exasperated his colleagues in a discussion of some issue charged with socialist principles. He had argued strongly and convincingly against what was their common belief. Finally, Attlee had broken up the debate saying, 'get off it Stafford, you know you don't believe a word of what you're saying'. Cripps had smiled and replied, 'Of course you are right Clem, but your arguments were so feeble that I just couldn't let them pass unchallenged.' The spirit of the Cabinet seemed to be pretty harmonious. The only minister toward whom I felt there was reticence was Harold Wilson. I sensed that he was seen as a young upstart and they were unsure how far he was to be trusted.

The more I saw of Herbert, the more I respected him. He wore the party dogma lightly and was always open to argument (e.g., on nationalization). I recall one occasion towards the end of our association, when the then Chairman of the Labour Party, Harold Laski, the left-wing economist from LSE, sought an interview to persuade him to set up a Social Science Research Council to parallel those for medicine and agriculture. I was very concerned at the time to ensure that research in the behavioural sciences should be built up. However, I knew the personalities involved and was certain that if a Social Science Council were to be established at that moment, its leadership would fall to a dominating group of experimental psychologists with little interest in sociological and anthropological matters. So I advised the minister to make very friendly noises and then say 'good idea, but premature'. As I sat beside him facing Laski, this was exactly what he did. Indeed, Herbert Morrison was convinced that the social sciences should have an important role in the evolution of both economy and society; he realised that their substance was still too slight to contribute much in the short term and resources would have to be allocated for their development, but he feared there was already a tendency for research in sociology to become too trivially theoretical. For this reason he wanted more serious investigations of real situations, such as those in industry. He pleased me by saying that I, with a background in the physical sciences, might have a better chance of marshalling support for the behavioural sciences than an 'inbred' academic sociologist.

The hectic load of legislation of the time laid a heavy burden on the ministers, and I was convinced that strong physical endurance was the first quality demanded of them. Indeed, exhaustion, heart attacks and death finally brought down the government. I was able to observe Morrison's work schedule from close at hand. He would wake at 6 a.m., read his official papers in bed before breakfast, then to the office to reply to letters, discussion with his secretaries on the day's business, phone calls with other ministers, and off in his car to speak at a lunch. In the afternoon he was usually in the Commons; debates and private discussions often continuing to late night divisions and usually with a dinner and another speech in between. Finally, he would reach home by midnight and rest for the next busy day. Frequently my only possibility of a talk with him would be to accompany him in his car as he sped to the luncheon speech of the day. On several occasions he tried to persuade me to stand for Parliament and offered to find me a convenient seat, but I was not tempted. The sort of life he led seemed to me a heavy price to pay for fame, influence and ephemeral power, while the utter boredom of the existence of back-benchers seemed even less tolerable. In the end, Morrison accepted my reluctance and agreed that my present work might be more useful and was certainly more satisfying.

I did have one direct brush with parliamentary life, trivial but amusing. The then chairman of the unofficial Parliamentary and Scientific Committee, the Socialist member for the Forest of Dean, Phillips Price, won the members' lottery enabling him to introduce a Friday morning debate. He selected Science Policy as his theme and spent some time with me mastering the details of the ACSP's new structural proposals. Then the member of the Opposition who had been given the task of responding to Phillips Price's introduction, invited me to lunch to find out about science policy and what its implications were. The debate was good-natured but pretty shallow. I sat in the official's box in the House. As it neared its end, Herbert Morrison slipped up to the box to be briefed on his reply. I had orchestrated this useful if quite minor debate rather well.

During all these months I worked closely with Tizard on day-to-day details of ACSP activity, but there was one subject on which I had virtual independence, international scientific relations. My old interest in international contacts, sparked off long ago in the Ministry of Supply, had followed me to my new job, as part of which I chaired an ACSP Committee on International Scientific Relations. This pleased me greatly as it enabled me to take part in the British Commonwealth Scientific Office and was able to continue old friendships. At this time many other countries were also considering how to develop post-war structures for science and were exceedingly interested in our plans. Accordingly I had to receive a flow of visitors; leading science planners from Finland, Germany, Greece, Italy, the Netherlands, Norway and the United States amongst others, as well as Israelis from the mandated territory of Palestine to be briefed about what we were up to. This entailed some missions abroad, including my vain visit to Delhi to advise Bhatnagar. It was about this time too that my old professor from Munich, Casimir Fajans turned up in London after having spent the war as a professor at Ann Arbor, Michigan. I took him to lunch at the Athenaeum and was amused by his deference to my new status.

It was also necessary to establish relationships with the scientists of UNESCO. This entailed a visit to Paris for talks with Julian Huxley who had become Director General, and its Science Director, my friend Joe Needham. The latter, true to his enthusiasm, was dressed in silk Mandarin's robes and sat in a Chinese decorated office, served by a Chinese secretary (or friend). We discussed UNESCO's plans to create a series of world science projects; initially the Hylean Amazon scheme and an international computation centre in Rome. These and other plans were regarded with suspicion by most Western countries as over-ambitious and costly.

More interesting to me was a more precisely detailed project to create a European laboratory for high-energy physics. This seemed, *a priori*, to be sensible for both scientific and financial reasons. The success of the Manhattan Project had given a big boost to nuclear physics and there seemed to be the prospect of unveiling the fundamentals of the nature of matter. Some half-dozen British research schools had plans to build high-energy particle accelerators. But these were astronomically expensive and if the government was to support this type of research, difficult decisions would have to be made on which of the teams to back. I felt that to carry out this work within any single European nation might be shortsighted in view of the huge resources required to ensure success. Even rich America recognised this problem and decided to concentrate skills and resources in a single institution, rather than leave them dispersed in many research schools. A corporation formed by a group of universities with the necessary skills and commitment enabled knowledge of the most relevant skills and experience of the country to be focused on a single enterprise, the Brookhaven Laboratory. The project of a group of European scientists, mainly French and Italian to create the Centre Europeean de Research Nucleaire (CERN) seemed a useful approach to a solution. Inevitably, the question of British adherence to the scheme came to our office. Instinctively I felt we should support it. To my disappointment British scientists were dead against it. Each felt he had to have such work under his own control. Even Pat (Lord) Blackett, the most respected of the physicists of the time, who had no personal ambitions in this field was against it. The only support I could find was from my friend John Cockroft, whose prestige did help to soften the position. Not being a physicist, or even a scientist of note, I had little say in the matter, but I decided to overstress the financial aspects and the political importance of such a European initiative. The Treasury and Cabinet Office staff had little understanding of the matter, but my Minister was open to persuasion. I worked hard on these people; indeed, reading my own letters and minutes of the time, declassified and open to public scrutiny years later, I realised how biased my action had been. Eventually we did join CERN. It has proved highly successful and is still active. I make no claim that my somewhat unethical tactics had achieved our participation, but I may have helped to tip the balance.

Among the more important aspects of our work was responding to questions posed by ministers. One of these came from the Chancellor of the Exchequer. Life in England in the few years after the war was grim. Rations seemed more restrictive than during hostilities when the public were tolerant of hardship, and food, although dreary, was sufficient. Now, in a world at peace, quality of life had to be sacrificed temporally in order to rebuild the country. This was much more difficult to bear. The Chancellor posed this problem: 'We are desperately short of hard currency for reconstruction, and much of what we do have has to be spent on the import of food. Can I save by reducing the rations still further without lowering the productivity of labour?' We responded to this by assembling a small study group led by a nutrition expert, Professor Jack Drummond, who was later murdered with his wife and daughter when camping in Provence. This group of nutritionists, ergonomics experts and economists reported that the savings could be achieved, but only by introducing a differential rationing system with workers in heavy industries receiving additional food in contrast to lowered rations of sedentary workers, housewives etc. This solution was regarded as politically unacceptable and food levels were maintained.

The Lord President himself posed a further and more important question. He asked for a review of possible contributions of science economic planning in the longer term. In private he explained that while a declaration of the general long-term influence of science on the economy was desirable, he urgently needed advice on how science could best help the economy during the next few years. At present the situation was abnormal; capital was acutely short, and reconstruction needs swallowed up what there was. Little was available for innovative new developments. This led to lively discussion on the ACSP about means of increasing labour productivity, as a result of which they formed a powerful Research and Productivity Committee to provide a detailed answer for the Lord President. This eventually sparked off a substantial productivity movement whose influence in Britain was significant, spreading rapidly to Europe and beyond. This is the subject of the next chapter.

17

The Productivity Saga

Sir Claud Gibbs, an energetic and friendly man, was chairman of this temporary Research and Productivity Committee and took to the task with enthusiasm. A member of ACSP, he was head of Parsons, an important engineering firm based in Newcastle. I visited his plant and discovered immediately that he was known affectionately by his workforce as 'the little bastard'. He had an uncanny memory for the names and personal circumstances of his numerous workers and made a point of forging relationships with each by showing interest in their personal affairs as he made his round of the plant each morning.

The committee soon warmed to its task. The trade union members were active and there was a good deal of discussion of what, if anything, they were likely to get from increasing productivity. Before the war there had been much worker opposition to Taylorism, regarded as a management tool for the exploitation of workers. By 1947 a few progressive firms such as ICI had managed to change this perception. Time and motion study had changed its name to *work study* and participation of the workers was seen as profitable. The committee stressed the centrality of worker participation and equity in any national crusade to increase productivity. Furthermore, human relations in industry should be a major research theme.

The committee duly advised that technological development based on scientific discovery was likely to become a main moving force of the economy and hence basic scientific research should be supported strongly. However, the inevitably long lead time from discovery in the laboratory to a marketable product meant little or no benefit to the economy from new research could be expected till much later. Efforts should be made to apply scientific approaches to existing systems, such as by exploring how wartime successes of operational research techniques could be applied in industry. Similarly, human industrial relations should be studied to establish increased labour productivity as a common aim for both managers and workers. To pursue these aims the creation of a Committee on Industrial Productivity (CIP) was recommended to advise on efforts to achieve productivity improvements within five years.

Normally such a recommendation, even if accepted by ministers, would be subjected to detailed scrutiny by civil servants and particularly by those departments most closely concerned, and would eventually emerge with a mandate carefully negotiated to avoid modification of the existing power structure. But no ordinary recommendation this; it resulted from the personal initiative of Morrison and Cripps, who had now moved from the Board of Trade to become Chancellor of the Exchequer, and had the approval of the Prime Minister. It was put into effect, mandate unchanged, thus bypassing consultations with departments, much to their discomfiture.

I shall return to the mixed successes of this committee later but now turn to other productivity initiatives of the time. As President of the Board of Trade, Cripps, with a scientific education, had set up a Special Research Unit (SRU) in his department. It was headed by Cecil Gordon, a South African Jewish Rhodes Scholar, a brilliant if eccentric biologist and a Marxist, whose wartime operational studies for Coastal Command, 'Planned Flying, Planned Maintenance', had contributed greatly to the closing of the Bay of Biscay to German submarines. Gordon was the most intelligent, warm and genuine person I encountered during this period. SRU was a personal move by Cripps, without consultation with middle-level civil servants who were fiercely *status quo*, resentful, and hence unlikely to sympathise either with its interventionist ideas or its operational style that had little respect for hierarchies. Its initial work concerned consumer needs, variety reduction and the structure of the cotton industry. Unnecessary product variety in a number of industries was seen as wasteful and inefficient. Its reduction could lead to longer production runs, reducing costs and adding significantly to productivity. Gordon's work on cotton spinning, on which he co-operated with L. H. C. Tippett of the Shirley Institute (the British Cotton Industry Research Association), involved productivity concepts even more directly. In industries where firms operate a common technology, productivity measurement should be able to identify gross disparities between the productivity of the best and the worst, suggesting how the latter might be encouraged to reach the level of the former. We shall return to Gordon later, but I should add that he and the SRU did not long survive the replacement of Cripps by Harold Wilson at the Board of Trade.

Then came a bombshell in the form of a report comparing the productivity levels of UK and US firms in upwards of 20 industrial sectors. This study by Dr Rostas, a statistician in the Board of Trade, appeared to show that in all but two of the sectors, the productivity of US firms outstripped that of UK firms. Naturally, this was unpalatable to the Federation of British Industry (FBI) and several attempts were made to discredit its methodology. It was gradually accepted, though, that it gave a true picture of the situation, striking a deadly blow at British industry's complacency. Cripps was impressed and depressed by the Anglo-American comparisons. The Marshall Plan had just been announced and Cripps, who had initially been suspicious of it as a possible threat to national sovereignty, quickly began to consider how best to use it to reduce the productivity disparities.

At this moment a new actor appeared on the scene in the person of James Silberman from the Bureau of Labor Statistics (BLS) in Washington. Dr Rostas brought him into my office one morning. I have no idea who invited him or sent him to London. Jim was an extremely active if somewhat brash young man with an unassailable faith in the virtues of productivity measurement as key to both industrial success and, incidentally, towards the American way of life. I saw him as a valuable ferment in the London scene. Rostas took him to see Cripps who was impressed by his message and began to consider how it could be used. Silberman's missionary zeal was irrepressible. He stormed the FBI and the highest levels of the Board of Trade and delivered his message to everyone even half willing to listen. The success of his mission paved the way for the creation of an Anglo-American Productivity Council (AAPC).

I decided that it would be useful at this stage, to visit a number of American firms to witness productivity measurement in operation and sound out the attitudes of both workers and managers. Once in Washington, I duly checked in at BLS whose director, Duane Evans, was outstandingly helpful, describing their methods and making arrangements for visits to firms. I started with a visit to Paul Hoffman, former President of the Studebaker Corporation, and now head of the Economic Co-operation Administration as the Marshall Plan was formally known. I was impressed by his acute alertness; he spoke exceedingly clearly and listened well. My intention had been to advocate productivity as a central feature of the Plan, but to my surprise he started by stressing just that. 'Europe', he said, 'has about double the population of the United States and its industry yields about half the product. This gross disparity tells us where American aid can be critically helpful.' He then outlined his hopes for the success of the AAPC, stressing that the performance gaps between the two industries were due to cultural rather than technological factors.

By this time Jim Silberman had returned from London and joined me in visits to a number of firms in Pennsylvania manufacturing men's shirts, which were participating at that time in productivity comparison. I had previously visited many US factories and was well aware of the easy shop floor relationships. Now, seeing a series of medium-sized firms in the same sector, highly competitive but at the same time co-operative, I understood how different the industrial cultures of the two countries were. In the US plants, hierarchical discipline was well respected but needed no enforcement; there was no caste or class distinction; no 'them and us'. Workers, shop stewards, foremen and middle management chatted together, dressed similarly and lunched together without distinction. The unions were strong and bargaining was tough, but the impression was of company success being a shared aim of management and labour on the assumption of a just distribution of profits.

I met the head of the Lady's Garment Workers' Union and was amazed to learn they operated a production advisory unit, available to any ailing firm in the industry, in the hope of raising its efficiency and hence profitability in order to protect jobs. The unit advised on plant layout, undertook method studies, suggested technical improvements and advised on financial management. Apparently the system was well used and successful. It is difficult to imagine struggling companies in Britain swallowing their pride by asking their workers' union to save them from bankruptcy!

I then moved to Detroit, accompanied again by Jim, to have a look at productivity attitudes in the line-production of automobiles. At the GM Chevrolet plant I had illuminating discussions with line workers, supervisors and middle management, talking about work satisfaction for those on mindless repetitive production, incentives and attitudes to change. I remember a lively discussion of the Elton Mayo studies that indicated slight gains in productivity seemed to follow each of a series of changes, a final modification bringing the situation back to its original state provided yet another small improvement. The rosy picture of US industrial relations contrasted with the more rigid two sides of the table situation in England. There, unquestioned acceptance of the inevitability of class hostility had been given a spurious virtue by being dignified as 'creative friction', a euphemism for 'creating friction'. We also play this confrontational charade in the House of Commons. I wondered how the American easy relations approach would stand up in a major power struggle. I was to have the answer that very day.

I had been invited to dinner in a Detroit hotel by one of the vice-presidents of GM. When in Washington just before, I had stayed with my old friend and neighbour from the BCSO days, Paul Sifton, the Washington representative of the Automobile Workers Union. He suggested that when in Detroit I should have a discussion about Anglo-American productivity matters with the head of the Union, Walter Reuther. Reuther, an intelligent, articulate and charismatic man was probably the most outstanding personality in the American trade union movement at the time. He was very popular with members and had built up strong educational and leisure programmes in the union. It would be useful to hear his views on productivity. I had therefore agreed to meet Sifton in the lobby of the hotel after the dinner so that he could drive me to Reuther's home in an outer suburb.

The dinner started very amicably. There were about a dozen guests, mostly GM executives. The host was warmly welcoming and we had useful talks about how the AAPC might work. Then as the main dish appeared, conversation became darker. They began to talk about their difficulties with the union and how they suspected infiltration by communists. There was an arrogance in their remarks. This was the period when they promulgated the slogan 'What is good for General Motors is good for America', and this included the hounding of communists. They seemed to forget that I, a foreigner, was present. Arrogance had become hubris. One of the group attacked my friend Sifton by name as the union's communist in Washington. I knew his political history well, he never had been a communist, but it would have been imprudent to enter into the argument. My embarrassment grew; the dinner was running late and the time approached when Paul would be waiting for me outside the door. I just had to get out and away before they caught me greeting the man they had just denounced. At that moment the waiters arrived bearing exotic ice-cream confections. Their consumption would take some time, so this was the moment to make an excuse and withdraw. But what to say? I had it. I made some vague hint of an assignation, 'Cherchez la femme' would at least be understood by these people and seen as respectable. They laughed and joked, wished me well and said goodbye at the door. I staggered out into the arms of Paul Sifton.

We drove some 20 miles northwest, urban sprawl all the way, and eventually entered a quiet street in a leafy suburb, to be stopped immediately at a barrier by the police. The whole block was illuminated by electric lamps as at a night football game. Apparently there had been an assassination attempt on Walter Reuther a couple of days earlier. He had been in his kitchen with an uncurtained window and was opening the refrigerator when a gunshot shattered the window, hitting the fridge door as he was opening it. A second sooner or later, he would have been killed. The union people were unsure whether the assailant was a communist enemy or a company hit man. After inspecting our papers the police let us through the barrier and we drove a few yards to draw up in front of the Reuther door. There we were handed over to two armed union guards. One was a large man who I imagined looked exactly like a Chicago gangster, the other a small, stocky person with a long chin that seemed to reach almost to the ground. Their examination was much less perfunctory than that of the police, but soon we were let in, entering the sitting room to find Walter and his wife listening to classical music. We had a long and useful talk about productivity that gradually turned into more general discussion about the state of the world and eventually to vaguely philosophical ramblings appropriate to the end of a night of good conversation. I was deposited at my hotel about 2 a.m. Jim was waiting up for me, relieved we had avoided what might have been a tricky little international incident. I returned to England on the Queen Elizabeth. Among the passengers were the American members of the AAPC, on their way to London for the first meeting of the Council. The voyage gave me ample time for discussion with both managers and union representatives of the team. I could not resist dropping some discrete hints that the few show firms they would be visiting could hardly be seen as typical of British industry.

Returning now to the CIP; by the time of its creation, action on productivity had passed to industry itself. The joint management/labour visits to America organised by the AAPC were already in full swing. What was left to the new committee, therefore, was an exploration of new approaches and of better methods of using existing experience and services. The Committee was an anomaly, being seen as a direct initiative of Morrison and Cripps with the backing of the Prime Minister, with little consultation with the executive departments; inevitably therefore the bureaucracy regarded it with suspicion, if not hostility. As it stemmed directly from ACSP discussions, Tizard was named as its chairman. I sensed he was uncomfortable with social issues and mistrusted the social sciences, leading him to conceive the new body as an umbrella, supervising work devolved to panels under more specialised leadership. The main Committee thus became more or less a façade. The four panels were: Technology and Operational Research chaired by Sir William Stannier, my old colleague of Ministry of Production days; Import Substitution under Solly Zuckerman; Human Factors in Industry chaired by Sir George Schuster; and Technical Information Services in my charge. The Import Substitution Panel had a modest success and, unlike the others, worked closely with the Board of Trade. I shall mention the work of my own panel in a later chapter, considering here only the other two.

Little thought had been given to the servicing of this heterogeneous grouping of bodies. My own small secretariat was obviously inadequate, having little or no expertise in the new fields. No suggestion was made that it should be augmented. Instead, the CIP at its first meeting created still another grouping, its 'Working Party', with a vague remit to rove around any subject matter it felt might interest the Committee or its panels. I was designated to lead this group of individuals mainly selected by myself from within government service. I attracted a fluid group of young innovative people, the most vocal being Cecil Gordon; it also included Tom Easterfield and Ronald Stansfield from SRU, Rostas, Willie Robertson from DSIR, Dr Womersley from the NPL and Tippet from the Shirley Institute. This strange creation proved to be an exciting group, ready to discuss any subject 'under the sun', a brains trust on every aspect of science in industry. Although having no formal status, it not only served but also gave tone to the Committee and its panels. Apart from its committee-servicing activities, the Working Party, considered many topics. I mention a few: the use of ordinance factories to replace manufacturing capacity destroyed by war; the scientific validity of time and motion study; management and trade union education; and management/labour relations.

Operational research (OR) had an important influence on the outcome of the war. It was critical in defeating the Luftwaffe at the Battle of Britain, and played an important part in mastering the submarine menace as in many other operations of defence and attack. Yet it had no place in the general public's image of the wartime successes of science despite efforts to secure recognition. This was partly due to its technical nature, but also because it had been almost wholly secret, and its methods so closely interwoven with the subject matter that even after the war ended, its work could not be published or discussed openly. Although wartime OR was barely known to the general public, there were many people in prominent positions who had contact with it in action, or knew at least something of its nature and successes. It should be realised that, in the first decade of peace, many academics like me, caught up in the war machine, were still involved in government activity, while the Treasury, and bureaucracy in general, had not yet completely regained control of government machinery. Thus there were many individuals in positions of influence who felt that OR, so successful in war, should have a useful role in rebuilding an efficient civil economy.

This conviction was highlighted in the panel's title 'Technology *and Operational Research*'. In fact it was little concerned with technology, but began to explore means to improve productivity in the cotton industry, using Tippet's comparisons of productivity achieved by individual firms that varied considerably between the best and worst. It postponed consideration 'of how operational research might be best extended to civilian industry in general' until it received a report on the subject from the Working Party, which quite quickly was produced. It was a lengthy document, written mainly by Gordon and Easterfield, entitled *Principles and Practice of Operational Research*. It sketched the wartime development and achievements of OR, explained (not too clearly) what it was, gave examples of OR-like work in civilian industry, described its various techniques and organisational methods and made a number of suggestions of areas of industry seen as propitious for its application. Amongst its recommendations, the report included several that went beyond its original mandate of industrial application, including suggestions for improving the efficiency of government administrative machinery and some wider national issues.

The Panel received the report with some enthusiasm, recommending that it was of 'such importance that it should be published and made widely known among both government departments and industrial executives'. The panel members did note, however, that it might need revision of its recommendations concerning the establishment of operational research groups throughout the administration. One member of the panel, a treasury civil servant, circulated the draft report around Whitehall for comment and it proved to be a veritable cat among the pigeons. Some dozen Departments responded. A few, including Education, were in favour of its publication, but the Cabinet Office, the Board of Trade, The Ministry of Labour and the Central Planning Staff of the Treasury were all strongly opposed to its publication in anything like its initial form. Objections in detail came from many sources. Most significant was the criticism that many policy issues raised were essentially matters for Ministers and hence could not be raised in a report available to the public. The report was seen as dangerous in that it touched on many sensitive areas and suggested interventionist thinking, diametrically in opposition to the *laissez-faire* philosophy of the bureaucracy. In fact, the criticisms of the report were being used by the power centre of the civil service as an excuse to take a swipe at the heretical thinking emanating generally from the Lord President's Office.

The Working Party recognised that there was no hope of publication of *Principles and Practice* without radical changes, and suggested that separate versions should be prepared for submission to Ministers and for general distribution. The redrafts took account of most of the objections and, after endorsement by the Panel, were discussed by CIP. It decided to send the new version once again to Departments for comment and that meanwhile, publication should be deferred. This effectively killed it. The comments received on the new draft were on the whole favourable. The working party had tried to save the report by removing the politically objectionable, but in the process making it less interesting. There were still criticisms of the document's length, verbosity and over-reliance on military examples. The Economic Information Unit prepared a new version, about a sixth of the length of the original. This was compiled by an economist rather than a scientist, on the assumption it would make it more intelligible! This trivialised document was sent to the Working Party with a covering note and almost completely derogatory marginal notes. The covering note gave the *coup de grâce*: 'The very simplicity of the language brings out very well the weaknesses of the original report and therein lies its value – it is difficult to believe that this paper really says anything at all.' The working party was understandably indignant. Evidently the distribution of this paper would do serious damage to the spread of OR in industry, so it disappeared from view. Thus ended the concern or rather lack of it by the government machine for OR. Meanwhile, however, the proponents and practitioners of OR had founded the Operational Research Club (ORC) that discussed in useful detail the various techniques of their metier. Sir Charles Goodeve, a prominent wartime leader in the field who had become director of the Iron and Steel Research Association, was successful in applying OR in that industry.

This was undoubtedly the least successful episode of my career. The cat (in this case the bureaucracy) did not budge an inch. Had I been less naive and less idealistic they might have moved a little – but I doubt it. Nevertheless I learned a lot.

The Human Factors Panel was vastly different in its approach. Its Chairman, Sir George Schuster, the last Governor of Bombay under the British Raj, was dynamic, self-assured and the apotheosis of urbanity. Forceful and knowledgeable, he had a deep conviction that too little was understood about the reality of human relations in industry, that the social sciences were essential in establishing harmony at the workplace, but that they were as yet insufficiently developed to be of much use. The core of his panel's work therefore was the stimulation of social research in actual situations in industry, a 'clinical' approach.

This innovation from within the government machine did not knit well with the *status quo* philosophy of the relevant Department, the Board of Trade. Furthermore, Schuster, in a paper presented to the Economic Planning Board, raised suspicions of his intention to interpret the Panel's terms of reference quite widely and embark on sensitive matters of policy. The bluff and powerful Permanent Secretary, John Henry Woods, who knew me from my Ministry of Production days, was warned of these dangers by his subordinates, who urged him to talk to Schuster in self-defence. Woods agreed and in a memorandum wrote, 'I am wholly sceptical about the likelihood of the CIP or the panels keeping off the grass of policy. In particular and *expereto crede*, watch Dr King. I will see Sir G. Schuster, but I shall need a witness.' What transpired, we shall never know, but I doubt that Schuster was much impressed by the warning. In fact, neither I nor my working party had much influence on the work of the Human Factors Panel, mainly because, in addition to having a secretary, it had the full-time services of a scientist, J. G. W. Davies as assistant to the Chairman. This relieved me of what would have been an intolerable burden. Nevertheless, the Working Party participated in and strongly supported the detailed working of the Panel and its research advisory group of scientists.

The Panel decided at the outset to design a programme of research projects to be carried out in industrial situations. The intention was to meet simultaneously three main needs, to: add to existing knowledge, then far from adequate; train research workers on the job to add to the numbers of the very few trained research workers then available; bring such research to the attention of industrialists to encourage them to appreciate its usefulness. The Panel worked closely with relevant private groups such as the Tavistock Institute, the Institute of Industrial Psychology and the Sociology Department of Liverpool University. Considering the short life of the Panel, its output of books and papers was impressive and its influence on industry considerable. As we shall see later, its existence marked an important growth point in the development of behavioral sciences in Britain.

In the meantime the main Committee, discouraged by the OR debacle, had gradually accepted Tizard's view that, bearing in mind the long lead time of research and development and the current shortage of capital, research had little to offer short term-productivity. It therefore recommended that it should be discharged and its dissolution was announced in the House of Commons in July 1950. However, arrangements were made for its residual activities to be absorbed in the traditional structure of the civil service. The work of the Import Substitution Panel went to the Board of Trade, that of the Technical Information Panel to DSIR. Sir George Schuster was able to announce that arrangements were nearing completion for the work of his panel to be continued by two committees set up jointly by the Medical Research Council and DSIR. In the event, negotiations were protracted and terms of reference proved difficult, so that the new committees on Human Relations in Industry and Individual Efficiency in Industry saw life only in early 1953. I shall discuss this in the next chapter.

A further aspect of the productivity movement had aroused my interest – its place in the rebuilding of industry throughout Europe. Ever since my visit to Paul Hoffman in Washington, I had followed the development of the Marshall Plan's activities in Europe and was fully aware of the progress of the Organisation for European Economic Cooperation (OEEC) in Paris. Rapid progress was being made throughout its member countries in making available the necessary investments and collectively to create the European Payments Union that did so much to increase inter-European trade. But it seemed to me that too little attention was being given to technological aspects. I felt there was a danger that industry would be rebuilt with plant and machinery 'bought off the shelf'. While there had been little technological development in civil industry in Europe during the war years since it was fully engaged in defence production, nevertheless there had been some interesting innovations in America, Switzerland and elsewhere. Also, prospects were bright for some radical developments by adapting some spectacular wartime technology to civil use. Surely it would be wise to scan these potentialities before conventional systems were rebuilt, otherwise European industry could well be re-equipping itself for obsolescence. I discussed this with Herbert Morrison, suggesting there was a need for OEEC to have a committee on science and technology. He understood the importance of bringing technology into the European discussions and arranged a lunch at which he and Stafford Cripps might raise the matter with Bill Batt of the SKF Company, who was the Marshall Plan administrator for Britain. The latter agreed that the Americans would support a British proposal to the OEEC Council.

The Council readily agreed to create this committee, even setting an early date for its first meeting to which I was sent. The French had afterthoughts and objected to the concept of such a body as mixing academics in the serious business of politics. As is usual, however, in such organisations that work on unanimity, avoiding majority decisions, a compromise had been agreed. Instead of a full committee on science and technology, they settled for a temporary working party on scientific and technical information. I heard this news only on arrival in Paris when I was met by Pat Young the OEEC Officer in Charge. I felt our plans had been diluted to near futility.

Young had arranged a dinner that evening for some of the working party delegates. We met on an ice-bound Sunday evening at a superb restaurant, Chez les Anges on the Tour Maubourg. The small group included French economist Professor Jean Fourastier, known for his historical studies of productivity changes since the Middle Ages; my friends Edy Velander, director of the Swedish Royal Academy of Engineering Sciences, and Robert Major of Norway; as well as a sharp-witted Greek, Professor Nicolides. Our disappointment was somewhat tempered by gastronomic excellence and calmed by good wine. Conversation flowed and the noise level rose as the evening progressed. Later, over a final cognac, someone announced that there was an obvious solution to our problem: 'After all, information arises from scientific research and industrial production is its utilisation. So, let whosoever is elected chairman of our working party tomorrow make such a declaration and rule that scientific and technical information must be seen to include everything from the laboratory to the production floor. That will give us the scope we hoped for'. In the slightly inebriate environment of the late evening, this was greeted with acclaim and we all departed happily through the frosty night to our various hotels

In the intense cold of the next morning, we met at the headquarters of the OEEC, feeling unsure of ourselves. The meeting was opened by the Secretary General, Robert Marjolan, who spoke a few anodyne words about the importance of scientific information and invited proposals for the chairmanship. Mine was the only name put forward and I was happy to accept this duty, but with horror I remembered our wild alcoholic pact of the previous evening. Marjolin withdrew, leaving me the chair. The first item on the agenda was 'discussion of the mandate'. Was I brave enough to implement our wise but irresponsible decision that could be perceived as running counter to the wishes of the 19 member governments? I decided to do so; let the cat turn round. I had little to lose; at worst they might sack me. I proposed that our mandate on scientific and technical information should cover the whole spectrum from research to production. There was some bewilderment but no dissent – and so it was. This was the beginning of a new chapter of the productivity saga.

18

Multidisciplinarity

When the Attlee government finally fell, our little empire collapsed. Churchill, once again in power, named Cherwell Paymaster General. The old rivalry caused Tizard to resign immediately. He was replaced as chairman of the ACSP by one of the most eminent scientists of the time, Alexander (later Lord) Todd, Nobel laureate, Professor of Chemistry at Cambridge. Tizard had worked full time as science adviser, but Todd, leading a large research school, could only devote minimal time to the job, meaning he had much less continuity of contact with ministers and the influence of the ACSP quickly declined. The defence segment of Tizard's work passed to Solly Zuckermann who retained it for many years through several administrations, becoming known to his cronies as the 'scientific Vicar of Bray'.

My Central Scientific Secretariat melted away. My reputation with key figures in the administration was such that I expected they would throw me out. To my surprise, I was kicked upstairs instead, promoted to Chief Scientific Officer in DSIR. I was appalled by the prospect of working within the constriction of a conventional government department; I could see no future. I went home that evening and told Sally that my career was at an end. In retrospect, I am convinced that these changes, diminishing the influence of science in government, had no basis in national politics, the new Conservative government was quite ignorant of the internal situation. Essentially what happened was that mandarins of the Treasury took the opportunity of the change of government to eliminate the still weak nucleus of a rival power centre around an ambitious, departing Lord President; a manoeuvre that passed unnoticed in the welter of administrative change. My successor was an administrator without a scientific background. I met him later as the British representative at an OEEC committee I presided over. I listened with horror at his introduction to a policy statement with: 'As my masters of the Treasury have directed me to say . . .' Obviously we were not in the same ball game.

Arriving in DSIR, I found the situation less depressing than I feared. My friend Appleton had just resigned to become Vice-Chancellor of Edinburgh University and been replaced by Sir Ben Lockspeiser whom I knew well. Apparently, a decision had been made to review the department's functions and structure, which had remained unchanged since before the war. Lockspeiser, the forceful Permanent Secretary of the wartime Ministry of Aircraft Production, had been moved as a new broom to DSIR, and asked that I be seconded to head a new Intelligence Division. This explanation of my transition did much to reassure my self-esteem for I knew that Lockspeiser was an open-minded, innovative man with whom I could work. A further inducement was that I could bring with me my long-held international co-operation activities – and also my very knowledgeable secretary. This enabled me to pursue my unfinished productivity business at OEEC. So I was only half-way back to the drawing board

DSIR had been created in the middle of World War I when it was realised that Germany had outstripped British industry in scientific and technical skills and products, but elements of the new organisation dated back to 1899 when the NPL was created. While formally a government department, NPL was under the scientific supervision of a committee initially set up for the purpose by The Royal Society. This concept of involving academics, industrialists and other users of research results in the decisions of an official body became the key feature in all DSIR's activities. Though remaining a government organ staffed by scientific civil servants and headed by a Permanent Secretary, DSIR had its programmes developed with the help of an advisory (and later executive) committee of academic scientists, industrialists and trade union representatives. In the same way, each of the National Laboratories of the department, on research fields such as chemistry, building, roads, hydraulics, forest projects and fuel, as well as the Geological Survey, were all advised by boards of academic scientists and representatives from the sector concerned.

A further function, assumed by DSIR as early as 1917, was to encourage and assist the creation of research associations for particular industries. These bodies, the number of which grew at one point to 50, were financed partly by government through DSIR and partly by their subscribing member firms. Each association attempted, sector by sector, to undertake co-operative research to upgrade the efficiency of its industry as a whole. It made no attempt to replace competitive development which remained the function of each individual firm. The role of DSIR was not merely one of passive financing, but of working with progressive industrialists for the promotion and setting up of appropriately structured associations. Into this well-disciplined, competent, but not too imaginative department it was now proposed to insert a new element, an intelligence division, under my leadership. There was no job description, no mandate. No one was able to explain what was intended by the word 'intelligence'. Certainly we were not being modelled on MI5 to indulge in industrial espionage, nor were we concerned merely with the acquisition and distribution of information. Lockspeiser agreed with me that the intention was to inject some modicum of economic and social thinking into the work of DSIR and to identify and encourage innovation.

I found my new colleagues competent and co-operative, but sharing with me a certain mystification as to what I was there for. The atmosphere was very low key and I could discern little passion in their work. The office seemed to be dominated by the chief administrative and finance officer, a loyal representative of HM Treasury. I had few real difficulties with him. He would warn me endlessly about the shortage of resources to launch my plans, but they always proved sufficient. All in all the environment was stiff and, although not hostile, vaguely discouraging. I sensed that Lockspeiser was also uncomfortable in his new job and hardly knew how to handle it. Years later, when I was back from Paris for a few days, I happened to run into him on the steps of the Athenaeum. He greeted me loudly, saying, 'the best day of our life, wasn't it, the day we left DSIR?' This was an exaggeration, but it was a rather uncongenial place.

Being a new division, my first task was to recruit assistants. I had inherited one well-established group from the old DSIR on scientific and technical information, a topic brought with me from the Lord President's Office. I already knew these colleagues well and they responded readily to my ideas. (Work with them is described in the next chapter.) I was given a deputy, John Knox, as steady as his name implies, who tolerated my nonconformity and protected me well against bureaucratic pitfalls. For newer approaches I looked to the members of the 'Working Party', managing to recruit Tom Easterfield and Ronald Stansfield, who had taken over as Schuster's research assistant, making him invaluable in follow-up work for the Human Factors Panel. I then approached Professor Simey of Liverpool for help in finding an imaginative industrial sociologist. He produced Ron Gass who had completed a useful research study of human relations in French industry. He was a person of many talents, with a nose for new ideas and a gift for leadership. He worked with me for nearly 30 years. I also found a good economist and an assortment of engineers and natural scientists, in this way building up an interdisclipinary group, not dissimilar to those of the 'Working Party'. Over the next few years we generated a good deal of intellectual excitement; although at a lower level in the hierarchy, we had fewer constraints if less influence. I was able to do pretty well what I wanted.

At this time I was somewhat obsessed by the concept of multidisciplinarity. I felt that many attempts to solve complex problems failed because one or more facets of the difficulty were undetected or ignored as insignificant. This was clearly the case in the neglect of the human factor in industry. It should be possible for people from the natural and behavioural sciences and engineering to learn to communicate, despite their stark differences of vocabulary and thought modes, as long as they are working together with a common aim. On the academic stage, co-operation and often even contact between the disciplines is trivial, while within the social sciences there can be suspicion and even hostility between various factions. While encouraging the involvement of sociologists in industry, I was seeking a suitable situation in which to test my ideas. That had to wait for a couple of years, but in the meantime my little group was finding its own multidisciplinary coherence.

Agreement on membership and detailed terms of reference of the DSIR and MRC committees took a considerable time. Two committees finally sprang into action just when substantial research resources became available through Counterpart Funds derived from American Conditional Aid. The committees had been planned in the expectation that they would be supported from normal Treasury funds. The unexpected appearance of American funding hardly influenced their activities, but it gave them a flying start. One favourable feature of the new structure was its acceptance of the customary DSIR and MRC practice of inviting committee members in an individual capacity rather than as representatives of their organisations as with the Human Factors Panel. There the representatives of the trade unions, employers and industrial federations inevitably inclined to a defensive 'two sides of the table' approach. The new Human Factors Committee initiated or selected a substantial nest of projects, useful both for the practical knowledge they would generate and the development of a cadre of research workers in applied sociology and social anthropology. They included a study by the Acton Society on management succession and a significant review by Burns at Edinburgh University on attitudes to technological change. A significant study was commissioned on the effects of incentives, and yet another on the effects of work study on relations within firms.

The Joint Committee on Human Relations was replaced by the DSIR Human Relations Committee in 1957 and in 1965 it was in turn superseded by the autonomous Social Sciences Research Council (SSRC) of today.

The Intelligence Division generated many new activities. One, directly operational, was through a small unit teaching and demonstrating work-study techniques, mainly to smaller research associations in the hope they would pass them on to their member firms. Another was a series of studies on research management. When in the Lord Presidents Office I had felt strongly that there was a need for improvement and training in industrial management. The success of American business schools indicated the desirability of establishing similar bodies in Britain. I had approached a few universities and industrialists, but found little interest. On the industrial front, 'managers are born, not made' was the claim, while in the universities, 'management has insufficient intellectual content to justify it as an academic subject'. There was obviously little hope of having this topic taken up through DSIR, but I discovered that little thought had been given to the special case of management in the research system, a subject of evident concern to the department. In general management, the unique element is economic success – dominance of the bottom line. Research has two goals, scientific success and useful application. Research workers thus have two motivations; to achieve economic success and to gain peer approbation for scientific quality. These two goals are often in conflict. A research project may succeed brilliantly on scientific grounds yet not be commercially viable. In which case the research manager may have to terminate the work just as it is giving creative satisfaction, much to the dismay of the scientist. Thus economic practicability, scientific quality and staff morale have to be finely balanced. This and other problems of research management intrigued me, leading me to spend much time, discussion and observation on the subject. I was soon invited to lecture on research organisation and management across Britain and also in the Netherlands, Sweden and Israel.

Somewhat later the intractable problem of economic evaluation of research programmes arose. Research was just beginning to be seen by both governments and firms as an area of investment. How could its yields be measured? The outcome of specific projects resulting in competitive production was easy enough to calculate, but how about the debit side of much uncompleted or failed research or the many small improvements to products and processes? This problem was important to DSIR in relation to its research associations which were continuously under criticism as dubious claimants for public funds and, with research aimed at general improvement, seldom ended in measurable monetary gain. We attempted assessment of the direct profitability of a couple of these bodies, but with pretty shaky results. Even in today's climate of cost benefit, this difficulty remains unresolved.

In addition to these internal activities, I had a number of extramural tasks. I sat on the research committee of the Federation of British Industry. I was on the Board of the Science Museum and was the Lord President's representative on the British Council. The discussions on these bodies, for which I was given no government brief, greatly pandered to my dilettante proclivities. The international co-operation side of my functions occupied a large proportion of my time. I attended as UK representative at meetings of the London BCSO, enjoying their evocative memories of my Washington years. I accompanied Ben Lockspeiser to the second British Commonwealth Science Conference in Australia, leaving London in advance and spending a few days lecturing in Greece and Israel before meeting up with Ben in Calcutta. The stop in Athens was memorable for one event, I was lunching at the Embassy when news came through of the death of George VI. The meeting in Australia was interesting but not significant. The membership had changed significantly with the arrival of Pakistan and there was no longer a representative from the Palestinian Mandated Territory, now Israel. We travelled by bus between meetings in Sydney, Melbourne and Canberra which gave an indelible impression of the back country with its vastness of open land and eucalyptus trees.

An interesting development was a semi-formal arrangement with the Foreign Office to keep an eye on the activities of scientists attached to the Control Commission for the British Zone of occupied Germany. This involved frequent visits to Germany during which I met most of its new leaders of science and became familiar with the evolving structures and policies. This stood me in good stead in later years. On a visit to Gottingen, I had a meeting with Otto Hahn, Germany's great nuclear physicist, who, had Hitler listened to scientific advice and become aware of its potentiality, might well have become the architect of a German atomic bomb.

A further contact with the Foreign Office arose in connection with a decision to appoint science attachés to our embassies in Paris and Stockholm. I had the job of appointing and supervising the incumbents. The attaché in Sweden had to cover developments in all Scandinavian countries. On one occasion I accompanied him on a visit to Copenhagen where we spent a pleasant afternoon at Niels Bohr's home – another of the great names associated with our growing understanding of the atom. Bohr had been smuggled to Sweden in a rowboat from Denmark during its German occupation and then brought to London where he was held secretly as Mr Baker. I was aware of all that and was happy to meet the old man. We sat, over cups of coffee by the fireside, listening to him gossiping vividly about his encounters with other famous scientists of the day. His mannerisms were amusing, if distracting. He was always pipe in hand, he would first knock out the ashes against the fireplace, ream it out with a pipe cleaner, stuff some tobacco in, tamp it down, light a match and attempt to light it. After a few puffs, engrossed in the conversation, he obviously forgot about the pipe and appeared to be surprised that it had gone out. Then he would relight, leaving the match in the bowl. This cycle was repeated several times until the pipe was full of semi-burnt matches. At this point the next lit match would set fire to the others. With alarm he would shake the contests into the fire and the whole performance would recommence. This mini-drama continued throughout the three hours of our discussion. I doubt if he ever actually puffed a breath of tobacco smoke, but it somehow lubricated his narrative.

In the mid-1950s there was considerable interest in the automation of industrial processes, the hope being it would revitalise repetitive and line manufacture – this hope was premature. Current mechanical and electronic devices were too clumsy to achieve significant automation, real advances would only come with the rise of microelectronics. Nevertheless, popular as well as professional imagination concerning automation was such that a study of the topic was warranted if economic as well as technical elements were considered, along with the human, cultural, skill and educational consequences involved. We had the right personnel for this task and the study was initiated under the leadership of Ron Gass. In addition, realising that automation would raise concerns for trade unions, we involved Ted Fletcher, head of the research department of the TUC. The study was duly completed and published by HM Stationery Office. By happy coincidence, publication of the report coincided with a strike at the Standard Motor Company, which the popular press widely announced as an automation strike. It was, in fact, not a strike against automation, but rather against the Company's methods of forcing technological change on its employees.

I was not in London on publication day. Ron Gass took a telephone call, which started, 'This is No. 10 calling, can you send us immediately by courier, some copies of the automation report?' Ron replied, laconically, 'No. 10 what?' Of course the Prime Minister's Office received the report immediately. This was Anthony Eden's administration. I doubt if they ever read the report, but it was politically useful that the people know the government had the foresight to commission an automation study and that the TUC were associated with it! Accordingly, our little effort was given massive publicity for the wrong reason. I was on TV and radio, and bombarded by journalists. In consequence it became a bestseller. It really was rather good and still largely valid now.

The reception of our work within DSIR was less enthusiastic. I must explain here that the situation within the department had changed and was considerably different from that described earlier. Ben Lockspeiser had retired a couple of years previously to be replaced as Permanent Secretary by Sir Harry Melville, a chemist of distinction but with little experience of working in government. Then it was decided, by whom I do not know, to promulgate a radical reform. The Advisory Council of the department was given executive powers under a new Chairman, Sir Harry Jephcott from Glaxo. Presumably he and the Permanent Secretary were expected, in some undefined way, to share decision-making – a prescription for confusion and conflict! The DSIR Council received the automation report with polite disdain. They had no criticism of its content, but questioned its appropriateness to the functions of the department. Was it science? It had little relationship to science in the sense of physics or chemistry. Even Blackett, a pioneer of OR doubted. At first I assumed the attitude was one of ivory towerism, but on mature reflection I concluded that it was perhaps suspicion of natural scientists and industrialists towards the behavioural sciences. The Council also rejected a request by the Ministry of Labour (which welcomed our report) that our group might be made available to advise them.

Jephcott and I had a serious discussion about my position. He was very kind but made it clear that my activities were a bit far out for the department. He suggested that it might be useful for my career, were I to settle down to the discipline of a more focused managerial function in science, say as director of one of the national laboratories such as Mechanical Engineering at East Kilbride near Glasgow. The idea was horrible, I had neither background nor interest in mechanical engineering. Obviously my wings had been clipped, and I decided to walk out as soon as I could see an interesting opening elsewhere. Opportunity was to appear quite soon, but more of that later.

Before turning to another subject, just a few words about the demise of DSIR. Ron Gass stayed on after my departure, but found life there quite unexciting and later joined me in new ventures. Later a 'Trend Committee' report was accepted by the government, abolishing DSIR and distributing its functions elsewhere – mainly to the Board of Trade and the Ministry of Education. It was interesting that the only political personality protesting was my old boss, Herbert Morrison, by that time elevated to the Lords. Despite my criticisms of DSIR, I greatly regretted its disappearance as removing a powerful and coherent centre of government science that then became diffused throughout official activity.

What I criticised DSIR for was its proliferation of national laboratories. These seemed doomed to cultivate mediocrity and not designed in symbiotic relation with the economic sector they served. Also, one could expect little mobility of the workforce. Research is a job for the young. Its optimal creative period is probably the twenties and early thirties. In the universities the research worker whose creative activities have become stale can still be brilliant and self-fulfilled as a teacher; in industry there are alternative openings in production, technical sales or management. The scientific civil servant with his secure post continues indefinitely in research; very little new young blood comes in. Were I consulted, I would advise against the creation of national laboratories of uniquely governmental character except where circumstances (such as security) allow no other option.

19

Some Gambols with Parsifal and Hagen

This chapter is about my interests in the dissemination of scientific information, despite its bizarre title relating to two singular and contrasting personalities in this domain. The advent of the computer has revolutionised the information scene, hence many of my observations may be of mainly historical interest. I do, however, take some credit for having foreseen the development.

When first embarking on research, the richness and complexity of the system of scientific publication immediately becomes apparent. There will be many papers by earlier workers on the chosen topic to be scanned, the indexes of the abstract publications must be consulted to identify them, and the difficulties nare just beginning. There will be many abstracts, appearing to have some relation, but it is soon apparent that various papers differ greatly in quality and relevance. Indeed, I soon realised that many investigations had multiple publication of the results – the letter to 'Nature', the definitive paper in a learned journal, a popularisation or an oriented version for readers with a specific interest, there may be review or annual state-of-the art articles to be scanned. Gradually I understood that it was the number, as well as the quality, of publications that counted for esteem and promotion. Until the rise of Thatcherism, the career research scientist had two motivations: to earn a decent livelihood and to succeed and thereby earn the esteem of peers. The latter has great weight. I remember Tizard telling me that a group of Canadian universities had requested him to interview on their behalf a number of bright young British researchers in electronics, as candidates for chairs in their research schools. He found the ambitious young men unwilling to move, despite very large financial incentives. The standard reply was, 'Working in Canada, we would have much less chance of getting into the Royal [Society].' Scientific recognition was more important than money, which is seldom the case today when funding of research is so often determined by its potentiality to produce products or processes of monetary value. Extension of knowledge becomes merely a useful by-product.

By the time the war abruptly terminated my research career, I had learned a good deal about the science information system, both by doing some abstracting and, as Honorary Secretary of the Chemical Society, in distributing papers submitted for publication in our journal to suitable peers for refereeing. This latter experience taught me much about the difference in quality to be expected from papers in scientific journals that referee all papers submitted, from those that do not.

Having a keen interest in the system of communicating scientific discovery, I was thus happy to be appointed chairman of the CIP Panel on Scientific and Technical Information. An immediate task was to facilitate the translation and distribution of German scientific publications, from which we had been cut off since before the war. One of my first visitors was a Captain Robert Maxwell who had linked up with Springer Verlag, the German academic publisher. With them he was making a business republishing and selling important German scientific and technical papers and journals, eagerly sought by scientists and industrialists in the West. These he sold directly to libraries, research institutes and industrial enterprises. He wanted our support, or at least encouragement. Maxwell was large, authoritative, and somehow rather frightening. Uninvited, he told me about his escape to the East from occupied Czechoslovakia, his arrival in England with £5 and a typewriter and of joining the British army. Later I heard several other and very different versions of this story. Whatever the reality, it is truly amazing that a poor Ruthenian Jewish peasant could have reached the West in mid-war, managed to be accepted into the British army and risen to the rank of captain. He won the Military Cross for deeds of exceptional bravery.

About that time I became friendly with Desmond Bernall, a professor at Birkbeck College and an outstanding crystallographer. Bernall, a Marxist for some reason not acknowledged by the Soviets, has an important place in the history of science due to his book, *The Social Functions of Science* (1939). I knew him during the war when he was a key figure in Combined Operations, and was now surprised to find that he was greatly interested in reforming the system of scientific literature, about which he held radical ideas. Together we visited Paris to discuss UNESCO plans and in particular to organise a world meeting on the abstracting services of all scientific disciplines. This conference eventually took place, electing me chairman to my great surprise, as I was virtually unknown to the assembled experts. I had never before chaired a large international conference, but again to my surprise I found it quite easy to guide the direction of discussions without appearing to do so, dismissing crazy or irrelevant interventions without offence and steering towards constructive conclusions.

Partly due to Bernal's enthusiasm, the Royal Society organized a conference in 1948 on scientific information that attracted not only those concerned with publication, dissemination, abstraction, classification and storage of information, but also many eminent natural scientists. This enabled it to examine the problems more fundamentally than in normal professional discussions. For example, it considered thoroughly Bernal's proposal that primary publication of scientific research should no longer be in the form of papers in journals, but as off-prints, refereed and published by the societies issuing journals, and distributed internationally. The idea was too radical to be followed up, but still had considerable support. The conference did much to raise concern for the growing problems of scientific information. Meanwhile, my committee in London had started a review of the national system of scientific libraries. The Science Museum Library in South Kensington, which also served as the library of Imperial College, had the major collection of works in pure science, while the Patent Office Library was a vast repository of technical information. The professional societies for the various disciplines also had rich collections of a specialised nature, and the central libraries of provincial cities also had useful science sections, but no one knew what was where. Dr Urquhart of DSIR had made a British survey of the number of times per annum a wide sample of journals from all over the world were consulted. Results indicated that overwhelming use was made of some 100 journals, mainly British, American and German, making it obvious they should be widely available. A further range of journals were consulted sufficiently frequently to necessitate their retention in some four or five regions, while the others were called for rarely or not at all. It seemed sensible that the latter should be held only at a single centre. This led to a recommendation in the 'King Report' for the creation of a National Lending Library for Science and Industry, built partly on the Science Museum library collections. This recommendation was soon implemented and the library built at Boston Spa, Lincolnshire, with an overnight postal service to all major industrial and scientific centres. It was suggested that the Patent Office collection should be extended and housed in a new library in London with open access, to be known as the National Reference Library of Science and Technology.

When it came to technical rather than scientific information, I came to appreciate the depth of my own ignorance. I had shared the naive assumption of many basic science researchers that technological advances arise from an entrepreneur becoming aware of the potentiality of a discovery in pure research and transforming it into a commercial product through rather pedestrian applied research. I was quite unaware of the thousands of small technical advances that contribute to ever advancing technologies. I had given little heed to the importance of know-how and still less to how it was disseminated. Once again the indomitable Dr Urquart came to my rescue. He made an investigation within large, medium and small firms of the channels through which they received new information that they incorporated into their processes. The results were surprising. Some major innovations could indeed be traced directly to the literature of pure science, but there are innumerable channels for the myriad of small technical changes that constitute industrial progress. These include articles in technical and trade magazines, advertisements and the personal communications that are important in the transfer of know-how. In view of these multiple sources of information and the importance of assessing the quality of individual items, many firms found it essential to have information units or at least an information officer. For medium-size firms and those too small to justify such a post, communal information services would be necessary. This was a function assumed by the co-operative Research Associations in the DSIR system. In the years since my report, there has been a steady spreading of information units within firms and in sectoral associations, libraries and elsewhere. The information officer has become a profession in its own right, distinct from the librarian and with its own training facilities.

Soon my attention turned from domestic to international aspects of information, much as was the case with productivity. One day I received a telephone call from a Mr Lemaistre, Director of the British Standards Institute (BSI), wanting to talk to me about The International Federation of Documentation (FID). I had never heard of it, but agreed to listen. Mr Lemaistre had just completed his term as President of FID and explained the potential importance of the organisation and his dissatisfaction with its methods and lack of vision. He feared that unless it was deeply reformed, its opportunities would fade away. Then, coming to his point, he said that the British should now state that unless FID 'pulled up its socks' Britain would withdraw. He pleaded with me that with my central function in the Lord President's office, I was the obvious person to go to the forthcoming FID conference in Rome to deliver the ultimatum. I hesitated to accept this doubtful assignment, particularly because I detest pouring cold water on struggling worthy causes, even if they may be inefficient. However, my interest at the time in problems of information was such as to make me say yes.

Before plunging into a description of this new (to me) world of international enthusiasts, I must say something of its history as narrated by Mr Lemaistre. In 1895, two Belgians, Paul Otlet and Senator Henri La Fontaine, concerned with the rapidly growing volume of world contributions to knowledge, scientific and otherwise, and fearing that it would soon get out of control, had called a Bibliographic Conference in Brussels. That led to the creation of an *Institut International de Bibliographie*, aimed at establishing a *Repertoire Bibliographique Universel*, classified according to a decimal system and to form a bibliographical union between governments. Otlet and La Fontaine became joint Secretaries-General of the IIB, that later went through several name changes to become FID in 1938. The two pioneers had worked steadily to build up the Repertoire and to elaborate the Universal Decimal Classification (UDC) until war broke out in

1914. After the war the repertory objective was abandoned as being no longer feasible. However, by this time UDC was well advanced, becoming the main activity. Its membership, national information centres, institutions and individuals had extended greatly and FID became recognised as the main NGO in its field. FID continued to be situated in Brussels until 1938, run from 1928 by a troika of Secretaries-General, the two founders and Frits Donker Duyvis, acolyte of Otlet, and my Parsifal. In 1938 it moved to the Hague with Donker Duyvis in sole charge. For him it was an all-consuming labour of love. His professional job was Deputy Director of the Octroi, the Dutch Patent Office. During the German occupation of the Netherlands in World War II, the Director of Octroi was removed and Donker Duyvis remained in charge of the small residue considered necessary by the Nazis for legal reasons, while the Gestapo occupied the rest of the building. Within this virtual incarceration Donker Duyvis had the impossible task of maintaining FID by the few possible contacts he was permitted.

As the Rome Conference was to be held in September 1951, Sally and I decided to take a short vacation in Umbria before reaching the Eternal City. We were then welcomed in Rome by the Council members who were mostly devotees of the UDC. They oozed an enthusiasm that seemed to me to border on fanaticism. I quickly discovered, however, that the US delegate, Harvard University's Librarian, had come with an identical brief. To have an ally in this gathering of the converted was comforting, the more so as I soon discovered he would exert pressure sensitively. I then met Donker Duyvis. My first impressions were disappointing. He seemed passive and subdued; he mumbled his words and spoke in such a soft voice as to be practically inaudible. The meetings began with the inauguration of the new President. This was Arne Moeller from Copenhagen, a clear-headed, scholarly young man. We then started on a long and not very interesting agenda, the Council sessions being interspersed with specialist meetings open to all participants, a couple of hundred in all. The titles of these meetings as well as the Council agenda gave a fair indication of the Federation's programme, which appeared broader than Lemaistre had suggested, but still very UDC-centred and not very exciting. I decided not to issue a crude ultimatum, but to try a softer approach. I mentioned that the British viewed FID as being too exclusively centred on UDC and that we were disappointed it was slow in looking at the important problems of selection and quality arising from rapid increases in world publications and technical advances. I added that we felt the time had come to look at and reform FID structure and membership before it could be sure of maintaining our support. My American colleague then spoke in much the same sense. A few intervened defensively, but there was agreement that reform was necessary and a decision reached to invite a small panel to put forward proposals for change at the next annual meeting. To my horror I was invited to lead it and to suggest colleagues to assist. I suppose it is routine for the person demanding reforms, should his views be accepted, to be given the task of suggesting the changes, but I had not bargained for this new task.

Over the following year I had several small meetings on how to reshape FID, helped especially by the common sense of the Swiss Council member who was the chief information officer of the firm CIBA. I was able, too, to question Donker Duyvis in detail about the reasons for existing practices, policies and priorities. With every discussion my admiration for his integrity and wisdom grew. This old man, so long the embodiment of FID, put up absolutely no obstacles to its change of structure or emphasis; indeed he welcomed change whenever convinced of its value. Our report presented a balanced view of FID's field of interest, a foretaste of developments in information science and a number of modest improvements to structure and membership. After a quite heated debate at the 1952 meeting in Copenhagen, it was accepted. I closed my files and thought to myself, 'this is my goodbye to FID, I don't expect I shall meet them again'. How wrong I was.

In the following months I received occasional packets of papers from FID that I scanned perfunctorily and passed to my secretary to file. Then about a year later she came to me with a paper in her hands and said, 'Do you know that you are a vice-President of this outfit?' 'Nonsense', I said, but it was there in print all right, apparently the Council had awarded me this honour in recognition of the work I had done to help them. I immediately wrote an angry letter to Donker Duyvis upbraiding him for the discourtesy (more likely inefficiency) of not asking me if I was willing to accept this position. Then I tore up the letter. I could not bring myself to hurt the devoted old man, so I sent a very weak protest with a caveat to say that my interest in FID would be purely passive. I received a rather pathetic *mea culpa* in response. Time passed; FID papers came to me regularly; I attended no meetings. I noted their intention to organise a world conference on libraries, documentation and information in Brussels in 1954 jointly with the International Federation of Library Associations (IFLA). 'Good luck to them', I felt in my remoteness. But alas, poor Arne Moeller died suddenly a few weeks before the conference. I was swiftly informed that with the passage of time I had become Senior Vice-President and, according to their statutes, the presidency fell to me *ad interim*. I would have to preside over the world conference together with the President of IFLA, Pierre Bourgeois, the National Librarian of Switzerland. I had a couple of weeks to read up on the various themes of the meeting, which proved to be a mammoth affair in terms of participant numbers. It took place without incident and everyone seemed happy, which is the best that can be said of any large meeting.

I was now established as President of FID and remained so until 1962. It was a heavy burden to add to my already overcharged efforts, but I determined to take it seriously. I had to go into all the nitty gritty of the operation of an NGO. Inevitably, its members were all enthusiasts, many sensible but quite a few were cranks. I had to visit the office in the Hague every few weeks for discussions with Donker Duyvis or to chair panels. He kindly invited me to overnight in his house on these flying visits; indeed most of our business was conducted on the evening of my arrival, so that I could return home early the next day. Although simple and rather sparsely furnished, his house was in Wasseraar, a woody upper-class suburb outside The Hague, attractively situated just behind the coastal sand dunes. During these evenings in their home, I became familiar with his family. He had one son, a serious student at the Delft Technical High School. His wife was a delightful lady with a huge sense of fun. She reminded me of a traditional Dutch doll, with smooth polished rosy cheeks. It was difficult to believe she was not wearing wooden clogs. Her hospitality was warm and generous, but, alas, the gastronomy was dreadful, to the extent that I soon began to dread my sojourns in the Donker Duyvis ménage, this greatly troubled my conscience in view of their generosity.

Donker Duyvis was a remarkable man. White-haired with fresh rosy cheeks; he was uncommonly modest for a scholar. He was totally self-centred, but the 'self' was the ideal that he shared with Otlet and La Fontaine, namely to demonstrate the essential unity of all knowledge. I deduced little about his private life, if indeed there was much to discover. Gossip had it that during the occupation when his small unit was located in a building used by the Gestapo, Donker Duyvus had a transmitter in the chimney of his office, for communication with the Dutch Government in exile, but I cannot vouch for the veracity of this. I have already mentioned his unclear and almost inaudible speech. He had a masterly knowledge of languages, fluently speaking English, French, German and Italian as well as his mother tongue, but equally inaudibly. I remember at an OEEC meeting in Paris, for which he was the Netherlands representative, he made an intervention in English. The British delegate, in despair of understanding, demanded an interpretation into French, whereupon, the old man, helpful as always, switched to an equally inaudible French before the interpreters could catch on. This provoked the French delegate to abandon a century old taboo, by asking for a translation in English.

My presidency of FID was part of a long transition. The early FID was a grouping of a few internationally minded people of fierce integrity, bound together by a common enthusiasm for documentation and a particular interest in the classification of knowledge. In fact FID was more like a club than an organisation. The members' warmth of friendship and mutual respect was remarkable and persistent. During the great congress of librarians and documentalists of 1954, we took part in a ceremony of remembrance in one of the Brussels cemeteries. Standing around Otlet's grave a number of mostly sentimental speeches of appreciation were pronounced. Donker Duyvis gave the eulogy to the founder, standing gaunt and red eyed with tears running down his cheeks. He was not alone in manifesting his emotion.

It was essentially amateur in the best sense of the word, but, although working with true scholarly skill, its efforts were dismissed by many as amateur in a pejorative sense. This period of amateur enthusiasm was appropriate to the needs of its time. There was still little interest in documentation matters. Such was the situation well into the Donker Duyvis era. During this time, however, documentation began to be seen as a separate discipline, allied to librarianship, but with its own reality and techniques. It was evolving to become information science.

FID's transition to professionalism was slow; perhaps too slow. Clearly an information revolution was about to erupt. Wartime successes had led to an upsurge of scientific research, with publications growing exponentially; industry had become more sophisticated and needed rapid identification and selection of new items of information from many sources. Public demand for information seemed insatiable; news events became known immediately throughout the world. In addition to this leap in the quantity of new information, questions about its quality and reliability were arising, calling for new techniques and devices for storage, selection and retrieval. Punch-card methods of selection had long been in common use, and soon after the war more complicated devices like the Bush Rapid Selector began to appear. It was soon clear that a whole new information technology was expected. We were unaware at the time how very soon this technological eruption would engulf us. It was only with the appearance of microelectronics with its microprocessors a few years later, that the force of this explosion could be foreseen.

During the amateur years, most of the national members were rather poor, small groups of enthusiasts. Increasing recognition of the practical value of information led to the growth and reform of these bodies, many of whom became part of, or were associated with, National Research Councils or Academies of Science, enjoying indirectly a modicum of government financial support. This in turn helped FID in its struggle to secure a stable financial basis. In my attempt to evolve FID to a new level of significance I had three main objectives: to ensure financial viability, increase the number of member countries so its functions would be truly global, and greatly widen its area of activity. In addition to its contribution to classification of knowledge, it would have to deal with the presentation of information, including initial, secondary (abstracts) and tertiary publications, reproduction processing, selection and retrieval, storage and transmission, and take in its stride new needs and techniques as they arose. Donker

Duyvis, despite his identification with the initial FID preoccupations, agreed on the need for his organisation to achieve a clearer professional image with the strengthened infrastructure this would entail and to focus on new horizons.

The Board members accepted my general approach after some initial hesitation, but implementation was frustratingly slow, due mainly to poverty. Later in my presidency, I wrote a 20-year strategy for FID development. A turning point in FID history came at their Warsaw meeting of 1959 at which proposals for new levels of financing were tentatively agreed. The American member had taken the lead to increase subscriptions by an order of magnitude, to break the vicious circle of financial weakness and limited effectiveness. At Warsaw also, my long-term policy document was accepted. Its outline was followed for many years. We had several highly successful conferences during these years: Brazil in 1960 did much to extend FID recognition beyond the industrialised countries, while Washington in 1958 had elevated its image and led to sustained intellectual inputs. I relinquished my office in 1962, disappointed that as I left, the transition was so far from completion, but I have the satisfaction today of observing the important role FID finally achieved and sustained.

If Donker Duyvis was the Parsifal, the pure fool of my information drama, then Robert Maxwell must be its Hagen, child of the Nieberlungs. Maxwell visited me occasionally during the years that followed. I vaguely sensed he was making use of me, quite legitimately, through my many contacts in the information world, but our relationship was quite friendly and correct. I was well aware of his founding of Pergamon Press, also of an investigation by the Board of Trade that decided that his dubious commercial behaviour disqualified him from holding any directorships in British companies. As a consequence he lost his place as chairman of Pergamon, but very soon cleverly wormed his way back. Then one day while on mission in Washington, I had a telephone call from him. He explained that by chance he was in town and suggested we might meet to discuss events in the information world, inviting me to visit him in his hotel. I accepted. Half an hour later the telephone rang again. A gruff male voice said, 'I understand that Robert Maxwell has invited you to visit him. I must warn you to be cautious. His room is bugged and we shall register the conversation. We in FBI suspect that he is a double agent of the British and Soviets. My conversation with Maxwell was anodyne and I heard no more from the mysterious caller.

Some time later Maxwell, or Bob as he was now, invited me to visit him at his house in Oxford. I had a pleasant lunch with him and his wife who had an endearingly charming French accent. Conversation over lunch was trivial and general and I began to wonder what he wanted from me. He had recently been surprisingly elected to Harold Wilson's government as Labour MP for Buckingham, a 'safe' Conservative seat. Equally surprising to me was that Maxwell should pose as a Socialist. Shortly before my visit I had read in the newspapers that the new member for Buckingham had been fined for speeding on the M1 in his Rolls Royce at 90mph, at the same time shaving himself with his electric razor – a surprising image for a new Labour MP. He talked to me about his plans to reform the operations of the Labour Party, which he was about to submit to Wilson. He was soon to learn that his ideas were not welcome. Maxwell waxed eloquent about life as an MP and worked hard on me to stand for Parliament, with his sponsorship. After lunch we walked around his gardens and sat talking beside his swimming pool. He related how he had created Pergamon Press and provided an outlet for scholars who would otherwise have had difficulty having their works published. He talked about the excitement of making money and mentioned the number of millions he had amassed. 'It has all been great fun', he said 'and I could easily double it; and much, much more, but that would become boring.' He added, 'I want to be remembered for having done good for humanity.' This he said with great conviction. To this day I am convinced that he believed what he said – at least at the time he said it. I think he regarded me as a sort of father confessor. Had I become *his* Parsifal?

At the beginning of the 1970s Maxwell came in contact with Aurelio Peccei, President of the Club of Rome (about whom more later). He expressed great interest in the Club and offered to help in obtaining financial support and in founding a British Association for it. I remember warning Aurelio not to rely on Maxwell's promises and that his public image was incompatible with that of the Club, but he was so captivated by the man's exuberance that he did not listen. To see Maxwell amd Peccei trying to charm each other was unbearable. In 1972 I accompanied Peccei to a dinner at the Ritz that Maxwell had arranged to entice notables, including Sir James Goldsmith, to support the Club. Nothing came of it. Indeed Maxwell's reputation by then was such that his advocacy was harmful. In 1978, very much against my advice, Peccei invited Maxwell to become an active member of the Club of Rome. He attended the Club's conference in Rome that year, where he made lavish offers of support by publishing newsletters and assisting our publication efforts. None of his promises were fulfilled.

In 1980 I had written a book, *The State of the Planet* for the International Federation of Institutes for Advanced Study (IFIAS) and was looking for a publisher. Erroneously I thought Pergamon was a suitable vehicle. Maxwell accepted it without demur and it was published the next year in a beautifully presented volume. But it was impossible to purchase; it was not stocked by booksellers, there were no copies sent out for review and hence no publicity. I began to understand the strategy of Pergamon. From his early success in distributing German wartime documents, Maxwell had built up an impressive network of client libraries. There was little call for these documents from the general public and thus no point in using normal distribution channels. When he started Pergamon he had an ideal network of libraries, automatic takers for the type of scholarly volumes that Pergamon offered; books published at high cost in small editions with little appeal to other than a few specialists. Thus his overheads were minimal – no need for reviews or other publicity, and only a primitive distribution system. Such methods were hopeless when it came to books with any pretension to appeal to the informed general public. My book was killed. Colleagues worldwide ordered the book from Pergamon, but there were no replies to their letters. I was informed the Chinese had asked for the translation rights, but again no response. I wrote and telephoned to buy 50 extra copies at author's discount for my own distribution, but they never arrived.

The next year a report to the Club of Rome, *Microelectronics and Society* by Friedrichs and Schaff, in which I had two chapters was published by Pergamnon. Again it was well presented but unobtainable. During a meeting in Tokyo, infuriated members of the Club protested to Maxwell by telegram but it had absolutely no effect. Fortunately a German edition of the book sold well. When, on the death of Peccei, I became President of the Club of Rome, I decided that his name on our list of members was an embarrassment and that I would have to ask him to withdraw. He had taken no part in the Club's activities, had attended no meetings since Rome and had broken all his specious promises of help. He had been friendly enough with me and had telephoned me in Helsinki to apologise for his inability to come to the meeting when I was confirmed as President. But enough was enough; he had to go. I called him, asking for a meeting. He was very warm and invited me to lunch at his office. I did not relish this encounter, knowing his reputation for fury and violence when thwarted: to sack Maxwell was unthinkable. I arrived at the Daily Mirror offices with trepidation; all was noise and bustle. Maxwell was continually demanding the presence of this person or that and he sounded angry with all of them. They were uniformly obsequious. Eventually we sat down at his dining table and I put my case to him. Suddenly he was quiet. He accepted my arguments and even apologised for his shortcomings. The rest of our lunchtime conversation was general and friendly. He shook hands warmly and wished me well. I departed and never saw him again.

In later years I had a considerable information system under my control, but my contact with it was indirect and uneventful. I am Honorary President of FID for life and follow its successes with satisfaction – from my armchair.

20

The Productivity Saga (and I Move to Paris)

The end of Chapter 17 left me presiding over a working party of the Organisation for European Economic Co-operation (OEEC). This organisation, created under the Marshall Plan, had a vital role in the reconstruction of a Europe ravaged and impoverished by World War II. Its seat, Château de la Muette on the verge of the Bois de Boulogne in Paris, originally a hunting lodge of the French kings, had belonged to the Rothschild family before it was taken over by the Nazis. Now belonging to the French Government, it had been made available as OEEC headquarters. The organisation was quite small and governed by a Council of Ambassadors from each of the countries participating in the Marshall Plan, who met each Friday, but it convened occasionally at ministerial level to decide important policy issues. Custom being never to put an issue to the vote; it was thus a nest of compromise. The international secretariat was headed by a Secretary General, at that time the French economist Robert Marjolin. The Council chairman was Sir Hugh Ellice Reece of Britain.

So here was I, sitting in a salon with magnificent *boiserie*, overlooking a classical French formal garden and presiding over an international meeting proceeding under dubious, contrived circumstances. I was excited, but rather worried our subterfuge would be unmasked, preventing us from launching a movement I felt was essential, namely spreading the productivity concept throughout Europe. I fervently hoped 'the cat would once again turn round'. I need not have bothered: our activities were so unimportant to the Council members with their political and economic concerns as to pass unnoticed – at least for the moment.

My introductory speech was the opportunity to suggest overall objectives and strategy for the group's work based on the conclusions of the Gibbs committee in London a few years earlier, and developed in the light of my experience. I painted a picture of the reconstruction period as one of transition, with the paramount need being to rebuild and improve the performance of industry and agriculture. Wartime successes suggested that science was likely to be a dominant factor in future development and, although capital shortage would temporarily slow down technological development, research should be supported as a long-term economic investment. I suggested the following as appropriate lines for the Working Party's agenda:

- Priority should be given to understanding how higher productivity could be achieved in the agricultural, industrial and service sectors of the economy.
- In particular, means should be sought to ensure full participation of the workers and their unions in the productivity drive by just distribution of its fruits between employers and labour.
- Basic research should be encouraged as a long-term economic investment.
- Stimulation of co-operation in applied research between the OEEC countries particularly in service areas in which governments have a research function (e.g. water and fire control, standards and energy).

These lines of attack were quickly accepted and it was agreed that productivity was the group's essential concern, with scientific and technical information added to conform to our title and mandate. There was a discussion of disparities between each country's best and worst performing firms in particular sectors of industry; hence inter-firm comparisons might help to identify reasons for low performance and provide incentives for improvement.

The working party quickly established coherence and an unusual level of enthusiasm amongst its members. At the end of the meeting they even decided to return to the Muette after dinner to draft the minutes to be sure of the accuracy of their expressed intentions. Thus began a year of extraordinary activity.

Pat Young of the international secretariat was extremely active in implementing our programme and I was often in Paris working with him. We also had co-operation and understanding from the Marshall Plan mission in Paris. We established an operational programme: visits to America of European teams based on those of the Anglo-American Productivity Council; seminars in Paris on practical ways of measuring productivity in cotton spinning and in the manufacture of boots and shoes; creation of a network of national centres of technical information; start up of a scheme of co-operative research, the first case being the European low shaft blast furnace.

As year end approached we had to come clean and report our work to the OEEC Council. By this time we were convinced that each country needed a productivity centre where all factors and interests on the subject could be brought together and experiences shared. Having worked on this idea with the Americans and having their promise of Marshall Plan counterpart funds, we were ready to propose the creation of a system of National Productivity Centres (NPCs), governed and operated on a tripartite basis by management, labour and government. The idea of productivity centres was obvious enough, but the concept of shared benefits as well as joint control by all parties concerned was the novel basis of the scheme.

I duly made my report to the Council. They were shocked and dumbfounded, many of the ambassadors saying that operational projects such as these were inappropriate in an inter-governmental body. It is significant that they seemed unaware that most of the work reported was outside our mandate! The storm subsided and they began to discuss our recommendation. I doubt any of them were interested in its substance or understood its significance. Nevertheless it was unanimously accepted. Why? Because it carried the promise of extra Marshall Funds for each member country. The Council agreed to continue this work by converting the working party into a permanent committee of the organisation, the Committee for Productivity and Applied Research (PRA). However, to stress the policy function, operational activities were delegated to a technical assistance group. Back in London, I was in some disfavour because of all the 'unnecessary' activity I had generated without their specific agreement. I was not reprimanded, but when a little later I was elected chairman of PRA, they were somewhat tardy and ungracious in giving consent.

The first task of the new Committee was to help in the establishment of the NPCs, which differed greatly from country to country as a consequence of history and social structure. In a few cases they became agencies of government with some participation from the two sides of industry, while in others, the government's intervention was very low key. In yet others it was difficult to secure co-operation from trade unions with little or no experience of working with management towards a common objective. I remember well a hard night in Athens convincing Greek trade union leaders it was in their interest to work with the employers in a Productivity Centre, and in Italy means to increase productivity were initially seen as crypto-Taylorism. In general, however, all the centres operated with management, labour and government pulling reasonably well together.

NPC leaders met frequently in Paris and it was soon clear they had much to gain from each others' experience and much to learn about the multifarious elements involved. Gradually PRA members and NPC directors came to the conclusion that some sort of European central body was required to deepen contacts as well as to provide the necessary information and skills. Experts from the Marshall Plan mission were involved in all these talks, in the end agreeing that OEEC should bud off a European productivity centre with sufficient funds to undertake practical work, assist in training and make available specialised consultants. When put to the Council this plan met with a mixed reception, but after a second meeting they decided to create the European Productivity Agency (EPA), thanks again to a financial offer by the Americans. The Foreign Operations Administration made a grant of $2.5 million to be matched by $7.5 million in the various European currencies by participating countries through conditional aid funds. The agency would work under the general guidance of the PRA Committee.

The next step was appointing a director to head the agency. The Americans had rather grandiose ideas and wanted a prestigious name at the helm who was British, as the Secretary-General of the parent organisation, OEEC, was French. They approached the British Government asking that Harold Wilson be made available. This was rejected with veiled ridicule; it was unthinkable for them to try to persuade the leader of Her Majesty's loyal Opposition to take on such a job. A second try suggested Sir Oliver Franks, well-known scholar and banker, but this was again turned down. In the corridors my name was mentioned as a possibility, but the British apparently had no interest in the job. The Americans then began looking for a German and, on Minister Erhard's proposal, secured the appointment of Karl Harten, formerly Director of an industrial association, the Verein Deutscher Eisenhuttenleute. This had some political interest as the first appointment of a German to a senior international position since the war.

Harten, though an agreeable, co-operative man, and no doubt a successful technical administrator, without experience of governmental or international activities, he never quite understood what his new job was about. He decided it would be useful to publish a series of cases of productivity increase in different industries; a good idea indeed. So he submitted as a first example, his own work on standardisation in aircraft construction in Germany during the war, his manuscript going for translation into English and French. Its preamble addressed to Von Speer, offered it as a contribution to the defeat of 'our enemies', the English and French! I cannot remember whether the original document began or ended with 'Heil Hitler'. A strike of the Muette staff was narrowly averted. Reasonable discussion between Harten and the Committee was very difficult and in the end I had to write the first year's programme on their behalf. He was supposed to take it to Washington for discussion with the Marshall Plan people, but to our great surprise he was refused a visa on account of his wartime activities; curious, as the Americans had sponsored his appointment. I went to Washington in his place and encountered no difficulties.

The committee easily appreciated that this situation could only end in disaster, so they asked me to take it up with OEEC's Secretary-General. The next Saturday morning I met Marjolin and his deputy Prince Colonna over a cup of coffee in the gardens of the Muette. I described the situation. At first they were politely hostile. Staff problems were entirely a matter for them and neither the national delegations nor committee members could be allowed to interfere. They did in the end realise that the malaise would eventually harm their organisation. A week later

Colonna was in London on other business; over lunch together I was able to discuss the situation in depth. Harten quietly slipped away and I never heard of him again. I knew nothing of the political discussions that took place leading to Harten's disappearance or about the choice of his French successor, Roger Gregoire, a member of the Conseil d'Etat. He was a very different type, quiet, low key, effective and exceedingly intelligent. Ted Fletcher of the British TUC was lent for two years to act as his deputy. Gregoire had to work on the programme I had designed, consulting with me in detail on its implementation. I found him friendly and reasonable, so there were no problems.

For the next two years Gregoire and I worked in harmony and indeed we both confessed surprise to the other that, with our very different backgrounds, we agreed on so many things. At the end of that period Fletcher had to return to the TUC. Gregoire asked me if I would come to Paris as his replacement. As it happened this was just the moment of my discouraging talks with Jephcott and my decision to leave DSIR. I decided to take the job, but only if it was upgraded from deputy to co-director, equivalent to my rank in the British Service. Gregoire negotiated this successfully with the Secretary-General. So off I went to Paris on 1 January 1957 for a trial year with no premonition that I would be living in that city for the next 37 years. I found a room near the Muette for a few months until Sally was able to join me. She remained in Clavering Avenue until we were confident that satisfactory arrangements were made for the children. By this time Jane was working in the Russian department of the Foreign Office and was a very responsible person; Catherine was a medical student and continued to live on in our house with a couple of friends; Margaret was to move to Paris when we had found a flat if she could gain a place in the Conservatoire. The girls commented, 'It's natural the birds should leave the nest at our ages, but it seems to us that the nest is now leaving the birds.' I shall touch on our life in Paris later, but must now describe EPA and its work.

The Agency concerned itself with productivity in the widest sense as was clearly stated in its first programme:

> Productivity means getting the best results out of any of the factors of production – capital, raw materials, plant and

> machinery, land, labour, etc. Emphasis on one or other of these depends on which of them, at a given time and in a given sector, is a limiting factor. Productivity is a means to an end. The final purpose of the campaign is the achievement of a higher European standard of living.

While the labour factor continued to be the main preoccupation of EPA, we were concerned with the productivity of other factors, especially energy and materials utilisation.

A basic question troubled me throughout my interest in productivity. How far was it possible to influence practice on the shopfloor from outside the system? In EPA's case, how could a group of technicians sitting in Paris and led by two disparate intellectuals hope to penetrate to the grassroots of production? This was a doubt inevitably shared by people in industry, a doubt that added to their general suspicion of government attempts to interfere. The agency, therefore, encountered a great deal of scepticism from its inception. I think we solved this problem surprisingly well by placing EPA at the centre of a European network of productivity organisations and activities, never approaching the individual enterprise except to highlight exceptional developments or experience as case material. Thus we never attempted to provide training to groups of enterprises. Our basic technique was 'training trainers'. Our work with the trade unions and on management education illustrates this well. The EPA trade unions section was led by Vermeulen, a burly Dutch union leader who, like many of his colleagues, had been in the underground during the Nazi occupation, sharing danger and action with companions from 'the other side of industry'. This long association in a common cause had generated a high degree of trust and understanding that persisted when normal conditions returned. Vermeulen was well placed to argue the tripartite case for just treatment for the workers in productivity increases. He organized innumerable round tables and seminars with the trade unions of member countries,

I had been interested in management education and development while working in London, had read much about US business schools and argued in favour of trying the system out in England. There was very little attempt to train managers in Europe; only low-grade commercial colleges, yet US industrial success was now generating a sense that something of the sort was needed with regard not only to techniques but attitudes. EPA made this a high priority. We sent many young potential teachers of management to attend courses at the Harvard Business School and elsewhere, and organised courses given by American business academics in European universities showing some interest in entering the field. Gregoire and I visited the business schools of Harvard, Columbia, Pittsburg, Chicago and Stanford, participating in courses, discussing methods and curricula and particularly weighing up the merits and disadvantages for Europe of adopting the case approach rather than some others. The report we wrote about our impressions was widely discussed in our member countries, helping to firm up proposals to create a number of new European management schools with Harvard's Dean Fox as our main adviser. Our work at EPA on the subject was intensified, achieving rapid success. In all this we worked very closely with the NPCs and also with the Ford Foundation which had an important European programme of business education.

There had been some pressures on EPA to create a European management institute, but this did not find much favour among the European countries or the NPCs. They preferred to create or strengthen management training foci in their own countries, using EPA for exchange of views, discussion of curricula or sending students abroad for training. However, an initiative for the creation of a European management school came from outside official circles. It came from General George Doriot, a Frenchman who became an American, rising to Quartermaster General in the US Army, while remaining very, very French. I met him in Washington when I had to visit the Quartermaster's Laboratory on some technical matter during the war. Gregoire also knew him, so, when we were in Boston visiting the Business School, we visited him and listened to his opinions on the trends of business education in America. This was very useful in enabling us to write a more balanced report. He also explained his ambitious plans to create a European management school in France.

Gregoire, very impressed, promised to help him as far as possible. We had, of course, no capital resources available for such projects, but EPA was able to give some peripheral help through our experts and Gregoire was very useful in encouraging support from his many friends and colleagues in French academic and bureaucratic circles. Doriot was finally successful and, with great support from the Ford Foundation, the Institut Européen d'Administration des Affaires (INSEAD) opened its doors at Fontainebleau and is highly effective to this day. EPA and later OECD retained a formal but not very active relationship with INSEAD. On my retirement from OECD in 1974, I became the representative of that organisation on the board of INSEAD, remaining so for many years.

Our approach to the introduction of improvement of existing or new techniques on the shopfloor was different and done entirely through the NPCs, which had direct access to firms and could bring together people from different enterprises for demonstrations of new methods. We acted essentially as their agents in making available the appropriate consultants. At a meeting with the representatives of the NPC, they or we would suggest particular subjects demanding attention. We would then approach the Americans or possibly a European group, requesting the temporary services of technicians we required. Eventually, the consultants would appear and be dispatched to the country or countries that had requested them. On many occasions several of the NPCs asked for help on the same technique, in which case our consultants would spend a considerable length of time in Europe, moving from country to country. This proved more economical than responding separately to requests by several countries for more or less the same type of aid. A very large number of such missions were accomplished during the lifetime of EPA; work-study, quality control, plant layout, materials handling, welding techniques, powder metallurgy, treatment of low grade ores, industrial drying, low cost automation, component standardisation, simplification and variety reduction, etc.

The agency programme on *agriculture* was complementary to that of other international and national organisations. We concluded that the main factors responsible for unsatisfactory productivity in European agriculture calling for EPA attention were poor management of farms (including farm accounting), slow application of new methods, and inefficient marketing and distribution. Our programme therefore concentrated on improving the economic organisation of farms, land development, agricultural training, advisory services and marketing, application of new methods and efficient marketing and distribution.

Of the other sectors of the economy in which we were active, building construction, with its high labour intensity and traditional methods seemed ripe for productivity gains. Some of the NPCs did useful work in introducing improved working methods, new materials and modular co-ordination. We even ventured into distribution with an interesting 'cold chain' experiment for the introduction of deep-frozen fish to two inland cities in Europe, Turin and Munich. The fish, freshly caught off the coast of Iceland, was filleted and deep frozen on the trawlers, delivered to a refrigerated depot in Dieppe, then shipped immediately by special refrigerated rail wagon to deep freeze outlets in the two cities. As a pilot effort the experiment was a success, but initially the product was barely acceptable to consumers who, accustomed to the taste of stale fish, found the deep-freeze product tasteless. But with the help of a touch of gastronomic art, this difficulty was soon overcome.

Inevitably, we had an important information function and our unit constantly fed its network with reports of new processes, know-how and suggestions. One useful EPA information vehicle was the *Productivity Measurement Review* presenting progress and experience in the measurement of productivity and inter-firm comparison wherever it might arise. The Americans offered to provide, through us, copies of a Department of Commerce publication called *Technical Digests*, whose aim was to describe small innovations, gadgetry and useful tips arising in one sector of industry but likely to be helpful in others. We were asked to approach European national industrial bodies to ascertain how many copies they might want to distribute to their firms. I remember approaching the FBI (Federation of British Industries) on this. A slightly censorious voice on the telephone responded, 'We really can't insult our firms by feeding them with such material, let us show willing by asking for six copies.' Reply to the same question from the corresponding German federation was for 50,000!

In that eventful year of the working party, we had already gained some experience of how to stimulate co-operation in research and development on the international scale. There was certainly much duplication between the research efforts of the member countries on subjects within the government domain. Not that I deprecate duplication in subjects on the frontiers of knowledge where the creative intuition of several scientists may throw new light on seemingly intractable problems. But in the public utility sectors, applied research is so close to objectives common to all countries that duplication of effort has little intellectual value. EPA took up research on roads, including materials and methods, durability of surfaces, accident prevention and economic problems, by bringing together representatives of the various road research laboratories for exchange of experience, planning of programmes and day-by-day signalling of results to each other. This proved to be a successful enterprise that continued indefinitely. Looking back from the present, obsessed by problems of the environment, it is interesting to remember that the agency had groups working on air pollution and water pollution as early as 1954. Our co-operative research programmes met with mixed success. Had we been able to provide even moderate financial inducements, the results would have been much more positive. As it was, we were able to provide a meeting place, interpretation and identify expert consultants who would help us to formulate likely themes for co-operation, visit the main laboratories having good reason to work together in each case, set up meetings and do follow-up work in between.

An interesting if somewhat trivial instance of this was a co-operative attack on the fouling of ship's hulls. At first sight this seems economically unimportant, but when one estimates the time spent in dry dock as well as the slower speeds and greater fuel consumption of the world's merchant ships due to friction caused by driving barnacles through the oceans, the cost is very high. Tackling this problem were only a handful of scientists at work in the main maritime countries, too few in each for significant progress to be expected. By bringing them together a critical mass of effort was assured. This project was understandably regarded with amused disdain by the macro-economists of OEEC and the national delegates who disliked being associated in any way with barnacles.

By this time, discussions with so many scientists and industrialists had strengthened my conviction that science just had to be regarded as an investment by both governments and enterprises. I foresaw that in the not-too-distant future, technological innovation based on research would become the dominating factor in increasing productivity. Yet what were governments doing about it? How far were they aware of the issues involved? I resolved to uncover the actual situation and invited Dana Wilgriss, the retiring Canadian Ambassador to OEEC to visit the scientifically advanced countries of our grouping and report on the government/science situation of each, but more on this later.

EPA's working atmosphere was impressively harmonious and enthusiastic. My relationship with the staff was as warm as it had been when I was chairman of their governing committee. My replacement as chairman was a Swedish economist of decidedly non-interventionist tendencies, who had, fortunately, little influence on our day-to-day work. He was rather suspicious of the sense of some of our projects. I remember, for instance, he questioned the usefulness of trying to make handicraft activities viable in some of the Mediterranean countries and insisted on inspecting such a project in Sardinia. On return he reported to his committee, 'I must confess I found it less bad than I had *hoped*.' He did not occupy the post very long and made little impact on our spirit.

Then there was the Advisory Council, unmentioned till now. This group of prominent industrialists and trade union leaders had been set up when EPA was founded, in order to give a bit of tone to the enterprise. Chaired by Axel Iveroth of the Swedish cement industry, its members included Vittorio Valletta, managing director of Fiat; Clarence Francis, President of General Food Corporation; Sir Wallace Akers of ICI; and Jack Tanner from the British unions. It was supposed to advise on the general policy trends of our work and suggest points of emphasis, but most members took their involvement very lightly, while Iveroth, who was a pleasant youngish and very active man, began to question decisions rightfully the function of the PRA committee. Thankfully, the threat of dual and competitive management of our work disappeared with a little quiet negotiation and plenty of good will.

I must just add a word about relations between the staff with their multiplicity of nationalities. The culture of the office was undoubtedly French. On arrival at the office in the morning, one inevitably shook hands with colleagues and addressed them as Mademoiselle or Monsieur, with repetitions on return from lunch and on departure in the evening. When Ted Fletcher arrived, this ceremonial jarred on him and he insisted on dropping the formal designations and instead staff were to use first names, such as Marie Claude or Jean Jacques. They did so quite willingly, but two years later, the day after Ted left for home, they switched back to addressing each other as Mademoiselle or Monsieur with the ritual of shaking hands.

In 1957 a new situation arose suddenly. The world woke up one morning to learn that the Soviet Union had shot Sputnik 1 into space. The Cold War having already started, it was unsurprising this spectacular technological triumph should make people and governments suddenly aware of the explosive development of Soviet science and technology, including the size and quality of its scientific and technical personnel. For the European countries, the Sputnik shock was even more poignant than for America. We on this continent realised not only the menace of the Russian development, but also how puny our scientific and technical manpower was compared to that of America. It was imperative that Europe make strong efforts to expand its scientific research and technological development.

Naturally I responded instantly to this suggestion. I had been advocating it for years, and started a round of visits to the heads of the OEEC delegations to sound out the depths of their concerns and their ideas for action (if any). I found plenty of good will but only vagueness as to how the OEEC should react. It seemed to me that the most obvious move would be to open up a new chapter of EPS's programme on scientific and technical personnel. However, the Americans brought the matter to a head by an initial offer of half a million dollars to be matched by a similar amount by the European members of OEEC for a detailed programme of activities 'in regard to shortage of scientific and technical personnel'. The OEEC Council accepted this and it became the *raison d'être* for a decision to establish an Office of Scientific and Technical Personnel (OSTP). This was envisaged as an agency of OEEC parallel to EPA and, like it, with its own governing committee. Initially, I was disappointed by this decision to separate the new programme from EPA, it might easily result in the appointment of a director unfamiliar with, and not necessarily sympathetic to, the free and constructive working of our agency, but to my surprise I was invited to take the post in addition to my EPA functions. As events worked out, it was a wise decision to give OSTP a separate identity. When, three years later, OEEC was wound up and OECD created, OSTP was more sure of survival on merit than it would have been if buried within the EPA programmes.

My appointment caused a minor blip in my continuing good relationship with Gregoire, who understandably felt left out in the negotiations and disassociated from subjects that interested him. However, in practice the two agencies worked closely together and harmony between the directors continued. I started immediately to collect staff for the new agency. The Americans produced a friendly elderly administrator, who proved to be a good facilitator, but more importantly, knowing that Ron Gass was feeling bored at DSIR in London, I invited him to join me as my deputy in OSTP – and he accepted. This second association with Gass proved to be long lasting and fertile.

Before the Council took its final decision to create OSTP it appointed a temporary working party (designated no. 25) to examine the problems involved in increasing numbers, quality and utilisation of scientists and engineers, and to work out a programme of action to help the individual member countries to tackle them. The working party duly produced its programme which was accepted by the Council in May 1956 at the same meeting as established the OSTP. Over the next three to four years a vigorous action programme was put in motion. I shall hold this topic over, returning to it later when it will be seen as the start of a brilliant continuing venture.

By 1959 European industry had returned to full normality, institutions had been rebuilt and the Cold War was in full swing. Gregoire and I agreed that EPA had pretty well fulfilled its promise, now entering a zone of diminishing returns. The NPCs had developed with varying degrees of effectiveness and in most cases aspects of their work had been taken over by existing bodies or perpetuated by delegation to new specialist institutions. In the course of its short life, EPA had uncovered many issues of importance to the future of society. For instance, our work on management education had spawned business schools all over Europe with deep implications for the education system as a whole. This would be even clearer with regard to the new OSTP activities on scientific and technical personnel. The concept of science as an investment was opening up a new field of investigation that would obviously raise difficult questions about the relationship of science to both the economy and society. EPA, with its operational mechanisms for attacking the problems of reconstruction, was obviously not the body to engage these deeper issues. We decided, therefore, to advise the Council that EPA, having fulfilled its mission, should be abolished but adding that arrangements should be made for a more profound examination of many of the issues its existence had raised and which had strong economic and social implications for the future.

The Council's first reaction was astonishment that the two directors were proposing the abolition of their own jobs and a fear there could be skeletons in the cupboard. Very soon, however, they understood our arguments and EPA ceased to exist. Looking back on these years of struggle assisting Europe to reach competitive levels of manpower productivity, I am convinced the results justified the efforts. We had achieved the acceptance of the need for management education and development and catalysed the creation of business schools; also, by thousands of small technological improvements, we had speeded up the recovery of European industry. Our work in persuading the unions that it was to their advantage to support the reconstruction procedures may have been somewhat marginal, but it was important. It is interesting to note here that the EPA model was emulated outside Europe; Productivity Centres were created in Israel, India, Japan and some countries of Latin America.

The moment was now propitious for following up the issues we hoped would be considered in depth. OEEC, one of the most successful international organisations ever, was just reaching the same conclusion as EPA, that it was at the end of its useful life and would have to be reformed or replaced. The customary group of 'wise men' was invited to advise on the functions desirable of a new organisation, an indication of the structures required and the appropriate membership. In 1961 the Organisation for Economic Co-operation and Development (OECD) replaced OEEC and a smooth transition took place. The United States and Canada who had held observer status in OEEC adhered, later Spain, Japan, Australia and New Zealand joined. Some of the non-operational activities of EPA as well as the programmes of OSTP survived in a new Directorate of

Scientific Affairs. I was appointed director and later became Director General. Gregoire relinquished international affairs and returned to the French administration. I missed him very much.

21

Policies for Science: Science for Policy

The end of EPA was a moment of crisis for me, although only I was aware of this. My work at the agency had been successful and fulfilling and now I was challenged by even more interesting possibilities in the new organisation. Yet I had doubts. The hurly-burly life I led, the sterile diplomatic cocktails and dinners, left few spaces for reflection or meditation. Still worse, the excitement of attacking issues which pretended to be important, but were cosmically zero, always appealed to my suggestibility and pleaded with me to identify with them. In the end I decided to accept the OECD function. I convinced myself there was no incompatibility between my interior striving and external activities as long as I could be aware of myself and of what I was doing.

At first sight OECD was little different from its predecessor OEEC: the same weekly meetings of the Council; its occasional discussions at ministerial level; the same striving for consensus without voting. But the atmosphere had changed, as had the agenda. The OEEC Council had always been presided over by a senior ambassador, usually British, while the Secretary-General was the servant of the Council and always French; the last such, poor amiable René Sergeant, so obviously 'knew his place'. In the new organisation, the Secretary-General was chairman of the Council, a sign that initiative was expected from him, not just service. I assumed that this applied also to the directors.

The new Secretary-General was Thorkil Kristensen, recently Finance Minister of Denmark, where he was known as 'Skinflint Kristensen' because of his austerity policies. Thorkil had been a consultant to OSTP and was attending a seminar arranged by

Ron Gass at the Rockefeller Foundation at Bellagio when the news came of his election to head OECD. Suddenly my consultant became my boss – a happy event for me. Thorkil was a soft-spoken, quietly intelligent man with absolutely no charisma. He was so simple that the Heads of Delegation mistook him for a simpleton when he first appeared and I suspect that some of them tried to run rings round him. But there was no naivety about the man; his was the luminous simplicity that occasionally attaches to people of high integrity. The only other person in whom I sensed this rare quality was Jan Tinbergen, the Dutch economist.

The creation of OECD gave fresh impulse to those countries hoping to deepen European co-operation, particularly the French for whom bitter memories of three wars in a century in which Germany had invaded France gave it special importance. In the early days of OECD there were many heated debates on how this co-operation might be achieved. I listened to one such in the Council chamber when the French and Germans were arguing for new political initiatives to be taken towards a united Europe and the British arguments were exclusively mercantile. I have never before or since witnessed such a pointless dialogue of the deaf. In the end the other countries lost patience with this futility and 'the Six' went off to the Messina talks that led to the Treaty of Rome and the creation of the Common European Market. As a consequence OECD lost its role as a major focus for the creation of the New Europe, although it retained many valuable functions. With the addition of Japan, Australia, New Zealand, America and Canada as extra-European members, OECD became the Club of the rich industrialised countries, well justifying its survival. The D for Development in its title gave it a role for common discussion of policies and aid towards the Third World.

From the angle of my work, this loss of political clout was an advantage. Because OECD was politically weak, it could be professionally strong. It could permit long-term examination of issues, sheltered from the short-term political expediency that disturbs progress in the main international and interregional bodies. It might also enable the study of intersectoral problems with the new approach that my concept of multidisciplinarity suggested.

The new Directorate of Scientific Affairs thus inherited the ongoing programme of OSTP, hardly noticing the transition from OEEC to OECD. It acquired EPA's scientific information unit, its co-operative research work, and lots of untested ideas. The development of OSTP, that had already discovered how closely scientific and technical personnel issues impinged on the educational system in general, is described in the next chapter. To my great relief the indomitable Ron Gass remained as my deputy for the whole of the Directorate's work, with special attention to education.

It was obvious our budgets would be quite small and no question therefore of operational activities or research financing; such weighty responsibilities would obviously fall to Brussels. I decided our work would have to be catalytic, long term in its approach, and alert to new ideas. After a couple of years of development that showed some success, I outlined to Kristensen the guidelines I hoped to follow. These were:

- Financial restriction limits our role to stimulus of new ideas and development of policies.
- We should think five years ahead, but never appear to be more than, say, two, if we are to be allowed to explore new approaches.
- Staff numbers should be low and expert consultants employed on appropriate short contracts, drawing on universities, industry or wherever for specific skills.
- Whenever success was achieved, that particular project should be abandoned as soon as sufficiently assimilated in the member country, keeping us flexible and uncluttered by continuing work that would prevent newer initiatives.

Kristensen approved my approach but added wryly, 'do what you think is best, but don't get me into *too much* trouble.'

From EPA I had invited Ambassador Wilgress to find out as much as possible about the attitude of member governments towards science and to discover what policies they had on funding. Mr Wilgress presented his findings to the OEEC Council just a few months before the demise of that organisation. He had visited most of the member countries for discussion with ministers, senior science administrators and industrialists. He came away with a strong feeling that in most countries research had grown in a haphazard manner in response to specific needs and was seldom considered either in terms of its real potential or of policy. He concluded 'the full implications of the scientific revolution have not yet sunk into the consciousness of large sections of the population of Western European countries – they are loath to scrap their traditions. In particular they are reluctant to adapt their educational systems to the needs of science and technology.'

Mr Wilgress produced a series of confidential notes on the situation of each of the countries and a general report to OEEC, in which he stressed the economic importance of science and the need for coherent policies: 'The first thing should be for each country to draw up a national science policy.' By this he meant an investment policy taking into account the balance between fundamental and applied research needs in view of science as a basis for technological innovation and economic growth including international co-operation. For OEEC, he recommended that the consideration of scientific affairs should be consolidated and a high-level group of scientists constituted as a science policy group. The report was received favourably on the whole but with some doubts. This discussion was the last item on the day's agenda and the meeting was already running late. With lunch approaching, they were a bit fidgety. Roger Ockrent, Belgian Ambassador and chairman of the Council's Executive Committee, had already risen and was about to leave the room at the critical moment I was trying to explain the significance of science policy. Impertinently I appealed to him to wait one moment and proposed that OECD should sponsor the Science Policy theme. He quickly remarked that science policy was inherently part of educational policy and hence inappropriate for an economic organisation; it was a matter for UNESCO. I mention this to show how far politicians were from regarding science as a national investment or a factor in economic growth.

I am sure the Wilgress report convinced the 'wise men' of the shape of the OECD to come, and of science's prominence within it. The report triggered concern in European countries and worldwide on the need to introduce coherence and deliberate policy into scientific activities, towards a better understanding of how newly discovered knowledge might contribute to the broad sweep of national objectives. With the arrival of OECD, the report's implementation fell into the safe hands of Thorkil Kristensen, who appointed an *ad hoc* group of independent scientists and economists to advise him on science and technology policy issues with Pierre Piganiol, chief scientist of the French Government, as chairman. The group began its work by examining the problems and opportunities of harnessing science and technology to the broad economic objectives of OECD, its analysis going much deeper than that of Wilgress. The need to develop policies for the management of national research resources was quickly acknowledged. If, as was the case, the opportunities offered by expanding knowledge greatly exceed the finance and skills available to exploit them, the need for mechanisms to allocate resources is essential. This inevitably carries with it the need to establish priorities – a concept abhorrent to scientists implying bureaucratisation and loss of research creativity. How to approach the complex issues of resource allocation was the dominant task of the new science policy bodies set up in the following decade.

The group stressed that allocation of resources is but one aspect of policy to ensure a strong and healthy development of national scientific effort. It is necessary for governments, advised by scientists, to consider the many special conditions that must be met if an environment conducive to the maintenance of high levels of creativity is to arise. Management of a research laboratory demands very different methods and skills from those necessary in the management of business or public administration. Piganiol's group termed these desiderata 'Policies for Science'. Quite different issues emerged with regard to the application of research and the contributions a scientific approach can make to the attainment of national objectives in, for example, agriculture, industry, defence, social policy, education and foreign affairs. A main point of the report was to demonstrate the wide spectrum of sectors in which research had potential to contribute to the formation of policies. This the group named 'science for policy'.

After barely a year of discussion the working party published its report *Science and the Policy of Governments*. There were two main recommendations: first, each government should set up a central mechanism to discuss science policy in this broad sense and advise ministers accordingly; second, OECD should call a meeting of ministers responsible for science policy and organisation in the member countries to continue and deepen the debate. These recommendations were both implemented.

Policies in general require a quantitative background to measure their progress and guide their implementation, clearly so with the allocation of resources. The development of science policy thus requires data on research funding and on numbers of scientists and engineers. I had already made some preliminary attempts to locate such data when EPA still existed. It was a difficult task. I found that while governments generally were willing to provide their science statistics where such existed, the data was fractional, based on uncertain and different definitions. Moreover, industrial firms, responsible for a large part of research and development activity, were often hostile to providing the information on imagined competition grounds. One of the new directorate's first moves was therefore the formation of a unit for R&D measurement. I enlisted the leading pioneer in this field, Christopher Freeman, as consultant.

A decade of difficult, painstaking work followed. The practical difficulties of gathering information from various countries were severe; it often simply did not exist. However, as the years advanced and governments became concerned with science policy, most began collating relevant data for their own national purposes and making it known to OECD. International comparisons were not easy owing to different definitions of science and its classification into pure, applied, oriented, etc., as well as wide variations of education systems. Eventually we called a meeting at Frascati near Rome where the national experts, negotiating on the basis of an OECD draft, managed to reconcile difficulties, and in 1970 produced the *Frascati Manual* as the standard practitioner's guide. It was later modified and extended to include the social sciences.

In 1965 OECD published an experimental analysis of 1962 research expenditure and manpower in its member countries and the USSR. These initial international comparisons with all their uncertainties were seen as 'League Tables' of technological progress. They were undoubtedly useful in giving governments some indication of their position in science in relation to their neighbours, which stimulated provision of increased resources. But it encouraged speculation about the so-called technological gap between the performance of America and the industrial nations of Western Europe. This was largely diversionary owing to the absence of a rational basis for thinking about investment in research and technology. Much more will be written about this 'Gap' in Chapter 24.

A couple of years later the OECD published its second report on resources devoted to research. By this time data was more abundant, statistics more comparable and analysis increasingly sophisticated. It presented the national effort in terms of the percentage of gross national product devoted to R&D. This worked out as 3.4% for the leader, USA, while for the UK it was 2.3%. The report did, however, point out how international comparisons based on percentages of GNP could be misleading for national planning because of different research objectives. For example, a very large proportion of US R&D was devoted to defence, and space and atomic development, largely absent in the goals of many of the others. Comparisons based on the numbers of scientists and engineers engaged in R&D in relation to the population of each country were more useful. This indicated that one American in 280 was engaged in such work compared with one in 340 Britons, one in about 500 French or Germans and one in 6,700 Greeks. The report warned of the danger of using its figures simplistically in policy development; there were too many qualitative disparities. For instance the cost of R&D was much less in Europe than in America, whose huge expenditures on defence masked the fact that its technological efforts for the economy were much less than the gross R&D might have suggested.

The first meeting of science ministers as envisaged by the *Piganiol Report* took place in October 1963 with considerable organisational difficulties. Very few member countries had designated ministers specifically for this purpose and doubts were expressed in some countries of the wisdom of convening such a gathering in the dominantly economic environment of OECD. The Netherlands felt it was definitely wrong; the Dutch Minister of Education came to Paris expressly to persuade Thorkil Kristensen to abandon the project. I remember how at lunch that day, he denounced association with the economy as a 'prostitution of science', demanding that science be considered part of social policy. When the meeting finally took place, there were only four or five titular ministers of science; ministers of education or industry represented the other countries. Theo Lefevre, the

Belgian Prime Minister who also had the science portfolio, was elected chairman. The meeting, based mainly on science and the policy of governments, was in fact an educational occasion, a broad *tour d'horizon* of science policy and its issues, an exchange of national experience and an attempt to deepen understanding of some of the problems ministers would face. As such it had no dramatic recommendations; its communiqué was anodyne, but it had a deep influence on many countries in their understanding of the problems and in the creation of central science policy mechanisms, in some cases with ministers to head them.

The most vigorous debate was on the desirable form of the national policy mechanism. Some countries such as Belgium and France favoured a strongly centralist approach; others, including Britain and America, argued for separate sectoral budgets. The most direct and systematic approach was that of Belgium. Under the Prime Minister's authority its National Science Policy Council had cognisance over all aspects of the problem, and was empowered to propose the reorganisation or reorientation of science activities throughout government structures. There was then a central national science budget, the cake being cut into slices for the various sectors on advice from the Council. The success of this strict system obviously depended on the power of the Prime Minister over his Cabinet colleagues. The Americans and British rejected this approach, arguing that in each sectoral budget – health care for example – the size of the research element must be a matter of opportunities opening up at the time and considered in balance with capital, operational, administrative and other aspects of sectoral activity. Many other issues were discussed in detail. There was strong advocacy of fundamental research and the need for academic freedom, sensitive matters at a time when the encroachment of government intervention was feared. The scope of the term 'science' was discussed and it was agreed that this first ministerial meeting should restrict itself to the natural sciences and technology, but at subsequent meetings should include the social sciences and humanities.

Before they departed the ministers agreed to meet again in two years time. This took place in February 1966 under the chairmanship of Monsieur Peyrefitte, French Science Minister. In many ways it could be seen as a continuation of the first ministerial meeting, with one important difference, nearly every country was represented by a minister of *science*, many having been created since the first meeting. There was discussion of the place of fundamental research, seen both as a long-term national investment and as essential in giving vitality to the education system. Ministers also agreed that it was necessary for all member countries, even the least industrialised, to sustain sufficient research activity to ensure a threshold of awareness of world scientific progress and of advanced skills, without which transfer of technology is likely to be ineffectual. We had presented a paper on the place of the social and especially the behavioural sciences in national policies. This was not opposed, but created little interest and was obviously premature.

During this second ministerial meeting there was much greater understanding of the relationship between science and the economy and the need for governments to establish policies for the stimulation of technological innovation. The liveliest debate was that based on the 1965 (Freeman and Young) study, comparing national R&D efforts. It was clearly a controversial issue as the heat of the arguments between the Americans and industrialised European countries indicated. However, it was quickly realised that the data available were quite insufficient to provide a firm basis for comparison of national efforts. Furthermore, too little was known of the influence of R&D on the competitiveness of different sectors. OECD was invited to intensify its work on comparative statistics on this topic, to analyse them and study their influence. These studies would form the basis for discussion of the problems of disparity in R&D efforts between countries and their influence on trade and industry at a third meeting of science ministers. This launched a series of studies that soon became well known under the title, *The Technological Gap*. More of this later.

Enough of the ministers! We were convinced that in the stage of change and uncertainty regarding science policy, countries would welcome opportunities to profit from the experience of others. OECD had already established a reputation for annual examinations of the economic situation of each of its members. We felt that a modification of this to meet the needs of science policy would be a constructive device. Accordingly, a series of in-depth explorations of the science policies of individual countries was launched. A parallel series of examinations of educational policies was also initiated, the exercise voluntary and, in each case, undertaken at the request of the country to be examined. Experience showed that countries engaged in internal debate on their scientific system asked for such an examination as guidance before preparing new legislation, the process normally taking a year or more to complete and the secretariat able to carry out only three or four examinations each year.

The initial stage entailed sending an expert from the science directorate to the country under examination to prepare, after discussion with the various authorities involved, a background report describing the structure of scientific organisation, its statistics on research expenditure and manpower, the economic and social situation, and many other related data. This preparatory work could take many months to complete, especially in 'scientifically complex' countries such as America and Japan. The background paper, essentially descriptive and avoiding value judgements as far as possible while possibly trying to formulate some of the main issues facing policy-makers, was then forwarded to the country and modifications sought, relating to accuracy of data or emphasis.

The next stage comprised appointment of a group of two, three or, in exceptional cases, four independent examiners, generally including an economist, none from the country under examination. Having read the background paper, the country was visited to question a broad range of persons concerned in one way or another with science and technology, including relevant ministers, senior officials, heads of scientific institutions, university leaders, industrialists and prominent scientists. The people undertaking the examination were chosen by OECD after negotiation with the national authorities to provide a broad range of experience. As an example of the types of individuals serving as examiners, I cite the review of America in which Theo Lefevre (Belgian Minister of Science), Pierre Masse (former head of the French Economic Planning Agency), Professor Casimir (Science Vice-President of the Philips Corporation in the Netherlands) and Conrad Waddington of Edinburgh University. For Canada, the examiners were Pierre Piganiol, Saburo Okita (head of the Tokyo Economic Planning Bureau) and me. The reports analysed the main problem issues and frequently suggested points for consideration by the governments in question, or even proposals for reform of their systems. Each report was signed by the members of the examining team, was not the responsibility of OECD, and was frequently supplemented by a number of questions.

The final phase was a so-called confrontation meeting held either at OECD in Paris or in a city within the country being examined, during which the examiners questioned a team of representative science policy-makers from the country, often led by their science minister. Present also were science policy people from the other OECD countries having the right to raise further questions. A final report was then issued, consisting of background information and examiners' reports together with a summary of the confrontation.

This series of publications provided OECD member countries with a detailed understanding of the science policy issues, preoccupations and experiments of Western Europe as a whole. For the countries submitting to examination, its utility varied considerably. The first few cases, during which we were learning how to make the system work, were somewhat descriptive and superficial, but as our confidence and that of the countries grew, they became more profound, frank and constructively critical. In general their influence has been greater in the smaller countries; less visible in the larger, industrialised nations. In a number of cases major reform has resulted in the interval between presentation of the examiner's report and the confrontation.

There were some minor political impacts. I refer to the somewhat ludicrous history of the Italian review as illustrating the deviousness of the grinding machinery of government. I accompanied the examiners, led by Harvey Brooks of Harvard University, on their initial visit to Rome. We were warmly received by a series of ministers, all of whom expressed the need for reform of their country's science and technology and a willingness to bring this about. I then returned to Paris and the examiners started on their round of discussions. I joined them again in Rome at the end of their enquiries when they were agreed on a diagnosis and wished to present their conclusions to a meeting of scientists, industrialists and officials for preliminary reactions before starting to write their report. They were indeed very critical of the existing situation and suggested quite radical changes. The reactions were interesting. The industrialists said that the analysis was accurate and the proposals sensible; the scientists said 'of course they are right, but there isn't a hope in hell of their being accepted'; while the official's response was 'very interesting, very interesting'!

The report was shortly after sent to the government. After a brief pause to allow for its digestion, I wrote asking Rome to suggest a suitable time for the confrontation. There was no response. I repeated the request several times; complete silence. Months passed and then, one day, an Italian emissary arrived in my office with a completely rewritten examiner's report, elegant and anodyne; in fact a polite rejection of the examiners views. I explained that neither the Italian government nor the OECD had the right to rewrite the examiners report. They could, of course, refute it at the confrontation. The emissary retreated with his tail between his legs and the report rested in limbo for about a year. Then, one hot day in August, I was sitting at my desk when a call came from the Italian Foreign Office. An agitated voice blurted out that for political reasons they had to have a confrontation meeting before the middle of October. I replied nonchalantly that they'd already had nearly a couple of years to ask for one, so why this unseemly haste now? Then it all came out. This was the year of student unrest all over Europe. The staff of the National Research Council in Rome had taken over the building, locked the President out of his office, found the *Brooks Report* in a desk drawer, translated it into Italian and spread it abroad. *Il Rapporto Brooks* was on sale at street kiosks. One of the innumerable Italian elections had been called. Student grievances and institutional inadequacies were public issues that made our report a *cause célèbre*. On no account must the government be seen ignoring well-intentioned international criticism, hence an early confrontation was a necessity. We obliged and a relatively mild discussion took place in Paris. I doubt it brought about any immediate changes, but it was a brilliant report, undoubtedly of long-term influence.

One tiny vignette from a country review. The Spanish confrontation took place in Toledo. Prince Juan Carlos sat through the whole proceedings attentively, having been sent as an element in his preparations to become king. Contacts with him later proved his retaining a lively interest in the science, economy, and society complex of issues that fascinate me.

It seemed to me axiomatic that parliaments were involved deeply in science policy matters, although at the time few were aware of it. Parliamentarians have not only to ensure the accountability of taxpayers' money voted for R&D, they have to decide on many varied, complex and highly specialised issues such as nuclear energy, the environment or biotechnology not to mention military technology. Most parliamentarians had little understanding of scientist's thought processes. Generally shaped by classical or legal educations, yet frequently having to take decisions on matters in which scientific or technical judgements were determinative, meant that in practice they had to accept propositions on faith. While this was perhaps acceptable during the period of scientific euphoria, as the general public began to see that scientific innovations had negative as well as positive consequences, this incapacity to understand the technical basis for decisions was disturbing. It was important that scientists and politicians be encouraged to meet in an attempt to achieve some sort of communication.

A pioneering attempt towards this end had been made in Britain with the establishment of the Parliamentary and Scientific Committee, an unofficial body that met regularly in the House of Commons. Its members were leaders of the main scientific societies together with any MPs showing interest by paying a trivial subscription. It discussed issues of scientific or technological interest and national significance. When its members, encompassing all the political parties, made a recommendation, a small delegation waited on the minister concerned to press their views. In my London days I had attended meetings of this body and was impressed by its sensible, non-partisan attitudes. I had attended one of its dinners, sitting beside Mrs Thatcher, then barely known. This was the only time I ever met her.

It seemed to me that here was a possible model applicable on the European level, so I arranged with the Parliamentary and Scientific Committee to organise a conference at the House of Commons. It was quite a success and encouraged several of the OECD countries to form parliamentary and scientific bodies of their own. I observed at that first meeting how European politicians showed little inclination to discuss the economic impact of scientific discovery; remaining bewitched by the mystique of science and eager to hear about high voltage particle accelerators, radio telescopes and the like. The meeting recommended holding such meetings on a regular basis. We decided it would be more appropriate that they be organised by the Council of Europe with its parliamentary assembly rather than OECD. However, the second of the series, held in the Austrian Parliament in 1964, was sponsored jointly by the two organisations. It faced up to contemporary issues much more realistically. I attended several meetings of the series. At that of 1972 it was evident that parliamentarians had become increasingly interested in the newer and especially the social aspects of science policy, a trend I discuss in Chapter 23.

I look back on these activities of 30 years ago or more with mixed feelings. We certainly were the pioneers of science policy thinking in Europe, and helped to generate understanding of the articulation of science into the web of economic activity. For me it was a fascinating phase. I came to know intimately the personalities in this field throughout Europe, was aware of their ambitions, petty power games and opposition to change. But it was also a period of frustration, of misunderstandings, faulty communications and thinly veiled hostility. At one moment of near despair, I decided that between the naivety of the scientists, the arrogance of the economists, the ignorance of the politicians and the bloody-mindedness of the bureaucrats there was little possibility of rational advance. More seriously, I was disappointed that no generalised model of an effective science policy structure emerged or was envisaged. This uncertainty continues. Most countries, and Britain is no exception, have changed their institutional and science advisory arrangements many times during these last 30 years and the issues remain the same ones we sought to resolve.

With its roots deep in the educational system, its place in the contemporary culture undefined, its tentacles penetrating into every department and agency of the State and its products impinging, often dramatically on the life of the people, efforts to shape constructive policies for science and its uses will long persist despite the complexity of integrating science into national activity.

22

A House in Provence

These last chapters have been dense with accounts of my professional work, giving hardly a glimpse of personal life. To redress the balance, I offer an account of how we were living during this period. It will be so brief as to be prosaic.

When I left for Paris at the beginning of 1957, Sally stayed on at Clavering Avenue until arrangements for the children were completed, joining me at the beginning of the summer vacation. I found myself a room in a private home whose owner was amiable enough and did not threaten my privacy, and immediately got down to work. He had some acquaintances in my professional circle, asking one of them whether I minded being so ugly! I asked him what he thought of the English: 'Did he regard them with the usual Gallic suspicion?' 'I have never thought about it', he said, 'I suppose I tend to avoid them. You see I was returning from Dieppe early one day when hordes of your compatriots poured out of their packet boat and crowded into the Paris train. I entered the dining car hoping to enjoy a leisurely breakfast, but to my horror it was already full of Britons with large plates of bacon and eggs and drinking tea. I found the sight utterly disgusting and have steered clear of your countrymen ever since.'

In the autumn Sally and I settled into a rented flat in the Rue de Rivoli with a view over the Tuileries Gardens, dehumanised by the unceasing roar of traffic. It was furnished in a heavy Napoleonic style with brownish-red upholstery proclaiming a shabby but pretentious history. The concierge, a quiet depressed man whose wife seemed to be screaming and cursing someone whenever we entered the building, gave a final stamp of authenticity to our new French dwelling. After a few months of wallowing in our historical surroundings, the traffic noise got us down. This coincided with our landlord wanting rid of us. Apparently our apartment was one of many where the sitting tenant paid only a minimal pre-war rent and couldn't be evicted. There was therefore great temptation to sublet at post-war market rates, making a large profit. This was illegal, but common practice. At the time of our arrival, the authorities decided to clamp down and thus we proved to be highly embarrassing to the landlord who did not dare throw us out lest we report it to the police. He resorted to removing articles of furniture piece by piece. We did not take the hint because we had not yet found alternative accommodation and soon nothing remained but our bed! Luckily a spacious apartment in Rue Madame, just by the Luxembourg Gardens, became available and the problem vanished for all of us.

By this time settled in my job and expecting therefore to be in Paris for some years, it seemed only sensible to buy a flat given the high rents. Sally was set on a bijou place in an old building on the Left Bank or on the Isle St Louis and spent three months searching for one. Amongst places on offer, this tenant of the Rue Madame inspected flats on the Rues Monsieur and Mademoiselle, finding nothing to her taste since all were priced beyond our reach, decrepit or needing extensive repairs. In the end we settled on a new construction off the Rue de Grenelle in the 7th Arrondissement bought from its blueprint. This proved to be a considerable success and was to be our home for 34 years. It faced on to a large garden with lawns and a pond, was well insulated from street noise and we wakened in the morning to the singing of blackbirds and the swish of swifts. Overlooking the gardens was a balcony where we planted flowers and which was big enough to serve as the summer dining room.

The *quartier* proved to be delightful. Situated west of the baroque Paris of the embassies that makes the rue de Grenelle memorable, it lies between the two green lungs of the city – the Esplanade des Invalides on the east and the Champ de Mars on the west; it is an oasis of vitality and cheer in the midst of the peaceful, sedate, rather self-consciously upper-class 7th Arrondissement. A main feature is the Rue Cler (now pedestrianised), the best Paris food market, always cheerful and animated, but at its best on Sunday mornings. Crowds throng the street admiring the mountains of ripe strawberries or the many varieties of mushrooms arranged to enhance their subtle differences of colour. For families it is a weekly outing to purchase the ingredients for Sunday lunch. Father may buy a dozen oysters while mother insists on langoustines for herself and the children. Then they will queue up at the *cremerie* to choose cheese from the more than 100 varieties on display. There is much friendly banter between families and merchants, who fight strenuously to retain the loyalty of their regular customers. Then, midday past, the crowds melt away, stalls are cleared and shops shuttered, leaving the street drab and empty.

Our *quartier* had many of the characteristics of a village engulfed and isolated in the midst of a metropolis. People were on the whole friendly and communicative and during our years there we never experienced the morose rudeness that so many complain about in Parisians. Our social centre was a butcher's shop at the corner of Rue Cler. The butcher and his wife, hard-working and conversational, would recommend items as especially tasty, with useful tips on how to cook and serve them, and there was never any pressure to buy. Generally there were several customers sitting around, as in a barber shop, waiting to be served or, having completed their purchases, loath to leave the atmosphere of *bonhomie* and good conversation. Inevitably much of the talk was about politics with a general disdain for politicians of all parties. The butcher was extremely well informed on international affairs, with a highly developed capacity for analysis. He was a perfect example of the effectiveness of the French education system and the ability of the French to analyse. Each time I returned from a journey abroad he would question me in detail about conditions in the countries I had visited.

Our social life was rigidly constrained by my professional function, with innumerable cocktail parties welcoming members of OECD committees or to give us the opportunity to meet visiting personalities. Each country gave a reception to celebrate its 'National Day' at which an appearance was necessary. In addition, there were endless – and unrefusable – invitations to embassy dinners. Initially, these were exclusively from ambassadors to the organisation, but later as the Club of Rome became well known, representatives of countries like Bulgaria, Japan,

Poland, the Soviet Union and Venezuela welcomed us. If I was bored to extinction by this, Sally was horrified by each invitation with its prospect of spending yet another evening sitting beside another stuffy ambassador talking banalities that were made even more primitive by linguistic inadequacy. We made no friends as a result of these gatherings, although some of the acquaintances we made proved useful professionally.

All this left little time for normal social contact, but spontaneous friendships sprang up within the international community of the secretariat. We remained close to Ron Gass and his wife Collette, but on the whole tended to cultivate friendships with colleagues of nationalities other than our own as being more interesting; we had very good relations with some French colleagues and their families. Gradually our French connections spread beyond the OECD community. After living for so many years in France, we agreed that our experience had been mostly of friendliness, and occasionally of affection, though a distinct formality persisted even in the most intimate cases. It was just not the done thing to visit unannounced or to prolong a happy afternoon together with an improvised supper of scrambled eggs. Apparently the French language has no expression for our word 'casual'. Weekends provided a happy respite from the high pressure of work and professional socialising. Ever since leaving Washington more than a decade earlier, we had been unable to afford a car. Now we could enjoy the independence and mobility it confers. We explored the countryside within a radius of 100 miles around Paris; picnicked by running streams and in forests carpeted by bluebells or lilies of the valley. We visited cathedrals, abbeys and chateaux, sauntered in the cobbled streets of old towns or simply enjoyed the rolling countryside.

What with her sterile social life and the frequent absences of her workaholic husband, Sally's introduction to life in Paris was hardly satisfying. Something had to be done about it. During the spring of 1939, a few months before the outbreak of war, we had spent a couple of pleasant weeks in Provence, first revisiting the classical sights of Avignon, Arles and Les Baux, before walking and scrambling amongst the Esterel's red rocks. Sally was so enchanted by the landscape she declared that she intended to have a cottage in the region one day. This was no passing whim, she continued to bewail the probability of attaining this ambition throughout the dark days of the war. When in 1957 the prospect of a job in France came up, the magnet of Provence played a big role in Sally's decision to join me.

So now, we hoped, was an opportunity to create a new quality element in our lives. One of my colleagues at OECD, Jean Poignan, lent us his secondary house at Peymenade for a couple of weeks in the summer. This was situated in the valley separating the Esterel from the Alps and near Grasse. He introduced us to a young couple Marcelle and Hugette who were setting out on a career to restore semi-ruins. We moved around with them for a couple of days, looking at various sites. We rejected them all as being too near villages and likely to be swamped by the flimsy new houses that were springing up as a result of the wave of affluence now beginning to sweep Europe. Finally, they took us to a remote and lonely place near the little Gallo-Ligurian town of Callian. It was barely accessible; indeed, it was unreachable by car and we had to cut our way through brushwood to enter on foot. There we found a ruined cottage with solid walls and no roof; indeed, there was a sizeable tree growing in what should have been the main bedroom. This shell of a house was situated at the edge of a territory of about 7 acres of unreclaimed *maquis*. On its west, a series of terraces still retained some traces of vines, while towards the east were the remains of small fields not yet completely taken over by the rampant growth of Mediterranean thorns that had taken possession of the main area. There were also many trees, oak and pine, mostly young but with a number of mature specimens. In the bright summer afternoon the place had a certain attraction; it was perfumed with pine and lavender and its wildness gave an impression of being outside the world that appealed to both Sally and me. But as a potential country home it seemed hopeless. There was no water, no electricity, no telephone and not even a road.

We returned to Paris empty handed; we had seen no other place we could contemplate buying. In the months that followed we talked often about this abandoned enclave of the Western World until in November, Sally announced that she wanted to have another look at it and travelled south with a friend who was looking for a place to retire to. With Hugette and Marcel they visited the site on a day of lashing rain and discussed the possibilities of obtaining the necessary infrastructural amenities.

The road problem was not insoluble; a judicious use of dynamite to shatter rocks and some cutting of brushwood could make a rough track for a car. There was a large well not too far off to provide enough water for the building process and there was plenty of good stone on the land for restoration of the ruin. Our friend, listening to all this and looking through the rain at the drenched landscape said, 'you wouldn't be fool enough to buy this place.' Sally bridled for a second and replied, 'I intend to do so.' So that was that. We negotiated with the owner of the land and settled the price; it was really quite inexpensive. The ruin had been a *cabanon*, built about 400 years before and abandoned at the end of the past century, as were many houses in the area when a disease wiped out the vines and sent the inhabitants scuttling to Algeria to start a new life.

The conveyancing was simple and rapid. There was a substantial tax on the sale of land but nobody was naïve enough to admit the actual purchase price. The question then was how much to declare – too much would be foolish, too little suspicious. Our lawyer said, 'that is simple. I shall telephone the *Maire* and ask them what would be the appropriate amount.' This worked; it was an amusing case of officials in collusion with the public contravening the intention of the law. The sale completed, Sally had at least the shell of a cottage in Province. We discovered in the *cadastre* that our *quartier* was Les Coulettes de la Negronne. As it was the only building there we decided that our house should be simply *La Negronne*.

I have the impression that most English writers have homes in Provence or the Dordogne and that quite a few of them have written charming descriptions of life in these parts – perpetual struggles with the local bureaucracy, the fight for water, the quaint antics of colourful local eccentrics, gastronomic delights, and the perils of the Mistral. While I enjoy these books all the more because I have enjoyed and endured the same experiences, I have no intention of emulating them by recounting our sylvan delights. I merely intend to insert a few details of the physical reality between the turgid descriptions of my professional doings, to indicate other, satisfying aspects of our life in France.

Our intention was to use *La Negronne* for occasional holidays. The idea of developing it to become a retirement home never entered our minds, so we discussed with Marcel and

Hugette the simplest and cheapest way they could convert it. We decided to raise the outer walls to upgrade what had been a granary storage loft into an upper floor of two bedrooms that could be entered by way of an external stone staircase. We also made provision for a covered patio outside the front door. We left them with a small advance for the purchase of materials and they agreed to start work immediately.

We visited the site a few months later, finding that much work had been done. The tree in the bedroom had disappeared and the skeleton of the patio had risen. The young builders had been understandably diligent, as was to be expected on this, their first joint job. Their equipment was primitive, they had a mechanical cement mixer, but that was about all. Stones and cement had to be carried by hand, and water brought from the well in buckets. They had brought in a local firm to explode the larger rocks on our highway, a track that I shall refer to as the *chemin*. Actually this track, shown on nineteenth-century maps, had in practice consisted of a 400-metre-long ridge of stones, no doubt accumulated over centuries by peasants clearing their fields. Finally it had served as a mule track. We actually saw a few of the explosions which were carried out according to procedures laid down by law. When the charge was set and the workers had retreated, a man with a feather in his hat stood on a high rock and tootled a thin sound on a small trumpet as a signal that all was clear for the detonation.

On arrival at *La Negronne* early the next summer the work was well advanced. The roof was on and covered with the local pink Roman-style tiles. The ground floor was unusable with walls being plastered and floors covered with red earthenware tiles, so we used the smaller upstairs room as a kitchen with Primus stoves for cooking and the other as our bedroom. Two of the children were in tents in the grounds and Jane had brought a tropical hammock complete with mosquito net that she had found in an army surplus shop in New York. This she hung between two suitable trees about 200 meters from the house. When she decided to go to bed, she couldn't find it, so dense was the vegetation. In the light of the flashlamp, thorn trees lit up, ghostly and menacing all around, but no sign of the hammock. She had to retreat defeated, to seek shelter in the tents with her sisters.

This incident focused our attention on the thorn problem. They flourished in their tens of thousands in some parts of the territory, obliterating all other forms of vegetation. We started to clear them around the house, cutting them as near the ground as possible with large clippers. We were delighted to see pleasant meadowland uncovered, but when we returned the next year the thorns had reappeared. We realised these tenacious plants had to be cut down for three consecutive years before they gave up. It took us 35 years to eradicate the thorns from the central part of our estate. The effort to do so gave me 35 years of physical exercise to counteract my sedentary life during the rest of the year.

That first summer at *La Negronne* was sheer bliss. Quite a number of friends visited, camping near the house. Everyone seemed to enjoy the primitive life. Sally and the girls established an outpost by the well up the *chemin* where they and clothes were washed. At other times they took sheets to Callian to the municipal wash house – open to the street and fed by a wonderfully cold stream brought down from the Alps by an original Roman aqueduct. From the same stream we filled large plastic containers for drinking water. Our builders had made provision for normal water use. The plumbing included a shower, prepared lovingly with its base of smooth, coloured stones, and a WC. These appliances were for the future. In the meantime we dug a deep trench as a latrine, hidden away in a small clearing in the thorns. When occupied, a Union Jack was hoisted as a warning to keep away. When we returned the next year it was a prolific bed of ripening tomatoes.

Summer days at *La Negronne* were beautiful if often a little too hot, but the nights were always comfortably cool at that altitude. There were occasional summer storms with violent lightning and, on several occasions, fierce forest fires tore through the trees often within sight of our house; some dangerously close. The summer heat made us long to plunge into cool water. At the southern edge of our territory was the Riu Blanc, usually a dry stony riverbed in summer, but becoming a raging torrent after prolonged rain in the mountains. A rivulet cascaded into it from the hills opposite, flowing all summer. One of its rocky pools was big enough for swimming and thus became an important summer asset. A few years later an artificial lake, Lac de St Cassien, was created a few kilometres downstream to provide for the resorts on the Côte d'Azur deprived of water by the collapse of the dam which had devastated Frejus. For many years this was a magnet for children at *La Negronne*, for swimming and boating. Later, as the lake became well known, its hordes of summer tourists tended to make us avoid it.

We returned to *La Negronne* for a couple of weeks around Christmas. The weather at this season was superb. The nights were cold, the temperature descending to minus five degrees, but we sat warm in front of roaring wood fires, burning logs we had cut from our own trees. The days were bright, with clear skies and we lunched in the open, warmed by a surprisingly strong winter sun. This was the time of *La Chasse*. Men with rifles and hunting dogs roamed the countryside shooting at everything that moved. When we first came to *La Negronne* there was some wildlife – a few partridges, thrushes and tortoises. These are now long gone and in winter there are only a few small birds. We resented the *chasseurs* for disturbing our privacy and making noise, but we were studiously polite. Any indication of opposition to the hunt would have invited vandalism, so primitively passionate are the peasants of the Midi about their traditional sport.

The next five years saw continuous development of *La Negronne*. Much paving was laid, new terraces built, and a wood shed and a double garage that served as a workshop 100 metres from the house. Finally, after much difficulty, we had electricity. It arrived by landline, carried on pylons, each one of which had to be negotiated separately by Hugette as their erection on other people's land naturally required the owner's permission. Water remained a major concern. We built an underground cistern to collect rainwater from the roofs, but it was insufficient and porterage of drinking water from Callian was tedious.

Then, having saved a little money, we decided to extend *La Negronne* by building a single-storey house up the slope from the existing one, the two patios being connected by a covered flight of stone steps. The new building was conceived as a dower house, with its own kitchen and bathroom. It had one large bedroom *cum* study with a paved breakfast terrace and a large salon/dining room with rustic open fireplace. Much later we extended it further by building a corridor with two guest bedrooms with their own bathroom. Despite the current lack of running water, we designed the extension to have central heating with an oil furnace in an external boiler house. By this time Marcel and Hugette had departed on their own enterprises, so the new construction was given to a local firm. The result was not a complete success. The firm went bankrupt in the middle of construction, causing many delays and difficulties and we were dissatisfied with the workmanship. Nevertheless, the new *La Negronne* promised to become a comfortable home, accommodating many people, summer or winter.

Our daughter Catherine and her husband Michael, both doctors, had been with us at *La Negronne* since its beginning. One day, Sally and Catherine, with her first son Alexander, were walking in a then rather desolate part of our territory we had named the 'wintry wood' when they heard the sound of childish voices in the distance. On the look out for companions for Alexander, they followed the voices until they came upon another semi-ruin in the process of reconstruction. They introduced themselves and explained where *La Negronne* was situated. This was the Darnis family. The father, recently retired from the French army, was particularly welcoming. The setting was idyllic, the children mixed well and the adults were mutually compatible. They were living in Draguignan, but coming each weekend to work on their house, *Saint Esprit*, which appeared on ancient maps. They hoped to complete it as their main residence. A sudden storm blew up and M. Darnis drove them home. This was the beginning of a friendship that has now lasted 35 years. M. Darnis gradually assumed the role of *La Negronne* guardian, keeping an eye on it in our absence, replacing loose roof tiles after storms, undertaking maintenance and minor improvement jobs, and after we had abandoned Paris, ferrying us to and from Nice airport.

The water problem had to be solved once and for all. Not only would the new house function as planned if there were constant running water available, but we had already, in a fit of optimism, constructed a large swimming pool that stood empty as a reminder of our folly. There had been plans to extend Callian's piped water system as far as our house, but they never materialised. Then the builders suggested we should buy water from the municipality in tankers. This we rejected. Our main hope was to find water by drilling. M. Darnis had recently found a plentiful supply of water at *Saint Esprit*, from a source identified by a water diviner and employing a drilling engineer. We approached this man for advice about the possibilities on *La Negronne* land. He made what seemed a very equitable proposal. He worked regularly with an old peasant who had a high reputation as a dowser. Should the old man fail to discover a water source, we would pay the engineer only his out-of-pocket expenses involved in bringing and removing the drilling equipment. If he did find water, we would pay a reasonable amount. The diviner apparently having already worked for the engineer for three full years and never failing to find water, we accepted the proposition.

The day came to make the survey. Half the villagers of Callian turned up to witness the spectacle. A few of them walked around with forked sticks to try out their dowsing skills independently. The water diviner (*sourcier*), an elderly peasant with a wood and coal merchant's business in a village some 40 miles to the west, had no particular knowledge of the local geology. I suggested he start by surveying the land adjacent to the dry bed of the Blanc as the vegetation there always appeared lush. He produced a small bottle filled with colourless liquid, presumably water. He walked along with this device swinging in a light metal stirrup. From time to time it agitated and swung quite violently, whereupon he brought out his forked twig, that had to come from a fig tree, just as in England it must be of hazel. The twitching of this stick apparently gave a fine structure delineation of the watery area. He said, 'Yes, you would find water anywhere here, but it would be foolish to drill for it, because it is in discrete pockets. It would flow freely and then dry up in a few months.' Finally, after a long search, he chose a spot on the highest and driest part of our land and announced that there was a narrow vein of water about 60 metres below the surface that would yield about 100 litres of water per minute. I felt vaguely that I was a traitor to the culture of rationality in cavorting with a 'sourcier', having been conditioned by my physics professors that dowsing was a dark abode of charlatans and their gullible victims. Nevertheless I immediately accepted his advice to drill. The fee for the diviner was trivial.

Then came the critical business of drilling the borehole. The bit sank through the soft limestone quite quickly. After a few metres it emerged into emptiness, a cave of some 5 metres from roof to floor. Then it made contact with rock once more and ground slowly until the 60 metres of depth approached. Small amounts of water were pulled up and excitement grew. Drilling continued very slowly and real amounts of water were raised. The decision then had to be made. Were we to stop at this point, not far into the body of water, we might miss the main supply and thus risk interruptions in arid periods if the underground rivulet was low. On the other hand, if we continued too far, we might penetrate the bed and lose everything. After much hesitation, there was general agreement that the position was near optimum and the drill was turned off. The result turned out to be much as the dowser had promised. The new plenitude of water transformed *La Negronne*. It provided water for all domestic purposes, made use of the central heating system possible, allowed us to water what passed as our garden and brought to life the swimming pool. It has been with us for more than 20 years and never failed. The submersible pump has had to be replaced twice because of lime deposits, but that is all.

A number of other major developments took place about the same time. We bought two adjacent pieces of land so that *La Negronne* now has about 25 acres. We did not want more land as such but, with the spread of habitations from the Mediterranean, extending our boundaries seemed the only means of avoiding too-close neighbours, thus preserving our isolation and tranquillity. Extra acres meant thousands more thorns, delaying our total victory over these green horrors. Also, we were able at last to install a telephone, this and later a fax, was an enormous comfort to me as it obviated innumerable visits to Callian post office and the long waits inevitable in my world-wide professional communications.

La Negronne was now fully operational and remains so today. It became the family centre, where our far-flung members most conveniently met. Catherine and Michael moved to Paris for a couple of years while Michael was working at the Villejuif Cancer Centre. This enabled their three sons to spend months of the year at *La Negronne*, usually in Sally's care. Later Jane's daughters arrived and joined their cousins. Margaret and her husband Hirohiko were also frequent visitors, as it was an easy car run from Monte Carlo where he was first flute in the orchestra.

This was one of the high points of *La Negronne*'s existence. Our children delighted in it as a family home, their children flocking there whenever possible. It really was a children's paradise with its birds and insects, its secret paths through the woods, the lake nearby and an infinity of country walks. As they grew and became more adventurous, there were excursions in the mountains behind and further afield to the Gorges du Verdon or the Ile St Margerite. In the early years visits to the coast were frequent, but it became more and more difficult to find a quiet bay from which to bathe, the coast became so crowded and commercialised. We were always relieved to arrive back to the tranquillity of our own domain.

A few years later again and we were hosts to school friends invited by grandchildren. Raucous broken male voices and petty feuds intermittently broke the spell. Then, with the passage of time, girlfriends appeared and eventually fiancées. But the grandchildren had reached a period of life when individuals want to explore the world and spend their leisure in exotic places. *La Negronne* for a time became a place for mainly adult gatherings, although our children and their partners came for shorter visits, faithful to its magic. It was not long, however before great-grandchildren began to appear, so the cycle has been completed and a new cycle begun. This new generation is being introduced, or should I say conditioned, to the sights, sounds and perfumes of *La Negronne* as part of their parents' nostalgia to become an element of their own heritage.

In many ways *La Negronne* is very different from the primitive Eden we first identified. The war against the thorns has more or less succeeded. Over the years we upgraded our attack by improved technology; cutting shears gave way to an oil-driven bush-basher and then to a tractor. We still have to run over the cleared part each year to keep down the brushwood. All in all, the maquis has been transformed into parkland as a result of 35 years of hard work. This labour has been a great joy to me, it has relieved the tensions of my professional life and, I am sure, preserved my health and strength. It has also been a great joy to Sally throughout the years. Nevertheless I often wonder if we were right to domesticate it. The easy life there today and absence of the primitive joys of our early years means less effort and fewer challenges facing the new cycle growing up. The magic is intact, the pine and lavender perfume that attracted us initially is unchanged, the cicadas still chirp their unfinished symphony and the praying mantis with its camouflaged ferocity still abounds. The wild boar population is on the increase, a fine specimen was recently observed near our front door. In any case, any disquietude about my possible environmental vandalism is misplaced. Given a few years of neglect, *La Negronne* would be well on the way to maquis again, if speculative builders were to allow it.

23

Education

This is one of the few subjects on which everyone is prepared to express apparently informed views based on their own experiences, and I am no exception. I find myself making wild and strongly felt generalisations from my personal, not at all typical, education. So, when the Sputnik projected me unexpectedly into educational aspects of the OECD programmes, I arrived with heavy baggage in the form of little-thought-through ideas and prejudices. I mention here a few of these that influenced my attitude in the early evolution of the OECD.

I have a deep conviction of the seminal importance of the process of learning throughout life in the actualisation of the fully human individual's potential. I regard formal teaching as one of several means to feed the vital mechanism of learning. Implanting a thirst to learn is the second act in the creation of a child; the most precious heritage parents can bequeath. Within education in its fullest sense learning is more fundamental than teaching and its processes demand paramount attention. The inclination and capacity for learning must persist throughout life if the individual is to sustain his or her vitality and complete development. It is my own experience, now in my nineties, to discover every day with delight that I have found new elements of learning and new insights into what I already knew.

Recently, the term 'Lifelong Learning' has been bandied about as a necessary concept if individuals and societies are to maintain employment in face of the rapid changes that are taking place throughout the world. This is, of course, related to what I have just written, but is usually applied to retraining rather than education and concerned with the creation of mechanisms for the recycling of skills. While such measures are obviously essential, they cannot be fully successful unless accompanied by education to inculcate an understanding of the nature of the changes they are designed to meet.

Good teaching feeds the learning process; poor or uninspired teaching dulls the perception of the pupil and the desire to learn. Students, parents and the state all applaud attempts to improve the quality of teaching, hence questions of class size, regular assessment of the quality of individual teachers and league tables of the performance of different schools are much in vogue. At other times concern with the need to improve the education system may focus on the need for equality of opportunity for education, the provision of pre-school facilities, higher technical education or the effects of financial constraints. Thus demands for a variety of educational reforms are always with us. The basic issues of education are hardly changed from generation to generation, but the nature of the emphasis for reform varies considerably, responding to current public concerns, modifications of the social environment, government ideology, technological advances or weaknesses, real or imaginary, exaggerated by the media.

Educational reform is an exceedingly slow process. First of all its need has to be identified and measures formulated to meet challenges; time will pass for consensus to be reached; legislation may be delayed because of parliamentary programmes. The huge size and inertia of the education system makes implementation difficult and prolonged. Finally, a few cohorts of pupils must pass through the new process before it can be evaluated. Thus it may take decades for a reform to be embodied in the system. By that time emphasis may have changed so that when the reform comes into effect, it is no longer of immediate concern. This hysteresis renders education resistant to short-term manipulation.

The heritage function of education is to transmit the traditions and values of a society through the generations to ensure continuity and stability as well as instilling a sense of identity to each new citizen. If too rigidly applied, or too chauvinistically interpreted, such conditioning can confuse young minds facing the difficulties in reconciling it with the realities of the political, social and behavioural environment in which they live. Awareness of the past, especially the recent past, is necessary for all of us, but tradition alone is no sure guide for a full life in the new world in which we find ourselves. Of course good teaching can soften the dichotomy, replacing it with a perspective, but therein lies the trap. Even the best teachers tend to teach what they were taught themselves. Thus, no matter the awareness of the teacher, an element of perpetuation of vanished societies is embedded in the school system. It is too facile to advocate anticipatory teaching; this could be practised only by teachers fully apprehending both the complexities of the present and the consequences of probable trends as they extend into the future.

If this argument is accepted, the critical point lies in the teacher training colleges. I have no experience of these, indeed I am seriously prejudiced on the subject of pedagogy. When I was a young lecturer in command of my subject, I felt that I should learn how to impart it to the students, so enrolled in an evening course on how to teach. The first lecture was so banal and unreal that I never went to another and my intuitive teaching skills proved adequate. I do not denigrate the need for pedagogic skills; but they have to develop a new spirit of innovation now there is so much interest in participative education with its emphasis on learning rather than teaching. However, the training of those seeking excellence in preparing the young for life demands knowledge of many subjects at considerable depth in addition to pedagogy. This necessitates the recruitment of candidates of high intellectual quality, not always achieved today when so many alternative opportunities exist in research, industry and other professions. Teaching, with its modest remuneration, is often seen as a last resort career, when its importance to society should render it amongst the most respected. I am convinced of the need for a fundamental rethink concerning the place of the teacher in society. This necessitates encouragement of teaching as a prestigious vocation, prepared by quality training that ranks with the level of education of the other higher professions and with corresponding pay.

Equality of education – a noble but illusory objective – cannot be achieved through changes in educational structures, provision of resources to support the underprivileged or other external devices. The fundamental obstacle is not only social and financial disadvantage, but also disparities in the capacity to learn that cannot easily be corrected by teaching. I ride many other hobby horses in the education arena. For instance the desirability of including some scientific topics in teaching of the humanities and law, along with cultural subjects in the training of scientists. I am greatly concerned with the consequences of the expansion of higher education, with institutions of greatly differing quality and function springing up everywhere and calling themselves universities.

But enough of my foibles! I must return to describing how in OECD a small series of studies on scientific manpower evolved to become a major international programme on education in general. Three years of activity concerning scientific and technical personnel had been accomplished before OECD's creation and the incorporation of this work within the new organisation was so smooth as to be hardly noticeable. The objective of the programme was formulated as: 'to help to increase the number, quality and maximum use of scientists and engineers trained in member countries in order to meet the present and future needs of the economy of Western Europe.' Seemingly a somewhat narrow mandate, we realised full well that this new emphasis on scientific and technical training would necessarily impinge on the education system as a whole and, wisely exploited, could be used to stimulate new thinking in education generally. This did indeed happen through the gradual evolution of the OECD education programme, one of the most influential international activities in the field, during the second half of the century. It is chastening to reflect that the launching of Sputnik, and the reaction of the US Department of Defence in offering a substantial grant to be matched by the Europeans, triggered this major education programme. It is doubtful OECD would otherwise have become a major player in education.

The chief actor in the construction and implementation of this programme was undoubtedly Ron Gass. His vision and energy were simply phenomenal. He had a genius for pulling ideas out of the air and doing something creative about them. In fact, his genial character, enthusiasm and open mind enabled him to form a network of contacts with creative minds throughout the member countries and had the perspicacity to pick out relevant ideas and synthesise them into brilliant practical proposals. It is not for me to describe the educational work in detail. Indeed, this has already been done by Ron's competent lieutenant, George Papadopoulos, whose history *Education 1960–1990 – The OECD Perspective* gives a balanced and elegant account of this work. I simply sketch our work during the early transitional period from technical manpower to education generally, in which I was intimately involved.

During most of this time education was part of the Directorate of Scientific Affairs, which I was in charge of. I had to devote most of my energy to the scientific side of its work but, especially at the early stages, was deeply concerned with the thinking and decisions on the educational side. Frequently I had to represent it at external meetings such as those of the European Ministers of Education. In 1974, when I retired from OECD, the programmes on education were detached from the Science Directorate and amalgamated with the Manpower Directorate to form a new Directorate for Social Affairs, Manpower and Education under Ron Gass. This was entirely logical as the two programmes of the science directorate diverged to meet new needs, resulting in the links between them becoming increasingly tenuous. Incidentally, during a hiatus between the retirement of one director of manpower and the arrival of the next, Dr Kristiensen asked me to act in charge and I had written the programme of this directorate in addition to my own work. I drafted and supervised this meticulously since I increasingly realised our education and science activities would have to evolve in terms of social and cultural as well as economic needs.

From Scientists and Engineers to Education

Clearly an initial task for the organisation was to attempt to quantify the existing situation regarding scientists and engineers in the member countries, and if possible give some indications of future demands. Indeed OEEC had undertaken a couple of international surveys of supply and demand for scientists, engineers and technicians during the 1950s. These had proved useful in alerting countries to the magnitude of the problems confronting their educational and training systems, but it was fully recognised that in the absence of available data and the lack of analytical methodology these could not give an accurate picture of supply and demand. The primary task was an exploration of the methodological needs of manpower forecasting in this special field, of how meaningful numerical comparisons could be made of existing stocks and outputs as well as of future needs. The ensuing report stressed that if OECD were to carry out a useable international survey of supply and demand for scientists, it would have to collect wide-ranging data on structures and expenditure in the participating countries. This proved to be a long and complicated task; it became of necessity a study of the educational and training systems of all the member countries at secondary and higher level, thus providing OECD with a unique view of the educational landscape that served as basis for its subsequent programmes. In the absence of satisfactory international standards, classifications and statistics, negotiations had to be entered into with each country to ascertain the signification of the terminology, the expected attainment at each level, qualifications for entry into higher education and numerous other data before any meaningful comparisons could be attempted.

Eventually a broad comparative picture emerged of the performances of the education systems of member countries and of their probable future development. The survey, which was widely distributed and quoted, evoked great interest in the Ministries of Education as a vivid and at times disconcerting indication of their relative situations. The methodology developed in the exercise also laid the basis for continued and more refined work in both the OECD and its member countries.

The survey also aroused great interest amongst a wider public. Surprisingly enough this was not so much with regard to the Cold War considerations that had triggered it, but to comparisons between the situations of the various industrialised countries of Europe and still more between those of the Europeans on the one hand and of North America on the other. The educational disparities between the two parts of OECD were dramatically wide, whether measured by the proportion of students entering higher education, numbers of degrees awarded or numbers of engineers. Educational expansion in Europe had been rapid since the end of the war, so the demonstration of an educational gap was a shocking, if salutary, revelation. It was now generally assumed that the future economy would be powered by technology, so the provision of a greatly strengthened corpus of scientists and engineers was paramount – catching up with the Russians now switched to catching up with the Americans! A parallel analysis by the science part of my directorate comparing the scientific research and technological development capacities of the two regions indicated a still more glaring and politically more sensitive disparity that became publicly debated as 'the Technological Gap'. This is discussed in the next chapter.

Simultaneously with these quantitative studies of scientific education and technical training in the OECD region, a qualitative approach was undertaken through country reviews, similar to those on science policy described in Chapter 21. At first these were of quite limited scope, somewhat superficial, being descriptive rather than analytical, but as confidence in the system grew and the OECD programme moved towards education generally, they became a powerful instrument of OECD's educational work. They greatly facilitated exchange of experience, stimulated national action and at times led to the initiation of significant reforms. The fact that countries voluntarily submitted to these examinations, prepared substantial background papers and frequently acted on their recommendations indicated that OECD had become a real factor in national education policy-making. That the examinations were conducted by distinguished educational and political personalities rather than by mere bureaucrats, gave the process a cachet and credibility, while their results, seen as OECD pronouncements, gave legitimacy.

Ever since the creation of OSTP in 1958, an intensive effort had been focused on the teaching of science at secondary school as the most obvious prerequisite for producing the large numbers of scientists and engineers the post-war world would require. While by the 1950s national education systems had recognised that more extensive and higher quality instruction was necessary and much progress had already been made, there was, however, little coherent policy on how this might best be achieved or how the expanded needs for science education might fit harmoniously into the education system as a whole.

The science to be taught was very different from pre-war school science. The knowledge to be imparted had expanded even more than the future needs for scientists. I explained in Chapter 8 how in the 1930s, the exciting new concepts of chemistry and physics had only just begun to penetrate undergraduate teaching. In the 1950s this was still the situation in school science. I fancy the scientific community in its euphoria was more enthusiastic about curriculum reform than the ministries and pedagogues, and at times pressed unrealistically sophisticated inclusions. However, the two sides conveniently joined in the common aim of achieving rapid reform. Although reform of the curricula was the declared objective, successful incorporation obviously required many other changes including, above all, teachers capable of teaching them. The existing body of science teachers had been trained pre-war without the benefit of the new theories and would themselves have to be re-educated. Laboratories would have to be designed and constructed, new instruction demonstrations devised and teaching aids including film and television tried out and provided.

Our unit at OECD made a massive attack on the totality of school science reform. We started by sensitising the national authorities to the nature and magnitude of the problems, exploring with them various facets and finally tackling curricula development in mathematics, physics, chemistry and biology. In the course of a few years these efforts amounted to a campaign of school science reform throughout the OECD region. Our role here was not only that of catalyst as indeed it clearly was, but also as a major contributor to the substance of the new curricula. On the teaching of mathematics, the most radical of our cases, our '*New Thinking in School Mathematics*' was a watershed.

By this time all industrialised members of OECD possessing advanced systems of education had established their own mechanisms for the continuation of this work, and we decided to terminate the exercise in order to clear the decks for new activities. Our wish to abandon this work at the zenith of its success bewildered Council members, but was in line with my agreement with Thorkil Kristensen (see Chapter 21). We did later continue work on school science teaching for the less-developed countries of Southern Europe as an element of a Mediterranean Regional Project.

From its inception OECD became the High Tabernacle of the economic growth movement. The very first meeting of its Ministerial Council announced a target of 50 per cent in the growth of GNP of its member countries for 1960–70. We had already been discussing education as an investment, I had been making speeches about the plurality of benefits that such an investment would yield to a country – increased prosperity leading to social development, enrichment of a nation's human resources and greater possibilities for individual development and quality of life. Recent years had witnessed an evolution of economic thinking on these matters. We were particularly impressed by conversations with the American economist Theodore Schultz who, in his 1960 Presidential address to the American Economic Association, demonstrated that traditional economic factors accounted for only a part of actual growth of national product, suggesting an important additional factor could be the quality of a country's education.

Although convinced from the beginning that our work on scientific and technical personnel would inevitably force us to look at education generally, it would have been dangerous in the early days to say so in the economic environment. Now, as economic growth had become the major preoccupation of OECD, we had a duty to investigate the role of education generally and our work on growth in particular,. Thus education and not merely scientific personnel began to feature in OECD's programmes. We moved quickly to take advantage of the political interest sustained by the Cold War in raising technological levels. Having identified the best available experts on both sides of the Atlantic, with their help we prepared a series of analytical reports as material for discussion at a conference on *Economic Growth and Investment in Education* held in Washington in October 1961. The conference, attended by an impressive group of senior budget officials, educators and economists addressed two issues:

1 The nature and magnitude of the task facing education in the next decade to meet the needs of social and economic progress in the OECD region.
2 What should OECD countries do to respond effectively to the requests of developing countries whose needs for educational expansion were even greater than their own?

The key paper by Professors Svennilson, Edding and Elvin, *Targets for Education in Europe 1970*, was in reality the most sophisticated attempt so far to elucidate the relationship between education and the economy in all its ramifications. It stressed that although all the indicators showed that investment in education could be expected to produce high yields, social returns were also considerable, concluding that methodology for setting educational targets must be built on 'the belief and assumption that more and better education for more people is desirable in itself and is, at the same time, one of the most important factors in economic growth'. Based on these assumptions estimates were put forward of the needs for educational expansion in OECD countries over the next decade, in terms of pupil numbers at different levels, teachers, buildings and expenditure. These were indeed dramatic, indicating the need for at least doubling the educational expenditure. When the ten years had passed, it became clear that these targets were realistic, perhaps even modest.

I presented the outcome of the Washington conference to a meeting of European Education Ministers in Rome the next year, securing their full endorsement. The Washington conference had a considerable political impact, leading to acceptance of the concept of education as an investment. It gave professional legitimacy to the need for massive education expansion to meet national goals including economic growth. Above all it contributed greatly to persuading Finance Ministers to provide additional funds making expansion possible. For this reason especially, Education Ministers welcomed the Washington findings as giving them a better position in the budgetary queue and indirectly enhancing their power in Cabinet. Economic units sprang up in most education ministries. Amongst the educators, however, there was still suspicion about 'OECD dabbling in education'. At an education policy meeting in Oslo, I was attacked for attempting to drag education from its hallowed place in the cultural environment to the sordid tables of the moneychangers. There was, indeed, a shadow of such a danger, but education was safe with us, amongst other things because of our conviction of the power of education to bring about social progress. Within OECD, we sensed that some of our economist colleagues would have liked to see education drawn more tightly within the commands of the dismal science. All in all, suspicion of OECD's incursions into the field of education faded quite quickly as our fertile partnership with the national ministries evolved.

For our secretariat, the Washington conference was a watershed. We were able to plunge into many of the fundamental issues of the educational system. Work on the impact of education on economic growth was pursued, our already considerable compilation of educational statistics further enhanced, quantitative analysis techniques devised, and a successful investment and planning programme initiated. For the educationally less-developed member countries of southern Europe – Spain, Portugal, Greece, Turkey, Yugoslavia and the Italian Mezzogiorno, an intensive Mediterranean Regional Programme was developed and employed to good effect.

1968 was the year of student unrest, culminating in the outbreak of riots, 'les événements' of Paris in May of that year. These were an unforeseen side-effect of the rapid economic growth of the decade and the many other factors challenging the prevailing social structures. Characteristically, Ron was down in the streets with the students, enquiring and debating as to the cause of their actions. He even persuaded Dr Kristensen to talk with the students and publish an analysis of the phenomenon and its underlying causes. The insights gained had some influence on the shaping of our future programmes.

These programmes blossomed under the leadership of Ron Gass and George Papadopoulos throughout the decade of major educational expansion in Europe and the periods of recession and change that followed. It is a rare story of undoubted success in the annals of international intellectual activity.

24

Innovations Galore

It is not in the nature of governments to be innovative, although it should be one of their main functions if they are to face up to the exigencies of the world's changing political, economic and social climate. Short electoral cycles, the time it takes to achieve legislation on new approaches and the inevitable *status quo* mindset of the bureaucracy make radical change a lengthy process despite the brilliance of new processes that may be in gestation within a country outside its official circles. For international organisations, innovation is still more difficult owing to the widely different objectives of constituent governments. This is all too obvious in major new political structures such as the EU, where every step is contested and national objectives range from the creation of loose co-operative arrangements between sovereign states, to full federation.

For OECD the situation was somewhat different; it existed in the shadow of the EU, had no contentious political goals, nor any ambition to acquire them. It operated without any rigid time constraints and thus could pursue longer term issues beyond the reach of national parliaments. Its highly professional staff was respected by member governments so that it was potentially important to its members for the cultivation of new approaches and means of governance, useful to those seeking better means to perform in the unknowable world of the next century. Being politically weak and professionally strong, its suggestions were there to be picked up by member governments if attractive, but not thrust upon them.

But from where do the seeds of innovation come? Hardly from the national representatives sitting on the specialised committees of OECD. They are inevitably briefed with their governments' views of the moment and not the future. In a recent conversation, Ron Gass and I agreed that throughout our time in OECD, new initiatives seldom came from national delegations. In these circumstances it is up to an innovative international secretariat to work up promising ideas that might later be assimilated into the political system. I gave examples of this in the previous chapter. Of course we did not have the arrogance to suggest we could originate the necessary bright ideas, but collectively we had an impressive network of contacts with original thinkers throughout the OECD region. We were thus in a position to discuss with them, sift their ideas, select those appropriate and frequently to employ them as consultants.

I was persuaded of the potential value of such an approach when OECD was founded in 1961 and I had to devise tactics to actualise it. Some of it was latent in the agreement I had with Thorkil Kristensen (see Chapter 21), especially the concept of being at least five years ahead of government thinking but never appearing to be more than two ahead, if there was to be any chance of being allowed to explore new approaches. Next, I felt that we would obtain a fair hearing only if I could establish good and relaxed relations with the members of the OECD Council, especially Ockrent, chairman of its executive committee, responsible for initial examination of new projects. There was also a need for patience, persistence and acceptance that there would be a longish softening-up period after the ambassadors had suffered the initial shock of a novel idea. It worked somewhat like this: I would propose a new opening to the executive committee and not be surprised or discouraged with its summary rejection, at times with scorn. Then, over the next few months, I would try to explain the ideas quietly one by one to a number of the members. The proposal would be modified slightly in view of these conversations and represented. Ockrent, who had developed a friendly bantering relationship with me, would introduce the discussion saying, 'here comes our Scottish chemist again, let's hear what he has to say for himself this time.' Then there would be a short discussion which made clear that a few of the ambassadors had some interest in the topic, a few felt that it was too far out to be acceptable and the rest remained silent so as not to expose their ignorance or lack of interest. The chairman in turning down the proposal then for the second time, would probably address me saying something like: 'you can't expect to pull wool over our eyes every time.' A few months and many conversations later I would return with a third version and obtain approval with little difficulty.

This game fitted well within the OECD culture, but was somewhat mystifying to outsiders. After my retirement, my successor, Keichi Oshima, a brilliant Japanese scientist and my own choice for the job, would bring a perfectly good proposition to the executive committee to have it turned down flat. He felt that this loss of face should demand he commit *Harakiri*, or at least resign. It took him some time to recognise the necessity of 'softening up'.

While my relationship with the national delegations was harmonious, there was a short period in 1963 when I sensed a certain reticence toward my activities on the part of some senior colleagues of the secretariat. In 1957 when I first entered OECD, the Deputy Secretary-General had remarked, 'You come with the reputation of being an empire builder. There is no room for empires here.' Now, seven years later, my programmes were flourishing and while my permanent staff remained quite small, our consultants were swarming in the corridors and discussing their work excitedly in the canteen. In the planning of OECD, provision had been made for operational activities, but now it was felt that the number and variety of our operational projects was distorting OECD's image of its essential function in economic policy. Fair enough! We had already terminated our science curriculum work and were moving increasingly towards policy concerns in both education and science. But the background sounds were ominous. Were we straying dangerously beyond our mandate? Was our enthusiasm and exuberance out of place in the respectable inter-governmental context? How did we manage to persuade the notoriously stingy budget committee to grant us resources when other worthy directorates felt starved?

Just then the Deputy Secretary-General was replaced. His successor was Michael Harris who had headed the Ford Foundation's activities in Indonesia. He quickly sensed the situation and set up a detailed review of the *Operational Activities of the Organisation*. Each of our activities was subjected to in-depth scrutiny. Initially I had sensed Harris was hostile to our work, but he soon became interested, ended up a firm supporter and ally. It was agreed some operational activities were acceptable in generating innovation in the policies of member countries. However, we should not undertake experimental projects designed for individual countries or small groups of them. These conclusions applied mainly to our educational activities, some of which, including the Mediterranean Regional Programme, had to be abandoned. Harris was strongly convinced of the need for many exploratory operations to be initiated, complementary to our policy-oriented studies as long as they were not financed by the common OECD budget. His enthusiasm led to the creation of the Centre for Educational Research and Innovation (CERI) and he was instrumental in persuading the Ford Foundation to make a substantial grant to launch it. This semi-autonomous body, set up within OECD and directed by Ron Gass, proved highly innovative and still exists today. The review safely over and its findings put into effect, our work continued.

The Residual Factor in Economic Growth

While colleagues were engrossed with the concept of education as an investment, the other side of the Science Directorate had a parallel preoccupation with science as an investment. Much of the source material was common to both these studies, which soon became a combined operation. As mentioned in Chapter 15, not much thought was given to the role of science in the economy before World War II. Post-war interest in economic growth and the widespread view that peacetime science would have an influence commensurate with that exerted during war attracted a number of economists to investigate the relationship. In a study of the growth of the US economy during the first half of the twentieth century, Dennison concluded that only some 40 per cent of the increase in GNP could be explained by additions to capital and labour inputs. The greater part has to be due to a *residual factor,* presumably consisting of a complex of elements, amongst which education, training, science, technology, entrepreneurship and management skills have an important part. At a meeting on *the residual factor in economic growth*, the economic growth theorists hotly debated Dennison's presentation. Doubts mainly concerned with methodological aspects of the analysis were not fully assuaged, but a strong signal was sent that the residual factor is a matter of quality; the increasing quality of labour and capital utilisation had enabled the GNP to grow so rapidly. Manpower quality at all levels, from unskilled worker to top management, is enhanced by education and training. Capital utilisation becomes more effective through new technologies that create new products, materials and processes and the markets they open up, as well as through increases in productivity. Both education and science had to be regarded as national investments in addition to their other essential functions.

The Technology Gap

In Chapter 21, I explained how our comparative data on expenditures on education and research, and numbers of scientists and engineers, deflected attention from fears of Soviet technological supremacy to concern in Western Europe that its future prosperity demanded its having an honourable position in the international league table of R&D effort. The United States outstandingly topped this table causing acute jealousy of its technological success. Evidence was all around – Boeing aircraft, IBM computers, nuclear weaponry and steady encroachment of US capital and management into European science-based industries. The French policy-makers found these comparisons alarming, and they infected their Italian neighbours with this concern. Foreign Minister Fanfani raised the matter of the *technological gap* and proposed a sort of technological Marshall Plan. The Gap had become a *cause célèbre*. We at OECD continued our work of refining and analysing the data, much being used (without acknowledgement) by J. J. Servan Schreiber in his popular book *Le Défi Americain* sounding its alarm of America advancing to complete domination of the technological industries of Europe, and even the world.

Our studies of the 'technological gap' demonstrated how misleading gross R&D figures could be. We broke down research expenditure into three categories of national objectives – atomic, space and defence; economic goals; welfare and miscellaneous (including medical research and developing countries). This changed the comparative picture entirely. It appeared that nearly two-thirds of US effort went into 'atomic, space and defence', a very much larger proportion than the European countries with which it was being compared, while only 28 per cent was devoted to economically motivated research. Resources for R&D towards economic goals in America being about twice that of the industrialised European countries – a not too troubling amount. Furthermore, the costs of research are much higher in America than in Europe, so there is little disparity except in the field of atomic, space and defence.

Research is an essentially international activity with its practitioners moving freely between countries. Likewise the products of technology, apart from some advanced military items, are freely available to any country able to afford them. Thus there is a relatively free market in patents and know-how, in addition to hardware, that spreads technological advances throughout the world. As part of our gap studies we attempted to develop the concept of *technological balance of payments*, expecting that national payments and receipts for the transfer of patents and know-how might throw some light on technological success and shortcomings. Such information is not easy to extract from statistics, but some interesting, if crude, comparisons appeared. It transpired that US receipts were nearly eight times greater than its payments, while France and Germany paid three times more than they received. The less technologically developed a country, the more negative its technological balance of payments. For any given country the situation varied greatly from sector to sector. Thus Germany, known for its successful chemical industry, had a positive balance of payments *vis à vis* the US chemical industry, while for the electrical industry its receipts were a small fraction of its payments to the Americans.

Our conclusions concerning the technological gap formed the main theme for the third meeting of the Science Ministers in 1968 and were fully accepted. This meeting was a turning point in understanding the relationship between science and the economy. Our investigations showed that there was little if any direct correlation between a country's R&D effort and its economic growth or trade performance. It appeared that diffusion of technology across frontiers, mainly in the form of purchase of patents and know-how or comprehensive arrangements between industrial firms, was sufficiently rapid to compensate for inadequacies in either domestic research effort or indigenous innovation. The example of Japan with its highly successful industrial development based on imported technology was a dramatic case in point. The technological gap demystified thus proved to be not a question of technology at all, yet a gap in advanced industrial achievement between America and Europe undoubtedly existed. It now began to be seen as a management gap as I had thought at the outset, but even that was too simple. The process of technological innovation, originating from a discovery in the laboratory, is exceedingly complex, as we shall see below.

There were some interesting political whispers in the corridors of this meeting. The British were pushing for entry into the European Common Market against President de Gaulle's veto and the French began to argue that the quality of OECD work on the gap was such that the British should have all their needs met by its work.

The Process of Technological Development

In the months following, our work on technological innovation in industry and on conditions conducive to its success intensified. We continued to explore the topic of technological transfer: was it necessary to create the conditions and facilities necessary to generate indigenous innovations if these could be purchased abroad? To neglect to do so would be a very inefficient and risky policy. Experience shows that effective transfer of technology can be relied on only if the recipient country possesses educational and scientific activity above a certain threshold. Certainly the implantation of imported technology does not require a high level of R&D effort, unlike the generation of new technology. Nevertheless, it demands excellence in education and a sizeable spread of domestic research effort to ensure that processes imported from abroad are relevant to the needs of the country, that adjustments can be made to meet local conditions and materials and, above all, that technical and managerial skills exist. Unless the level of scientific awareness surmounts such a threshold, countries can easily import the wrong processes and operate them ineffectually – innovating obsolescence. OECD studies suggested that in countries aspiring to achieve an advanced industrial position, attempts should be made to ensure that a proportion of innovation flows from its own R&D if a sustained vitality is to be enjoyed.

Such a sequence took place in the Japanese model that built its economic strength magnificently on the basis of imported technology. This was a carefully devised policy that realised the need for simultaneous construction of advanced educational and research capacities that allowed easy assimilation of foreign processes and in many cases yielded products superior technically and in design to the originals. The Japanese research facilities, now matured, have attracted many creative scientists and have already provided a substantial array of technological innovations, many of which have been exploited elsewhere. Japan, traditionally the perfect imitator, is now receiving payments from the sale of its original creations. Many years later my friend Suburo Okita, onetime Head of the Japanese Planning Agency and later Foreign Minister, explained to me how deliberately such a strategy had been determined. Immediately after World War II ended, a meeting had been convened in the office of the Head of the Department of Economics at Tokyo University. This professor was a father figure to leading economic thinkers, who turned up in force. They decided there and then, on the outline of an economic development plan, to transform defeated Japan into a modern industrial power. They agreed that the first step had to be radical reform and extension of the education system, including many new universities with a range of research institutes. Meanwhile, they should encourage importation of the most effective advanced technologies available in the world, sending groups of experts everywhere to scan developments and future plans in the main industrial countries to choose the best. The plan was put into action immediately.

The assimilation of new knowledge from research into industry cannot be considered in isolation from elements that facilitate or discourage it. To be a serious starter in the innovation process it has to be judged viable on an economic as well as technical basis and some assessment will have been made of the market potential of its expected product. In the long chain of development from basic research through applied research, pilot plant or prototype through to scale production and market penetration, many factors operate. National fiscal policy with incentives or deterrents is obviously important, along with accessibility of risk capital – so too are entrepreneurship, management skills and attitudes, good labour relations and marketing. Important also are levels of education and the national psychology and culture. It is only by a full understanding of all variables that the innovation process can be comprehended and mastered.

All the OECD countries, being committed to policies of economic growth, are naturally concerned with the stimulation of innovation in their industries, but policies adopted by members varied greatly. This led us to make an analysis of national approaches, with a view to correlating their experience, success and failure. In the meantime we published a study on the role of governments in the innovation process, as seen by industry as well as by the government itself. It stressed the importance of internal communications, discussed problems of market size and sophistication, looked at university-industry co-operation and examined different means of direct intervention such as taxation strategies, standardisation and direct government procurement.

We also studied factors favouring innovation within the individual firm with special reference to the place of the research department within corporate structure. First, we enquired about their justification for maintaining a research unit in a broad sweep of enterprises. Answers were varied and rather surprising; some advanced firms clearly viewed research as essential for the development of new products and processes, but were in the minority. Many admitted that their research departments were important for company prestige, a sort of corporate 'keeping up with the Joneses', but that they were useful for testing, quality control and trouble shooting. Explicitly, many quite technologically sophisticated firms expected little in the way of original innovation to spring from their own research, but relied on their scientists to keep them abreast of technical advance in their special field of manufacture, advise on new advances which might be adopted and help in integrating new processes.

Our impression was that corporate management of innovation depends on entrepreneurial flair outside the research department. In too many cases the research department was found to be more or less isolated from corporate policy-making, seldom consulted on important technical decisions and could easily adopt a defensive attitude – the NIH complex discouraging the introduction of new innovations. Lack of communication between research and other departments seemed widespread at the time and the coupling of research and marketing particularly important. In due course these matters were reported to ministers to considerable interest and appreciation. However, some countries (especially America under pressure from the industrial lobby) felt these matters were inappropriate for an intergovernmental body and should be left to industry. Consequently, to my great disappointment, this activity was stopped. Our exceptionally bright group had built up a corpus of knowledge on industrial innovation and the team had to be disbanded. Fortunately three joined our consultant Christopher Freeman, leader of a newly established Science Policy Research Unit at the University of Sussex where the work continued successfully.

I turn now to a related subject, *technological forecasting*. Looking into the future has long captivated human imagination; in fiction, from Jules Verne and H. G. Wells to Aldous Huxley and George Orwell, visions of future societies have entertained or irritated us. Since the end of the war attempts at non-fictional forecasting had sprung up all over the world in attempts to respond to the needs of military analysts and industrial planners for informed guesses about future battlefields or markets. By the mid-1960s futures research or futurism had become widespread and respectable; academies and multinational corporations studied or practised it. Consultancy firms such as the Rand Corporation thrived by selling their forecasts to governments. I was convinced there was a place for systematic technological forecasting in the planning of industry and decided to make a compilation of related activities and methods, recruiting Erich Jantsch to the task. Jantsch, an eminent systems analyst, was brilliant, arrogant and scathing, but he settled down to the job with true Germanic thoroughness and published *Technological Forecasting in Perspective* in 1967. He identified more than 100 forecasting activities in the countries he visited and estimated US industry expenditure on it was about $60–65 million per annum. Until then few realised how widespread technological forecasting had become. Although Jantsch recounted several new techniques of forecasting, few had much influence so far in practice. While forecasting is still basically unscientific and pragmatic, relying on intuition together with projections of identified trends in research, it can nevertheless usefully provoke fertile dialogue, often leading to a clearer notion of the human and social consequences of technological change.

In 1968 Jantsch and I organised a symposium on long-term planning and forecasting, held in the Rockefeller Villa at Bellagio and attended by an outstanding group of planning philosophers. In their *Bellagio Declaration* the experts issued a warning that existing social and technological developments could exacerbate matters beyond political control. They advocated the adoption of a humane system of planning, fully sensitive to its consequences to individuals and society, with the greatest possible degree of public participation. A striking feature of this meeting was the contrasting and often mutually hostile views of the two most vociferous participants – Hassan Ozbekhan with his philosophical approach and J. Forrester of MIT, strictly technocratic and practical. Chairing the meeting, I found it difficult to dampen the heat of the debate. How much OECD profited from this symposium is debatable, but I know how important it was for me when, a couple of years later these two personalities replayed their rivalry in the councils of the Club of Rome.

Research Management

To the individual scientist management was an alien bureaucratic intrusion in the creative world of research. However, from the 1900s, as research institutes arose and flourished in industry and the public sector, their directors were inevitably faced with complex management problems for which they were completely untrained. In the 1950s the European Productivity agency was deeply involved in the promotion of management education, realising that the problems of research institutes were very different from those of general management, necessitating quite different approaches if their products were to be both creative and effective. To this need, a proposal was made in 1958 that a college be established somewhere in Europe at which scientists would be trained in *the art of managing*. Judged premature at the time, it nonetheless set in train a valuable fact-finding operation, leading in turn to a series of seminars on the main problems faced by laboratory directors and including criteria for the selection of research projects to be initiated, and those for deciding when to terminate projects – particularly difficult when results were scientifically successful, but not economically viable.

These meetings contributed greatly to the emergence in Europe of a more positive and articulate awareness of the place and problems of research management and how these problems are common to all those directing research programmes. These issues had for some years been discussed regularly in America in the National Conference on the Administration of Research (NCAR). Japan also founded a similar body. Europe now had sufficient awareness of the matter to profit from US experience. I had attended a conference of NCAR in Puerto Rico in 1961 at which the desirability of a conference of US and European research management was advocated. OECD picked up the challenge and decided to convene a European/North American conference which took place at Monte Carlo in February 1965 with Professor H. B. G. Casimir, head of the Philips Research Laboratories, as chairman. It had been decided the conference should concentrate its attention on large research-intensive firms engaged in large research projects in the private sector. Its theme was defined as *The Integration of Research Policy into the Overall Policy of the Firm.* Its participants were to include representatives of top management as well as research directors because of the importance of discussing the place of research within corporate structure and policy.

The conference attracted extremely important and authoritative participation from both sides of the Atlantic. Discussion was lively and profound, signalling for the Europeans a new understanding of the role and place of research in the growth of the corporate economy. A strong conviction was expressed that this type of activity had to be continued by establishing a body analogous to NCAR in Europe. OECD followed up immediately by assembling a working group of European industrial research leaders to sketch out an appropriate organisation. Thus the European Industrial Research Management Association (EIRMA) was born. OECD seconded a member of my staff as its first executive director and Dr Casimir was the initial board chairman.

A number of leading European firms became its first members and it organised a series of working groups on many detailed aspects of the research process and its integration with company policy. Now, after about 35 years of useful activity, practically every European firm with research capacity is a member. Not only has EIRMA contributed substantially to the quality of industrial research in Europe, but it has continuously brought the implications of changing political, economic, social and technical conditions to the attention of research leaders. Its discussions between colleagues from competing firms are a prime example of the fertile coexistence, co-operation and competition frequently advocated but seldom achieved.

Science, Growth and Society

Successful work on technological innovation having been abandoned for strictly political reasons, it was now back to the drawing board – an opportunity as well as a necessity. When we had started to discuss science policy two decades earlier, our intention had been to see how new knowledge derived from research could best be applied for the benefit of society. With Europe in ruins from the destruction of war, the overwhelming need was to rebuild an effective economy to generate resources for social renewal and wellbeing. Thus effective harnessing of science and technology to economic growth became the paramount objective. The Cold War made defence technology the other major priority, but that was overseen by other institutions. During this period, while well aware of the social concomitants of economic growth, the national science policies had given scant attention to other sectors of civil concern; there had been no mention of research for medicine and healthcare despite progress in that sector triumphantly contributing to human welfare through the conquest of diseases and significant lengthening of the average lifespan.

The nature, quality and attitudes of society had altered greatly in the industrialised countries since the end of the war: new aims, issues and ambitions had been generated, warranting a reassessment of science policy and its priorities in serving a changed society. Much of the change had resulted from economic growth itself. A proportion of the new prosperity was of course diverted through taxation to the provision of social good, but the larger quantity reached the citizens directly as material affluence; the luxuries of yesterday became the necessities of today before being claimed as rights tomorrow. Thus the Consumer Society was born and formed the dominant culture; science and technology were its progenitors. Denis Gabor, the Nobel laureate physicist expressed my feeling well when he said, 'Our present civilisation is based materially on an extraordinarily successful technology and spiritually on practically nothing.' Inevitably, this transition of society threw up a multitude of social problems, many of them susceptible to research attack. Then there was the need to help alleviate monstrous world poverty, hunger and disease. We had already done some work on the place of science in the process of development in Third World countries, but clearly this issue would have to be related in the nexus of science policy.

Thorkil Kristensen had just retired as Secretary-General of OECD and was replaced by Emil van Lennep, who I had no difficulty in persuading to form an *ad hoc* group of high level personalities to reassess the place of science policy. With Dean Harvey Brooks of Harvard as chairman, they set to work enthusiastically, meeting several times in Europe and America, publishing their report, *Science, Growth and Society*, within two years. Now a classic, it analysed the present situation of science policy and concluded that the boundaries of the subject might have to be extended considerably in the context of changing social structure. It postulated that policies for science and technology in the OECD countries would have to evolve in the light of two central assumptions:

1. economic growth *per se* is no longer a sufficient overall objective, and
2. further interventions in the operation of the market economy will become necessary.

There was considerable discussion on the management of technological innovation; society largely appreciates the important part technology has played in achieving its present prosperity and assumes that even more technological advances will be necessary to meet its aspirations. However, there were worries about undesirable side-effects and fears about the rising of 'evil technologies' such as nuclear weapons. Most technologies arise in the private sector; can we rely on market forces to select future developments beneficial to all rather than profitable to the few? The market cannot be expected to deliver certain types of need which must be supplied to all citizens – health care, education, social welfare and a suitable environment – and are largely met by taxation, although private enterprise is claiming a role here also. However, it is in industrial development that more difficult issues arise. It is often felt that problems of rapid change and growing complexity are due to the unbridled character of technological progress. Governments have sought means to exert more control over the processes of research, technological innovation, the balancing of risks and benefits and equilibrium between social and economic objectives – but with little conspicuous success. The Brooks Report discussed these matters in some detail and proposed instruments to deal with them. I return to these topics in the next chapter.

By this time I was nearing OECD retirement age and concerning myself more and more with problems of development in poorer countries, but I wanted to have a last fling with the innovation issue before I left it. My long experience with bureaucracies, both national and international, convinced me that however competent and intelligent they might be, they constituted an impregnable obstacle to change with their essential function being anchors of stability. It seemed therefore that there was a need for new attitudes towards change within the administrations themselves, so I proposed a project on *innovations in the structures and attitudes of government*. The Council members were kind and gave a green light. Working within an inter-governmental organisation and hence in contact with senior national bureaucrats whose ways I proposed to change, I knew the task was well nigh hopeless. Nevertheless, I hoped to track down some experimental structures and discover some non-linear thinking. I recruited a brilliantly innovative consultant, Martin Lees, who broached the problems in the centres of reflection that exist somewhere in most administrations, but his only moderate success was hardly a disappointment to me. When I retired from OECD, the topic was quietly dropped and Lees went on to do better things. I was, however, to return to the subject a few years later with the Club of Rome as an element of the perennial subject of Governance.

25

Economic Growth is Not Enough – The Launch of a Club

During my lifespan the world has changed out of all recognition. My first childhood awareness of an advertisement was for pen nibs: 'They come as a boon and a blessing to men, the Pickwick, the Owl and the Waverly pen.' Today we are deluged by publicity for computers and the internet. I remember riding in a horse-drawn tram, but now I have flown at supersonic speed from Paris to Rio on Concorde and witnessed corresponding shocks of change in most other areas of human activity. Today's world is utterly different from that of my infancy, largely as a result of technological developments and the affluence they have created, at least in the so-called developed countries. Indeed, the technology-based approach to development is penetrating into societies and cultures of all types, so that even the most rigid, fundamentalist cultures find it impossible to resist the promise of power and prosperity that it appears to offer.

This burgeoning growth has inevitably caused radical changes in lifestyles and the norms of social behaviour. Even now, 50 years after the new shape of things became recognised, adjustment is still incomplete and there are many contradictions. Many institutions and mechanisms are still more appropriate for the world as it was at the beginning of the last century than it is at the start of the new; lip-service is paid to many traditional values, while they are largely ignored in daily life. A troubling ambivalence has invaded society in public attitudes towards science and technology. On the one hand there is general fear of nuclear war, encouraging activists of anti-nuclear and peace movements to demonstrate against military installations and even nuclear power. Some feel technology is out of control and others that environmental deterioration is seriously threatening the life of the planet. On the other hand the majority were enjoying the affluence that technology was delivering and craved more – much more. Each novel appliance and device was enthusiastically welcomed, thus further stimulating consumerism and economic growth.

These contradictory attitudes, often sensed simultaneously by individuals, had a somewhat disruptive influence on society with many people, particularly the young, troubled by this situation. In all European countries the relevance of contemporary education to meet foreseen needs was hotly contested. Large numbers of people were appalled by the crass materialism of society, the domination of the machine and the clammy hand of the huge bureaucracies that seemed to have sprung up to dominate their lives – impersonal, faceless and heartless. It seemed as if quality was leaking out of life.

The significance of this dichotomy in European society was starkly illustrated by student unrest erupting throughout the continent, culminating in riots beaking out in Paris in May 1968 (*les Evenements*). Student anger was ostensibly about authority's failure to take seriously their dissatisfaction and often utter rejection of the existing structure, organisation and contents of the education system and demands for participation in decision-making. However, there was also underlying concern about the uncertain aims of society beyond material satisfaction. The resulting riots, violence and street battles with the police disrupted life in Paris as the students attempted to give their movement a wider political significance, inviting the workers to join them. Had they done so, a very ugly situation might have arisen. Although the topic was more or less taboo in official and international circles, my educational colleagues in OECD, as noted, inevitably entered into dialogue with the students and attempted to analyse their motivations. When time for the summer vacation arrived, radical manifestations gradually faded. There was no immediate follow up, but the student revolt was no aberration as the authorities hoped. The way was paved for deeper analysis and gradually many of the student demands were considered and some met.

While most people accepted the malaise of society passively, it generated a sense of alienation for a sizeable minority; this was the era of the hippies, the flower children and 'make love not war'. It was a period too of demonstrations and peace marches, of debates on the 'alternative society', of the first appearance of environmental activists and green politics. So here we were at the end of the 1960s. The Cold War was in full swing and dominated geopolitics; the First World was creakingly trying to adjust to its new-found affluence; post-colonial Third World countries were trying to establish their identities within their seas of poverty. Technology was gradually taking over the direction of human advancement and economic growth had become a main priority for most countries.

OECD was the high tabernacle of economic growth. Tables similar to football league tables were compiled, indicating the position of each member country in the struggle for maximum growth. Incidentally growth mania seemed to reach its peak in 1968, the year of the student riots. But there was little discussion on what the growth was for: growth for growth's sake? OECD as an international economic organisation could not answer this question. It was essentially a matter for individual countries. But there was good economic reason for the question to be posed. I had long been worried by the absence of discussion on the use to be made of the new wealth generated by successful economies. I was convinced economic growth was not enough and that a whole array of new problems had appeared, demanding much more attention in the context of resources becoming available.

I had discussed this dilemma on many occasions with Kristensen and he shared my concerns. He had come to the end of his term as OECD Secretary-General and would soon retire. A meeting of ministers was planned and the Council decided to reserve time for a discussion with him as a swan song. Thorkil decided to present the ministers with a paper on 'the problems of contemporary society' and invited my ideas. The meeting duly took place and he was given a couple of hours before lunch on the first day to expound his ideas. Much to the surprise of the ambassadors, some of whom had been reticent about 'wasting' precious time talking about non-economic topics, the ministers were interested, even excited with what he had to say, and an unusually lively debate ensued that spilled into the afternoon session. The problems of education provoked exceptional concern, hardly to be expected from ministers of economic and foreign affairs. When they went home and the Council took over at ambassadorial level, there was little in the way of follow up, except that they budded off an Environmental Committee from my services.

I resumed my talks with Kristensen. We agreed that potential dangers were very real and, if not tackled in time, might even threaten the survival of the species. Immediate action was required, but Governments could not be relied on to act quickly; they were insufficiently aware of what was at stake, and the mass of people on whose votes they depended were still less well informed. We perceived the need for an international group of independent people of integrity to study the state of the planet and its evolving trends and inform people about prospects for the long-term future. I reflected deeply on this idea which seemed to contain some sense. Having meagre personal resources and soon, on my retirement, to be without any supporting infrastructure, there seemed little I could do about it. I had no inkling that a new actor was about to enter the scene a few months later, fortifying my resolve to act.

The Launch of the Club

I first encountered Aurelio Peccei in 1967. At that time he was President *ad interim* of Olivetti, following the death of the founder. Peccei, in the 1930s a young entrepreneurial manager in Fiat, had been in the Italian underground during World War II and only just escaped assassination at the hands of Mussolini's agents. At the end of the war, the Allied Control Commission involved him in the reconstruction of Fiat as part of the national recovery plan, a task he accomplished successfully and quickly. When normal times returned, the Agnellis, traditional owners of Fiat, realised he had reshaped the company according to his own ideas, and were nervous about his ambitions and exuberant entrepreneurial zeal.

He was then sent, dare I say exiled, to Buenos Aires to manage Fiat activities in South America where, with the independence that distance from headquarters afforded, his exceptional energies soon manifested. He avoided replication of the conservative structures of the Turin headquarters in building up automobile, agricultural and heavy industry branches, and made Buenos Aires his home. However, he maintained close contact with Turin and also with the consultancy consortium Italconsult for developments in the Mediterranean region, which he led. All these activities necessitated an enormous amount of travel. He was shocked by the poverty, disparities and injustice he found in Latin America, alarmed by the explosive growth of population in Third World countries and disenchanted by the influence of technology on societies at different levels. He perceived a number of fault lines in the world system and was convinced that governments, immersed in day-to-day problems, were unaware that quite fundamental problems could be foreseen that demanded immediate attention.

In 1962, US Senator Javits and Vice-President Hubert Humphries, seeking means to alleviate appalling economic conditions in Latin America, persuaded a number of prominent Americans including David Rockefeller to make available industrial venture capital to innovative industrialists in the subcontinent. Peccei was obviously interested, and co-operated with the new investment company Atlantic Development of Latin America (ADELA). His by now extensive knowledge of industry throughout Latin America was immensely useful to them. ADELA organised a meeting of Latin American bankers and businessmen in Buenos Aires in September 1965, inviting Peccei to deliver the keynote speech, which he did in fluent Spanish. The speech, *The Challenge of the 1970s for the World of Today*, was very far ranging and raised issues of population explosion, environmental degradation, the North/South divide, the importance of electronics, and the need for a long-term global perspective. Apparently, a transcription in English was passed to the Department of State in Washington and reached the eyes of the Secretary, Dean Acheson, who was impressed and is said to have remarked that it was much more useful for not having been presented by a Yank.

Together with other papers and pamphlets as is common practice, copies of the transcript were placed on the side tables on some UN occasions, including at meetings of ECOSOC and the Advisory Council of Applied Science and Technology (ACAST). At the end of one ACAST meeting, the Soviet delegate, Academician Jermen Gvishiani, picked up a copy of this document and read it on his flight back to Moscow. Gvishiani was very excited by Peccei's ideas and determined to invite him to the Soviet Union for talks. But who and where was Aurelio Peccei? The document gave no clue; it had no heading and concluded 'Aurelio Peccei, the Military Academy, Buenos Aires' with the date. How could such a forward-looking message have emerged from a Latin American military academy? Gvishiani sent a copy to his American colleague at ACAST, my old friend Carrol Wilson, asking him to find out who this man Peccei was and put them in touch. Carol had not heard of Peccei either, so he sent the material on to me with a request that I should do my best to locate him and put him in touch with Gvishiani. I too had no idea who Peccei was, but I quickly found out and made contact.

The invitation to visit Russia soon followed and it is interesting to note that the two men met at Akademgorsk near Novosibirsk, just about as far as possible from the political intrigues of Moscow, where it was relatively safe for Gvishiani to talk and listen. When I visited Akademgorsk 30 years later, I met a couple of scientists who still remembered Aurelio's visit and the excitement it caused. The talks were very fruitful and led to Gvishiani's long association with the Club of Rome. In writing to Peccei to make the connection, I had mentioned that I had read his paper and shared many of his anxieties concerning the way the world was heading; I suggested that, should he find himself in Paris at any time, he might lunch with me to discuss the situation. Some ten days later my telephone rang: 'This is Aurelio Peccei. I am in town today, may I join you at lunch.' Aurelio was a handsome man of about my own age, highly charismatic, warm and friendly. He was articulate in English, French, Spanish and Russian as well as his own language. We talked at length and quickly established a bond that lasted until his death in 1984. It was an unlikely alliance – Aurelio, an exuberant Latin industrialist with an economics background, and I, a quiet Scotsman, international civil servant and scientist. A couple of weeks later I had to go to Rome where we dined together, thus initiating a series of intense, intermittent discussions over the next six months.

We found that we had many common concerns, sharing a vision of global threats such as over-population, environmental degradation, poverty and misuse of technology. The significance of these problems, coming on top of existing distortions, did not seem to be attracting sufficient recognition, nor did there seem to be any single body capable of analysing, let alone starting significant action against them. Neither the UN, nor any other official international body, seemed capable of quick response to demands for solutions to existing problems, how then could they be expected to take significant action against global threats just appearing?

While our main concerns were global, we could not avoid discussing the specific problems of Europe, including its disparities with America, a hot topic of the time. We felt Europe was lagging behind with traditional and static policies that took little account of rapid change and had little inclination to look ahead at the probable consequences, economic and social, of existing trends. By this time we were ready to share and enrich our thinking by including others in the discussions. A convenient approach might be to invite a few eminent Europeans of broad outlook to meet with us and then extend the group gradually, bringing in others from different parts of the world. So we sat down in my office in OECD and drew up a list of personalities who might be invited. Aurelio thought he could persuade the Agnelli Foundation to fund the meeting, while I would ask my consultant Erich Jantsch to write a keynote paper to set the tone of the meeting. It took place in April 1968 at the beautiful Villa Farnesina in Rome, seat of the Academia dei Lincei, the oldest academy of sciences still extant, the academy of Gallileo. Some 30 distinguished personalities attended, including the futurists, Bertrand de Juvenel and Dennis Gabor (Nobel prize physicist), bankers Guido Carlo and Jean Saint-Geours, Pierre Massay, head of the French planning organisation, Conrad Waddington the biologist and Hugo Thiemann, director of the Geneva Battelle Institute.

The meeting was a complete flop. Jantsch's background paper was a scholarly plea for long-term planning, leading to proposals for action partly based on the methodology and recent experience of the Californian think-tanks. While his paper was academically exciting, it failed to secure the undivided interest of many of the invitees, being too technical, too dense in the expression of ideas that were partly concealed in half-understood verbiage. Its very title, *A Tentative Framework for Initiating Systems-Wide*

Planning on a World Scale, was formidable enough to frighten off many participants. Discussion was distinctly irritable and often irrelevant. Many were deterred by the technical language and took issue with the use of words and terms; for example an endless semantic fight between English- and French-speakers on the meaning of the word 'systems'. All this detracted from the serious discussion of approaching world issues we had intended and even Aurelio's charismatic appeal was quite unable to bring the meeting back to coherent discussion. There were also political undertones, the Vietnam War was raging and a murmuring Anti-Americanism was noticeable. The Rand Corporation, frequently held up by us as a model of successful forward planning, was best known as consultant to the US Department of Defence and that was enough to discredit prospective advisory bodies in general in the eyes of some at the meeting. To some it was unthinkable to have the Vienna Opera House and the Rand Corporation on the same continent. I angrily rejoined that it was impossible to envisage a future Europe that did not have both.

The meeting growled to a close and its members parted, not exactly in disorder, but with no sign of enlightenment. Aurelio and I were bitterly disappointed, but not utterly discouraged. Closing ranks, Aurelio invited a few of us to dinner at his apartment for a post-mortem. We agreed that we had been naïve; we knew too little about international politics; our presentation had been too flamboyant and technical. However, we were more than ever convinced of the need for an independent international exposure of oncoming world problems. We were not yet able to undertake this role and agreed we would have to go through a phase of self-education before facing a sophisticated audience again.

The next morning our group of six – Peccei, King, Jantsch, Thiemann, Saint Geours and Max Konstamm (a close associate of Jean Monnet) – met and quickly reaffirmed the views that had emerged at Aurelio's convivial dinner the previous evening. There was unanimous agreement that we should continue to act, very modestly at first, but with long-term ambitions. We should not form an organisation, the term carried a smell of bureaucracy, but rather an informal Club. I proposed the name Club of Rome, with no other significance than the place of our first meeting. I have often regretted my suggestion, as it has frequently been hinted that *Rome* indicates some hidden association with the Vatican, the Treaty of Rome, or even Imperial Rome.

We then began to talk freely about what this Club should be and to design a framework for its functioning, before discussing the substance of its concerns. We all agreed that unless it could be independent and seen to be free from all political and vested interests, there would be little justification for its existence. It had to be free to criticise policies and actions of governments or institutions without external constraints. This meant we should not seek or accept any funding from governments or industrial corporations. We would have to rely on resources from foundations or other bodies. Also we should not invite persons holding political appointments to our membership. Should an existing member accept a political appointment, they would become a sleeping member while in office. The importance of stressing our independence was manifest a couple of years later when the Club became widely known. A number of people questioned our integrity because of Aurelio's well-known connections with large industrial corporations. We were suspected of being members of a hidden Mafia or of multinationals. Throughout my long connection with the Club, I never saw any sign of attempt by anyone to influence our work or publications. Indeed after the publication of *Limits to Growth,* industry tended to regard us as disruptive.

As far as structure was concerned, we decided that the Club should have as little as possible and preferably none. All of us knew of non-official bodies whose work was hindered by restrictive statutes or an excessive number of committees. We regarded the bureaucracy of governments and other large organisations as guardians of the *status quo* and hence enemies of change. We were determined not to fetter ourselves with an internal structure and in a fit of impossible idealism decided to have no president, secretariat or budget. A year or so later, the need to have a legal existence forced us to appoint Aurelio as President. Fortunately, his office was in a position to function as secretariat. It was many years before we had anything approaching a budget. Our small group was determined to recruit like-minded colleagues from all five continents, but in view of the absence of structure and resources, we decided to limit the membership to 100 as the maximum number we could cope with. We expected all members to be active, join in debates and develop a Club loyalty. We put no term on membership, but agreed that when individual members retired or became inactive owing to pressure of other interests, they would be politely asked to withdraw to make room for new, active members.

At last we got down to discussion of the main thrust of our work. We talked about the main threats to society and the planet as we saw them, and tried to draw them up in order of their relative importance. We soon abandoned this approach as we saw each was related to the others and agreed that our theme was intrinsically composite. We termed it *the Predicament of Mankind*. This sounded very general and a bit pretentious, so we decided to concentrate our concerns on three concepts. These were:

1 We should tackle longer-term issues. Governments normally attempt to remedy short-term difficulties. They are chained to these by the electoral cycle. Citizens are naturally concerned with problems immediately facing them and this means that government must seek to resolve them. In view of impending problems not yet generally recognised, and especially in periods of rapid change, it is essential that independent groups such as our Club should inform people and prod governments to look ahead.

2 The need to identify and explore problems that are essentially planetary in scope. Examples of these are the degradation of the ecosphere by man-generated macro-pollutants, international trade and finance, water resources, desertification and expansion of world population. These are problems that cannot be solved by individual countries in isolation.

3 The *problematique*. A term that has become the *leitmotiv* of the Club of Rome. It denotes not just a single, specific problem, but a cluster of problems. Thus the *World Problematique* is a tangled complex of a huge number of problems; how to maintain an environment propitious to human life; the population explosion; food growing capacity; just and effective trade; economic and financial systems; the maintenance of peace and a myriad of other

> factors that interact in unelucidated ways to maintain the dynamic equilibrium of the planet.

'Problematique' has certain normative and idealistic connotations. The policy of a government consists of the sum of its sectoral policies for security, health, education, industry agriculture, etc. Reform or change of policy in any of these sectors is generally considered exclusively within the confines of the sector itself. Yet these changes may have unforeseen influence, positive or negative, on the performance of other sectors. Thus the idea of the *problématique* suggests the need for a holistic approach in policy planning, or at least serious considerations of cross-impacts before major policy changes. Investigation of the many interactions within the *problématique* was to become the leading feature of the new club. The phrase 'predicament of mankind' soon faded from the Club's vocabulary, the predicament seen as mankind's failure to master the *problématique*. Without the concept of the *problématique*, the Club would have appeared little different from the many existing political pressure groups. The term has slipped into the common parlance and is still a major insignia of our identity.

How then should we begin to attack the *problématique*? One of our members suggested that we should do a pilot exploration of the interactions of major factors in management of a city, say Florence. Another felt we should select a small country and look at the interactions within it. Aurelio, Jantsch and I thought we should have the courage to tackle the world-*problématique* in its entirety, however primitive the first sketches might be; starting, perhaps by describing the main variables. The only realistic approach, we felt, was a top-down one, otherwise we would fall immediately into parochialism. Two of the group, Konstam and Saint-Geours, found this hopelessly unrealistic and withdrew. After the publication of *The Limits to Growth*, Konstam wrote a charming letter to Aurelio congratulating us and saying how wrong he had been to walk out. Saint-Geours never actually resigned, but had no further contact with us for two years, after which he crept back as if nothing had happened.

I looked for an historical precedent for the type of body we envisaged. The sole case I found was that of the Lunar Society of England active in the late eighteenth century (see Chapter 15) – a small group of eminent scientists, industrialists and philosophers who met in Birmingham once a month when the moon was full to light their safe return home in their carriages to various towns in the Midlands. They soon became known as 'the lunatics' and the poet William Blake wrote a lampoon thus entitled. They were esteemed members of society, independent and quite critical of the establishment on rational rather than party political grounds; they even sided with the American colonists against George III, enraging Birmingham merchants who paid a mob to burn down Joseph Priestly's house during one of their dinners. Voluminous records of their discussions survive. Themes covered a wide range of issues from science, the economy, politics and philosophy to religion. They were particularly interested in the influence that new inventions and new concepts of science might have on social wellbeing. This is clear from their vision of how the improvement of the steam engine by one of their members, James Watt, might lead to the arising of new industries and eventually to the abolition of poverty. They were, in fact, the Fathers of the Industrial Revolution. Independence, multidisciplinarity, foresight, imagination, ambition to influence and not for power – these characteristics of the Lunar Society seemed to fit the needs of the Club of Rome two centuries later.

During the next two years we met frequently, usually as guests of Hugo Thiemann in Geneva, mainly discussing possible methods of exploring the tangle of the *problématique*, but also recruiting a few new members. The first of these was Hassan Ozbekhan, the Turkish-American systems philosopher mentioned in Chapter 24, who contributed greatly to deepening our considerations.

Then a new approach to our work suddenly appeared. Every year a European 'Summer University' took place in the village of Alpbach, a well-known conference centre high in the Austrian Alps. Ozbekhan and Jantsch had been invited to lead a seminar for which they had prepared a paper on *long term forecasting and its possible application in decision-making*. Aurelio and I travelled to Alpbach to support them. As it happened, the Chancellor of Austria, Josef Klaus, had chosen that day to pay his annual visit to this European forum located on his national territory and he attended our seminar. We had dinner with him that evening in the Gasthof. Conversation was lively and wide-ranging. At the end Klaus said, 'all these things you have talked about should interest me, but we never discuss them. Why don't you come to Vienna, say one day next month, and spend a morning talking to my cabinet?' This we did, accompanied by Thorkil Kristiensen, who on retirement from OECD had become a member of the Club, and Jermen Gvishiani, a hidden member. The meeting was useful but there was no immediate follow up. We had a meeting with a group of Austrian industrialists that same afternoon – again much interest but no action.

A couple of months later, by chance Aurelio and I found ourselves in Canada on separate missions. I was fairly well acquainted with a couple of the federal ministers and arranged an informal meeting with them in Ottawa to discuss the Club of Rome. Also at the meeting were three of the bright young men of the new Prime Minister, Pierre Trudeau. At the Sunday morning meeting Aurelio was in his most expansive mood and his enthusiasm seemed to fire the Canadians. During the lunch break one of Trudeau's men telephoned the Prime Minister, summarising the drift of our discussion. Soon a message came through, inviting us to cocktails and dinner at the PM's residence the next evening. This we accepted and duly turned up at 6 p.m., and leaving past midnight after hours of exciting conversation. These Canadian and Austrian incidents demonstrated that it was relatively easy to make contact with top political personalities, for what it was worth. We felt that, with our deepening perception of the complex of the *problématique*, we had something to say to political leaders that was necessary for their vision of planning. We decided that a sympathetic hearing from Heads of State had to be a suitable complement to our main task of alerting people generally to the dangers ahead. Consequently, during 1969–70, Aurelio and I met 17 senior political figures.

Meanwhile our membership climbed to 30; at Alpbach we had recruited Eduard Pestel, Rector of Hanover University; Detlev Bronk, then President of the US Academy of Science; and Paul Weiss. Another important addition was Saburo Okita, then head of the Plan Commission of Japan, with whom I had become acquainted through my OECD work. We held the first Club of Rome conference at Berne in June 1970. Two of the main actors at this meeting were Ozbekhan and Forrester, whose altercations proved fertile for the direction of the Club's future.

26

My Flirtation with the Soviet Union

Prelude

I first became aware of the existence of Russia when, as a small boy during World War I, a family friend gave me a bizarre propaganda map of Europe, on the top right-hand corner of which sprawled a large bear, marked Russia. For some reason this drab brown colour associated itself in my mind with Russia and even today, 80 years later, whenever I think of that country, the brown image reappears. During my childhood I learned new facts (or myths) about Russia. For instance, my public school grammar book had a dramatic trick of illustrating figures of speech by slightly absurd rhymes so that you would never forget them. In order to dissuade us young rascals from slipping into the sloppy habit of saying 'different *to*,' it quoted: 'The use of the explosive bomb has rendered Russia different *from* the kindlier nations of the West'!

And then, when I was about 15 I discovered the Russian novel and devoured Turgenieff, Lermontov and the rest, but above all Dostoevsky. I had entered a new and utterly different cultural world in which introspection, moral wrestling, understatement as well as overstatement, insinuation and paradox crowded across huge canvases. I soon learned to appreciate the quiet subtlety of Chekov. I was aware of the political evolution of the Soviet Union. I had read some Marx, but that was enough. I had no intention of allowing my preconceptions of Russia to be destroyed by its current materialism. However, by the early 1930s I was politically more aware and no doubt influenced by a swell of public interest in goings on in that country. It was the period of:

To Russia, to Russia to have a quick look;
Home again, home again; write a fat book.

We had been present at a couple of *soirées* at which suave 'parlour pinks' had waxed eloquent on the achievements of Soviet society but I had not been impressed by the type of world they described and admired. Nevertheless, I felt that it would be interesting to see how this new and seemingly innovative society had assimilated what I had dreamed of as its qualities and weakness. Were there still vestiges of the myth of 'the pure Russian soul'?

Sally and I decided to visit the Soviet Union in the summer of 1933, the year of our marriage. This was no easy venture. We were almost penniless, my salary at that time only £275 per annum. Russia was far away and its rate of exchange made living costs exorbitant for foreigners. Another problem was that it was only possible for ordinary people to visit Russia through Intourist, its official tourist agency. As free-ranging travellers, we had an inbuilt snobbery concerning packaged tours and, particularly in Russia, we would need a degree of independence to wander around. Eventually we were reconciled to the reality that our means would permit only a short visit to the nearest point of Russia and that any dream of reaching Moscow was not for that year. We would take a boat from Hull to Helsinki, travelling steerage with Petroushka. Then we would sail and paddle to Finnish Rajajoki on the frontier, then take a train the few miles to Leningrad. We bought a two weeks stay in Leningrad from Intourist.

We were happy to be in Helsinki again and still more so to be sailing east, through the Archipelago. It was a long voyage for a small craft like Petroushka. We camped on uninhabited islands in generally fine weather. The islands looked much alike; rocky and covered with conifers. Some of them had accumulated soil between the rocks, providing a relatively soft, grassy base to pitch the tent. On one occasion we landed on a likely island to find it occupied by thousands of large black spiders. We evacuated quickly to a neighbouring one that proved virtually spider free. At this high latitude, the light has special qualities; a sort of friendly hardness, particularly striking during long summer evenings. We sailed from island to island under a moderate wind, the cloudless sky a pale enamel blue shading to green towards one horizon and a dramatic orange at the other. The solitude was accentuated rather than disturbed by the occasional plaintive call of a bird, or the distant chug of a fisherman's motorboat. At the beginning of our voyage we had passed several evenings sailing through this dreamland, reading a Taushnitz edition of the then censored *Lady Chatterly's Lover.*

Having to reach Leningrad by a certain date, after which our visa would expire, and appearing to be on schedule, we decided to stop off for a look at Viipuri or Wyborg, the chief town in Finland's eastern province, Karalia. This was an attractive place with many old wooden houses, but little of particular interest. However, a few years later I was glad to have visited Karalia in her final years. After Finland's brave defeat in the Finno-Russian 'Winter War' just before the outbreak of World War II, it was ceded to the Soviet Union, as was Finland's Arctic coast. The Karalians are a distinctive people speaking a language akin to Finnish, but with their own culture. Many of the inhabitants migrated west and became assimilated with the Finns. Stalin dispersed most of the rest to distant parts of his empire. We had, of course, no premonition of this as we strolled in the leafy streets of Viipuri.

Continuing eastwards, we sailed within sight of Hogland, the only large inhabited island of the archipelago. Soon we began to sense a subtle change in topography, impossible to describe. The islands, their distribution and vegetation were as before, but there was a feeling we were approaching the end of something; it was as if the Gulf that until now had seemed an open sea was being brutally barred by the huge land mass of Russia. The great fortress of Kronstadt was not yet visible in the distance, but we imagined we could see it, with the misty spires of Leningrad beyond.

Since entry into Russia was approaching, we decided to find a green island to camp on, not too far from the mainland and within an easy day's sailing from Rajajoki. We pitched our tent on a calm, luminous evening and went to sleep, only to be wakened in the morning by the rustling of trees in a high wind. The sea was angry and it would have been folly to attempt to reach the mainland until the storm abated. This was really a piece of vengeful ill luck, the first bad weather encountered since leaving Helsinki. In vain we waited for the wind to die down and spent a second night on the island, worried lest we should be marooned there, losing our visa and thus being barred from entering Russia when we were virtually at its door. Next morning the storm still raged and we became more desperate as the day advanced, we had to reach Rajajoki the next day. Later that evening, with the wind still roaring, we had to decide whether to launch into the storm, or abandon the whole enterprise. We selected the risky option and launched Petroushka. With only a shred of sail on the mast, we lurched out with our paddles towards the nearest point of the mainland. Immediately the sail tore away and the mast broke like a matchstick. There was no option but to paddle with all our strength against the wind or allow ourselves to be pushed back ignominiously to our island. We just made it, landing roughly on the rocks, pulling Petroushka to safety. We managed to erect the tent in the wet twilight and fell asleep exhausted, fearful of what the next day would bring.

We were awakened early by voices near the tent and, peering out, saw a group of men puzzling over the remains of Petroushka. It was a calm day with only a watery sun to remind us of yesterday's storm. I struggled into some clothes and went out. Apparently there was a village nearby and they had come across Petroushka on their way to the fields. I tried in vain to find words in common, so had to converse in sign language They did understand my pronunciation of the name Rajajoki and eventually to comprehend the general nature of our plight. Soon the men were joined by a few women who tried to communicate with Sally, and in no time cups of coffee were being passed around. Everyone was friendly and it was obvious our unexpected arrival was a welcome diversion in a village rarely visited by outsiders. We discovered that the main road from Wyborg to Leningrad ran a few kilometres inland from the village, so the problem was how to get there and how to find transport to Rajajoki.

Soon an old wooden cart was produced and a farm horse harnessed to it. We clambered on; Petroushka in her bags was handed up, we were joined aboard by several of the villagers and sauntered sedately to the road which was nearer than we had expected. We sat down by the roadside, somewhat disconsolately. It still seemed the back of beyond and we were pessimistic about the prospects of finding an early hitch. However, eventually a bus approached and stopped at our signal to allow our friends to explain the situation to the driver who agreed to take us. Again everyone was friendly and cheerful, and several of the passengers alighted to help the driver tie Petroushka to the spare wheel at the back of the bus. Having thanked our friends and their horse, we climbed aboard and in no time were at Rajajoki station.

Having left Petroushka in the Finnish *consigne*, we passed through immigration unchallenged with several hours to spare before the visa lapsed. A truculent customs officer tossed all our pathetic belongings out of our rucksacks with a gesture of distaste and scrutinised every piece of printed material including *Lady Chatterly's Lover*! Failing to discover anything incriminating, we were moved into a waiting room until the train arrived. There were some 40–50 passengers already there, talking and joking in many languages but predominantly English with a central European accent. Someone in a corner strummed on a guitar and occasionally sang a line or two. The atmosphere was somehow intense with apprehension at entering the unknown. At long last a train puffed in and a second group of passengers joined the already overcrowded room. These were people about to leave the Soviet Union and they exuded a quite different atmosphere; one of liberation, the relief of escape to the familiar world. In the course of this long wait, we found ourselves talking to a couple of Jews of Russian origin, from the East End of London. They were pleasant people who had relatives in Leningrad. We decided to meet up again when we had settled in our hotels. After a long delay we were ushered to the train and in no time reached Leningrad. We arrived (as had Lenin) at the Finland Station, carrying no banners of ideology or even idealism, but well padded with naïve prejudice.

Our Intourist guide (watcher), a prim, bossy young woman obviously experienced in dealing with stupid foreigners, duly met us. She didn't hide her disdain. We were taken by car the short distance to the city, lodged in a hotel of fading grandeur, given a room and meal tickets for the duration of our stay. The hotel staff were elderly and deferential, in some cases servile – survivors, I suppose, from tsarist Petersburg.

Having deposited our belongings, we strolled the streets for a couple of hours. The hotel was on Nevsky Prospect, which at that time had a bleak revolutionary name, although citizens apparently referred to it as 'the Nevsky'. We trod it so often during those two weeks that we felt it bonded us somehow with the Petersburg of the great Russian novels. In fact there was little overt attempt to control our movements. The proportion of our time devoted to formal sightseeing was quite small and we changed some money into roubles at an exorbitant exchange rate to give ourselves a modicum of independence, to use the tramways or stop at a street kiosk to quaff a glass of Kvass.

Leningrad, despite accretions of utilitarian communist architecture, remained one of Europe's most beautiful cities. Its most immediately obvious characteristic was a skyline of buildings with graceful domes surmounted by fine spires. The huge square in front of the Winter Palace was overwhelmingly beautiful, but the feature of Leningrad that affected me most was the intrusion of nature into the city itself, not in the form of its famous white nights, but in the sheer momentum of water gushing down the Neva to the sea. Ladoga, Europe's largest lake, lies not far behind the city. I had a strong impression of the vastness of the snowfields of Arctic Russia, draining into Ladoga, and then rushing through Leningrad. It endowed the city with a quality of nature I never felt before or since in a great city.

Our formal tourism was much as expected; we visited a hospital, a crèche and a workers' housing project. Our guide was touchily defensive. When visiting one of the worker's houses, I noticed a large crack in one of the exterior concrete walls. Curiosity prompted me to enquire its cause. There followed a flurried consultation between the guide and the housing manager. 'The children must have been throwing stones', was the answer! The guide obviously found us difficult. Clearly her tourists were normally either sycophants, seeing a virtue in everything communist, or else hostile critics. We were neither, so she was unsure how to react to our questions.

One of the visits interested me greatly. We entered St Isac's cathedral that had become the Anti-Religious Museum. The propaganda was in the form of pictures and tableaux purporting to expose scandalous cases of cruelty or exploitation by each of the main religions. I could see nothing in the exhibits that had anything to do with religion or the aspirations of man. What did intrigue me was an extremely long pendulum hung from the highest point of the cathedral dome. This was a superb demonstration of Foucault's pendulum. Its displacements as observed by its progression around a large circle on the floor gave a dramatic sense of the rotation of the earth – another blow, I suppose, at religion!

The real joy of the Leningrad visit lay in our human contacts. We had early on reconnected with our friends from the border waiting room. They could not have been warmer, introducing us to their friends and relatives who were equally welcoming. Thus we spent much of our time in a friendly Russian Jewish circle. Communication was established with some difficulty. The Russian families spoke Yiddish intermixed with a good deal of Russian, our friends from London, Yiddish, peppered with English and American slang. I spoke a German distorted to make it, what I imagined, sound Yiddishlike and incorporating as many Russian words as I could. There was much joking over this Babel at our first meeting, but we soon settled down, believing we could understand each other. As well as visits to their homes, they took us on excursions outside the city. We came to understand their attitudes to the world and their situation in the communist state. As the merest gesture of appreciation of their hospitality we were able to share our few hotel food coupons with them. It is interesting that they felt no danger in associating with foreigners – a few years later such associations would not have been possible.

1933 was one of the most dreadful years of Soviet history, all guns and no butter, with everything sacrificed for defence and economic dogma; a failed harvest in the Ukraine, combined with oppressive Stalinist policies, caused innumerable deaths from starvation. The cities were apparently better off than the countryside, but judging by our new friend's kitchen, living conditions were exceedingly hard. Despite all these difficulties, people we met were optimistic about the future, even euphoric. Conversations were seldom doctrinaire, but exuded pioneering pride, a sense that they were willingly enduring hardship for the creation of a better world. My impression, the most striking I brought back, was that this spirit was quite general, I felt it in the streets as a quasi-religious emanation. The only other occasion I have met this phenomenon was in Israel in its earliest days. We left Leningrad with regret, yet when we had passed back through that ever-crowded station waiting room and emerged into the fresh free air of Finland, I must admit that there was a feeling of relief, a sudden easing of an unsuspected tension.

Before leaving Karelia, we made one diversion, related to my search for the complete Russian experience. We went north to Sortevalla on the banks of Lake Ladoga in the hope of reaching its famous island, Valamo. Before World War I, when Finland was a Grand Duchy of Russia, it had become the Mount Athos of the Russian Orthodox Church. It was inhabited only by monks and covered with monasteries. It was possible to reach Valamo by a small steamer, run by the monks. We used it to cross to the island, delighted with the romantic image of the black-robed, black-bearded monk standing at the wheel, steering us through rocks to his Holy Island. We spent a couple of days on Valamo, lying on the grass in the sun, sleeping in a monastery cell and attending services at all hours, a marvellous antidote to our recent exposure to secular enlightenment.

The Time We Didn't Visit Stalin

After the Germans were defeated in the Battle of Britain, Hitler's offensive was mounted against the Soviet Union. From bases in Poland, the German juggernaut crashed towards Moscow, despite heroic Russian resistance, penetrating to within sight of the western suburbs. There is a monument marking the furthest point the Germans reached beside the road leading from the airport to the city. Stalin in desperation appealed to the Allies to open a second front to release the pressure on the eastern front, but this was rejected; given the state of preparation, landings in France would have been suicidal.

Nevertheless Churchill wanted to show willing. He remembered the Tizard mission to Washington in 1940, pondered whether a similar approach to offer new Anglo-American war technology might be helpful, and therefore invited Tizard to fly to Moscow to ascertain whether they would accept a mission of war technology specialists to help in upgrading their weaponry. It was not a matter of delivering another black box of secrets, that would be folly, as some of its items could easily have fallen into German hands. It was a question of selecting items – weapons or gadgets we knew were superior to existing Russian equipment, yet safe to communicate. The Germans would already have known most of these by capture.

Tizard accepted the challenge, asking me to be secretary of the mission and accompany him to Moscow. I was delighted by the prospect of a dramatic adventure as a break from London routine, and the opportunity to work for Tizard once more. The initial task was tricky. I had to explain the exercise to representatives of the three services and get them to propose items to give to Moscow. These lists had to be consolidated and argued with each service and its American counterpart, negotiations lasted several months. I remember at one stage thinking that the war would be over and Moscow in ruins before the Russians were given anything, but we did finally come up with a consolidated, rather impressive list.

The plan was that Tizard and I should fly to Moscow in a light (Mosquito) aircraft and hand a letter from Churchill to Stalin at an appointment already made through the British Ambassador. Should the reception be favourable, groups of experts on the main classes of weapons proposed for transfer would fly out to join their Soviet counterparts. As we had to fly over enemy territory with the danger of being shot down, Tizard and I had to be in uniform. Had we been shot down, landed by parachute and been captured in civilian clothes, we would have been shot as spies. It was important Tizard be seen in Moscow as an important person, so he was to wear a general's uniform. As for me, if shot down, I had to be in the same prison camp as Tizard so that we would be together for interrogation, so I was given an absurdly high rank. I don't remember what – Lt Colonel or Brigadier. This whole business struck me as highly theoretical, I saw no chance of us landing safely on earth if shot down, nobody even bothered to tell us how to open a parachute!

All was set for departure. We had been inoculated against countless diseases, Churchill's letter was in my pocket, and we were about to go to the airfield when the whole thing was called off. We were told that President Roosevelt had suddenly got cold feet; we were giving the Soviets too much. My hopes for adventure were shattered.

Cold War Contacts

After this failed visit to the Soviet Union I had no contact with Russian affairs for several years. However, my eldest daughter Jane went to the British Embassy in Moscow as a junior diplomat and I was served an endless stream of Moscow street stories, many of them hilarious, but I resist temptation to tell any of them here. I have no intention of poaching on material for her own eventual memoirs. My next encounters were during my OECD years in Paris.

In the early 1960s a mysterious Russian appeared sporadically in Paris. I first came across him at a UNESCO function, seeing him now and then at diplomatic cocktail parties. He spoke English impeccably, was dignified and well tailored, unlike other Soviet citizens at the time with their ill-cut suits and rough manners. His name was Jermen Gvishiani, obviously of Georgian extraction. I noticed he was treated with extreme deference by all the Soviets in Paris. This was no run of the mill apparatchik. But who and what was he? I learned Gvishiani was deputy director of the All-Union Council for Technology, gossip having it that he was the powerhead of Soviet technology policy. If that were so, why was he hanging around at diplomatic cocktails in Paris. He was also often in America and had given lectures at MIT on management and systems research. Then information came through that his wife was the daughter of Kosygin who was a widower and often took her with him on state visits as his first lady. Now, after having known him for 30 years I still do not know what Gvishiani was after in Paris, but I have some guesses.

Eventually Sir Ronald Walker, the Australian Ambassador introduced me to Gvishiani. I had a strong feeling that he had some special reason for cultivating the Georgian. I found Gvishiani easy to talk with; his mind-set enabled him to fit in immediately in social and professional conversations with Westerners. He was well aware of the problems of scientific and technical institutions in Europe and America and quite open in admitting similar problems in the Soviet Union. During the next couple of years we had further discussions that fired my interest in the Soviet approach to science and its implications for policy.

Then he suggested I should visit him in Moscow, meet his colleagues and see the system at work.

This was a shocking suggestion. The idea of a senior OECD official visiting the Soviet Union at this stage of the Cold War was bizarre, but I was intrigued by the proposal and consulted Dr Kristensen. He thought the idea interesting and agreed I should go in a personal capacity. He did not consult the Council, some of whose members would certainly have vetoed the visit. So off I went to Moscow alone. Gvishiani was welcoming and hospitable. I visited him in his office in Gorki Street and met several of his colleagues to discuss our approaches to science and government. I also went to his institute to speak with some of the younger scientists of his team. I also made a courtesy visit to the Academy Nauk (of Science) and to the organisation set up by the Warsaw Pact with a mission for those countries corresponding to that of OECD. There were no embarrassments, no prying. On my last evening, my host took me to dinner at a Georgian restaurant near his office to meet Genadi Dobrov, the main adviser on science policy matters, whom I had met previously at UNESCO making conversation easy. During an exotic meal I had an idea on how we might co-operate in a practical way. At OECD we had completed a round of examinations of science and technology policies of our member countries. Several members of our committee had suggested it might be useful to complete this series with an analysis, based on the available literature, of how the Soviet Union dealt with these matters. A number of consultants were already at work with us to this end. Why, I thought, should we not make the exercise more lively by a visit to Moscow by an expert to question their interpretation of the situation on the spot. I put the suggestion to my friends at the dinner table. The ever-enthusiastic Dobrov agreed immediately. Gvishiani promised to seek permission.

Returning to Paris I found easy agreement for the scheme and shortly after, a telegram from Moscow confirmed theirs. My next appearance in Russia was thus to be an official visit to arrange the collection of data on their scientific institutions: the Academy, the Council of Technology, the ministries and the universities. It had to be very discreet so as not to embarrass Gvishiani. Obviously I, with little understanding of either language or substance, should not attempt this alone, but be accompanied by the Russian-speaking consultant from my office who was co-ordinating the work of the British Sovietologists, Helgard Wienert, an East German whose family had escaped from the East and were now American. I therefore cabled Moscow for permission to bring an expert with me and we turned up in Moscow a couple of days later. Gvishiani could not be seen to be involved personally in what might be seen by party colleagues as a dubious exercise, so wisely arranged to be away from the capital during my visit. He had called Dobrov from Kiev to settle the arrangement.

Helgard came with some anxiety lest her escape with her family from behind the iron curtain be remembered and endanger her freedom when in Russia. Fortunately this did not arise. When we arrived at Moscow airport, we found a group of Gvishiani's colleagues, whom I had already met, waiting to receive us. You can imagine their faces when they realised that the 'expert' who accompanied me was not a sober male scholar, but an attractive young woman. Clearly, they assumed I had brought along my mistress, regrettably not the case, but they pulled themselves together to make the occasion a success. We drove off to the city, everyone in good humour and settled in at the monstrous Rossiya Hotel. I had a quiet night but Helgard was disturbed by several abusive telephone calls during the night.

The next day we settled down to business with Dobrov in Gvishiani's office. Everything was settled with ease. Dobrov already had all the arrangements in hand; Helgard would accompany him to Kiev and work in the science library where the books and papers were already being prepared. I could return to Paris immediately. Helgard and I spent the afternoon sightseeing and then met with our reception party that evening in the Georgian restaurant I had visited before. It was a hilarious occasion with communal warmth of the kind one only finds in Russia, fuelled by innumerable toasts in vodka and Dobrov's conjuring tricks. The whole group affectionately escorted us back to the hotel. There was only one small difficulty, Viktor, the interpreter who had worked with me previously, fell in love with Helgard, was very suspicious and treated me with veiled hostility throughout the visit.

Our report on science and technology in the Soviet Union was completed and eventually published in Paris. It proved to be a useful contrast to the very different approaches of the OECD countries. Naturally, I fully reported all my contacts with the Soviets to the OECD Secretary-General who remained interested and supportive. His deputy Michael Harris was as intrigued by the enigma of Gvishiani as I was. I promised to introduce them and eventually invited both to lunch at an Italian restaurant near my home. They got on famously and Harris invited us for dinner at his home, which made possible a hard-hitting discussion of the world situation. Gvishiani showed a deep understanding of the Western point of view, was clear about Soviet strengths and weakness and had many interesting insights. It was amazing to listen to two people of authority discussing so dispassionately in the middle of the Cold War.

Soon after this the Club of Rome was created, partly through Gvishiani acting unintentionally as an intermediary. Formally, he did not become a member until the end of the Cold War, but in practice was more active than many declared members. When our report, *The Limits to Growth*, was published in 1972 he invited members of the executive committee together with Dennis Meadows to hold a seminar on the book at his institute, quite an event in Moscow. Most of the Russians were system scientists interested in the growth model and appreciative of its arguments, but fundamentally opposed to its findings due to their conditioning, and able to confront reason only with ideological conviction. My daughter Jane came with us as interpreter and was able to give some feel of what the Russians were saying between themselves.

I was in the Soviet Union a few more times before it ceased to be. During one of these I happened to be walking across Red Square with one of Gvishiani's research workers when he suddenly remarked, 'If you wanted to protest or make a public statement, this would be the place to do it. But you would have only 20 seconds, because one person in six in this square is police or secret police.' This was my final impression of the Soviet Union, but my association with things Russian was about to flourish as the old order collapsed.

27

Perestroika and Glasnost

The end of the cold war led quickly to more frequent and more relaxed contacts with the Russians. It was no longer dangerous for Jermen Gvishiani to reveal his association with the Club of Rome, at last becoming a formal member. He invited me to a meeting of scientists and political people to discuss what should be the new attitude of the (still) Soviet Union to environmental issues then being discussed in the West. This was held in Tallin in Estonia. This location interested me as Sally and I had spent a few days in Tallin in 1933 on a short excursion from Helsinki when Estonia was, as the Soviets described it, a 'bourgeois democracy'. On the occasion of Jermen's meeting, I hardly glimpsed the Soviet Republic of Estonia as I was chained to the conference room participating in a fascinating experience of internal communist debate. I visited the city again some years later when the country was once again unqualifiedly independent and just as I had remembered it from 1933.

Another incursion, before the break up of the Union, was to Tbilisi. On that occasion I was interviewed for Russian television by Sergei Kapitsa, son of Peter Kapitsa, the nuclear physicist responsible for the development of the Russian atomic bomb. Sergei, also a physicist and director of one of the institutes of the Academy of Sciences, had a second career as a television personality. His regular broadcasts of discussions and interviews on science topics were a top feature, seen by millions, including, I was told, all members of the Politburo. He only just escaped assassination when attacked in his institute by an anti-science fanatic with a hatchet. Later he was elected a member of the Club of Rome and I have met him frequently since. I also met his father, then a crusty old man full of nostalgia for his great days in the Cavendish Laboratory of Cambridge. We chatted happily about common acquaintances there such as my friend John Cockroft. Kapitza was outspoken in his criticisms and even denunciations of the Soviet system, and had he not been one of the most important of its heroes, he would certainly have been segregated as a dissident. As it was, he circulated freely, a tolerated national hero.

In 1985, when Gorbachev had brought openness to the Soviet Union, relations with the West were only just beginning to be normalised from the frozen hostility of the Cold War. A first summit meeting between President Reagan and Gorbachev was planned for Geneva towards the end of that year. The Club discussed whether there was anything we could contribute during this critical period. I felt we might help to build a degree of mutual trust and confidence by persuading the two superpowers to work together on a few topics of self-interest. One of my closest colleagues, Eduard Pestel, then Rector Magnificus of Hanover University, suggested we should approach Reagan and Gorbachev with a memorandum proposing a joint project of the superpowers to stop the sale of arms to poor Third World countries. Such a possibility might be accepted by both parties as not jeopardising their national security. We agreed and Eduard and I drafted a closely argued plea. Among other points we wrote:

> In the West, the supply of arms to developing countries by the USSR and other Warsaw Pact countries is seen as a means for the establishment there by force, of communist governments and the acquisition of military bases against the USA and its allies. By stopping the supply of military equipment, the USSR would make it clear that it no longer intends to promote world revolution through *military means* thus removing the basic American fear *vis-à-vis* the USSR. On the other hand the USSR believes that the USA is trying to encircle it and to preserve its power in all parts of the world, denying the USSR any foothold that it might gain by replacing corrupt governments and military dictatorships by 'popular' movements A great deal of mutual distrust and suspicion would be removed if the two powers were to join hands in putting an end to this highly immoral arms trade which in the long run would be politically

> counter-productive for them. For example, Egypt, for many years armed solely by the USSR, is for the time being in the American camp, while Iran, after having been supplied for a decade with the most modern arms, turned abruptly hostile to the Americans.

The letters were signed on the 4 September 1985. Pestel was due to visit Moscow that week for a meeting at the Academy of Sciences and was confident that he could ensure the material would reach Gorbachev's desk rather than the waste-paper basket of the secretary of his third assistant. Similarly, we had means to deliver it directly to Reagan.

From Washington we received a perfunctory acknowledgement, but no response. But we learnt it had been discussed during briefings for the Geneva summit. From Moscow it was otherwise. A couple of weeks after the letter had been sent to Gorbachev, the Soviet Ambassador to France telephoned my flat, saying he had been instructed by the Kremlin to deliver personally the reply of the First Secretary. His three-page letter was distinctly defensive in describing in detail the virtuous Soviet attempts to reach agreement with the Americans on the reduction of both nuclear and conventional arms. He then went on to profess a positive attitude towards development of the poorer countries and the abolition of poverty. While registering the need of these countries to maintain their national security against attack by neighbouring states, he agreed on the need to reduce arms sales to them, but would only act in concert with the Americans.

Having easy access to Gorbachev through the Ambassador, we sent an additional note suggesting he might consider joining with the Americans in an experimental scheme to reduce armament imports in the African region. We had a quick and friendly reply stating their willingness to press for such reductions but only within a general agreement.

A few months later I was approached by the Soviet embassy in Paris, asking advice on a cultural initiative they hoped to launch. Apparently Gorbachev was anxious to try out a novel way to establish communication between scholars, artists and scientists in the West with colleagues in the Soviet Union. Gorbachev was well known for consorting with artists and intellectuals in

Moscow. He now proposed that his writer friend Chingiz Aitmatov should invite a number of Western personalities to the Soviet Union as his personal guests. This really was a revolutionary idea. Until then the possibility of an individual Russian citizen inviting a group of uncleared foreigners to his home was unthinkable. The embassy request was that I should receive and advise Aitmatov in Paris. They also wanted to contact my Club of Rome friend Federico Mayor, who was at that time Spanish Education Minister. I readily agreed and arranged that Mayor and I would receive Aitmatov in the UNESCO building.

Chingiz Aitmatov was a native of Kyrgyzia (now Kyrgyzstan), a small mountainous republic within the USSR, squeezed between Kazakhstan and China. The native language Kyrghys is a dialect of Turkish. Aitmatov is regarded by many as one of the few master writers of the late Soviet period to rank with Tolstoy. I have since read some of his novels and can vouch for his quality as a unique and bewitching artist.

Federico and I duly met him at UNESCO House in early 1986. He was a large friendly man in his sixties. Talking via an interpreter, communication was remarkably easy. Gorbachev had insisted that the planned meeting should take place in October. What he wanted from us were ideas of whom to invite and what they should discuss. I felt strongly that the proposed date was too early, as many of the people we hoped to attract would already be otherwise committed. There seemed no possibility of postponing the meeting so we got on with the discussion which was quite short as Mayor had to return to Madrid and, to my astonishment, Aitmatov proposed to spend the night in Fontainebleau with his family – a Soviet family permitted to leave the country for an uncontrolled weekend.

They arrived at my apartment on Sunday morning – Chingiz, his wife, their 10-year-old daughter and Rustem Khairov, Gorbachev's assistant as interpreter. We spent a pleasant day together discussing the project and suggesting personalities to invite. The little girl played with my granddaughter of about the same age although they had not a single word in common. In the evening I took the opportunity of Rustem's presence to write a letter to deliver to Gorbachev. With the arrival of *Perestroika*, we needed a couple of new-style Russians in the Club membership and who better than the initiator of the new style himself to make suitable suggestions. I received some names a couple of weeks later from which we chose Aitmatov and Academician Evgenie Primakov, director of the Institute of World Economy and International Affairs.

On 12 October 1986 a collection of colourful personalities assembled in the VIP lounge of Moscow Airport on the invitation of Chengiz Aitmatov. There were Arthur Miller and his wife, James Baldwin, Claude Simon the French Nobel laureate, Yasar Kemal the Turkish colossus, accompanied by his compatriot composer Zulfu Livanelli; novelist Lisandro Ortero from Cuba; Alvin Toffler and his wife; Narayana Menon an Indian musician; Afewerk Tekle an Ethiopian painter; Federico Mayor and Augusto Forte of UNESCO and me. Peter Ustinov joined us a day later. The atmosphere was of friendly anticipation. The threat of Babel was there, but somehow easy communication was established. We spent most of the evening in a shabby bus crossing Moscow to another airport far to the east, where we boarded an Aeroflot plane for the more than 3,000 mile flight to Frunze (now renamed Bishkek), the capital of Kyrgyzia and home of Chingiz.

We saw little of Frunze, a pleasant city overshadowed by snow-capped mountains. Sleepy after the long night flight, we washed and brushed up hastily and were ushered to a meeting of local writers and the Republic's dignitaries for a largely ceremonial invocation to the spirit of universal cultural brotherhood. It was obvious that a visit by celebrated people to this remote corner was a rare event, and the Mayor of Frunze declared us honorary citizens. I had to make a speech of thanks for this honour in which I asked if it was a prelude to obtaining a Soviet passport! After Christmas that year the Mayor of Frunze telephoned to wish his fellow citizen (home in the South of France) a happy New Year.

Early the next morning we climbed into a couple of small planes and flew a further 100 miles east to Lake Issyl-Kul. This is the most spectacular place I have ever visited. The very salty lake is about the size of Lake Geneva and overlooked by the majestic Tyan-Shan Mountains, some of whose peaks are higher than Mont Blanc. Its true magic is the sense of scenic magnificence in a lonely, detached enclave of the planet. It is the authentic Shangri-La, its border with China only 50 miles away. The region is very sparsely inhabited; it has no towns and not even a village within many miles. The few nomadic families live in leather tents. Some of us were entertained in one of them and ate food with the friendly, Mongol-featured, Turkish-speaking family. We were housed in a pleasant sanatorium built as a retreat for the top Soviet brass. It was because of this that it was possible to come by plane to this otherwise very remote place.

We were comfortably settled in this quiet outpost and magnificently fed. We sat in a circle in a beautiful room with splendid oriental carpets. Each had an interpreter assigned. These were all intelligent young women, research assistants in institutes of the Academy of Sciences. Each sat beside and slightly behind the person she worked for, whispering in his ear an interpretation of what was being said. We all spoke in our own language, but this system allowed continuity and was better than consecutive interpretation.

Discussions were quite unstructured, but well disciplined and frequently exciting. Chingis started by outlining his idealistic hope of mobilising creative minds and attempting to develop new thinking that might contribute to the understanding of contemporary social and human problems and, hopefully, their solution. He hoped this could be the first of several encounters and that there would be regular communication between us. All those present entered into the spirit of the meeting. There was no intrusion of political or ideological thinking; the names Marx, Engels and Lenin were never uttered. The only sour note came in a long speech by Claude Simon that appeared to castigate the Nobel Committee for having delayed too long in awarding him his prize. Peter Ustinov kept the meeting enlivened by his witty, wise and pertinent interventions.

Much of the talk touched on global problems that the Club of Rome is concerned with and it was particularly interesting to listen to the various slants of this artistic group towards familiar issues. At one stage they came near to advocating the formation of a group to tackle global problems with objectives similar to those of the Club. I was able to tell them about our activities in detail. Federico helped me persuade them of the need for specificity, namely that any new group should consist mainly of artists and writers aimed at generating creative and intuitive approaches that might suggest leads towards the solution of global problems rather than attack the problems as such.

Eventually the group decided to form *The Issyk-kul Forum*, named after the lake on whose shores we were meeting, as a network of contact between those present and others invited to subsequent meetings. Then came the inevitable statement which was an elegant piece of prose. It stated that the political system in all countries operated on linear thinking. But the future was no longer what it used to be. Global society was changing enormously as were its problems. There was thus a need for the political system to be complemented by new thought processes the best creative minds in the world might suggest. Telegrams describing our decisions and aspirations were sent to Reagan and Gorbachev with requests they might meet the group, or at least a representative.

There was an immediate response from the Kremlin. Gorbachev invited the group to meet him, explaining that as he had just returned that day from the Reykjavik Summit, he was unusually busy. However, if the members could stay over in Moscow on their way home, he would be happy to spend an hour with them on the Monday. There was, as expected, no reply from the White House.

The significance of the Issyk-kul meeting lies not in its achievements but in the fact that it could have happened at all. I asked Aitmatov where the money to support it had really come from. He told me that the main funding had been donated by the Russian 'Book of the Month Club', Aeroflot had provided the transportation and he had managed to persuade the government of Kyrgyzia to pay for hospitality. The Soviet government that had stimulated it sensitively desisted from funding it.

Half of our group was able to accept Gorbachev's invitation. We arrived back in Moscow on the Saturday to be met by a fleet of black-windowed limousines. We were driven swiftly to a house for important guests of the government. I was allotted a suite of six what appeared to be early seventeenth-century rooms and a cheerful old lady to look after me. Breakfast that Saturday and the two subsequent mornings consisted of coffee, brown bread and a huge pot of caviar – enough to justify three journeys to Moscow. After lunch we went to a reception by the Soviet Writers Club where we met the cream of the Moscow literary world. In the evening there was a visit to Swan Lake at the Bolshoi, a disappointment as I had already seen their Swan Lake and had hoped to see another piece from their repertoire, however a party with the ballerinas and musicians behind the scenes enlivened this particular occasion. Gorbachev had suggested we spend the Sunday at Zagorsk, the holy centre of the Russian Orthodox Church, some 60 km northeast of the capital. There we attended, with a couple of thousand Russians, a service in the twelfth-century cathedral, after which we lunched with the archbishop and his colleagues and had a long discussion on world affairs.

Monday came and the discussion with Gorbachev. This took place in the communist holy of holies, the office of the central committee of the party. The black limousines took us there and the General-Secretary was waiting at the top of the stairs to greet us. As the only member of the group he knew, if only by correspondence, he took me aside and we had a short talk before we joined the others in the boardroom where so many historic decisions had been made.

After Chingiz Aitmatov introduced us, Federico gave an account of the Issyk-kul meeting and its outcome. Gorbachev expressed satisfaction and hope that the contacts would continue, saying that his head was still buzzing with the images, issues and possibilities opened up by the Reykjavik meeting and his consequent inability to spend more time with us. He was very optimistic with regard to the summit. 'We came to Iceland in an attempt to break down prejudices. It was not a failure; the situation is far from hopeless. President Reagan showed willingness to reach agreement and we identified some of the roadblocks on the way to agreement.' He then added:

> propaganda interfered, what with the forthcoming US elections; but while elections come and go, the problems remain. Then the media is like a pack of wild horses rearing here and there; its distortions increasing our difficulties. The basic need is for both sides to find new methods of thinking, because our present linear approach from a dogmatic past leads nowhere. It is through bodies such as the Club of Rome and creative individuals such as you here, that we hope to find new thought processes. The immediate need is to protect the world from the

> possibility and fear of nuclear annihilation. Despite appearances to the contrary, this somehow unites us despite ideological differences. This is the supreme task; common values must dominate over ideological differences.

He related some anecdotes from the summit. He told us how Reagan had attacked him over the Soviet jamming of the *Voice of America*. Gorbachev had retorted that he would see to it if Reagan would lease him a TV and radio station in the Eastern United States to put across the Soviet position. Reagan had apparently reacted that this was an interesting idea that merited consideration. Then a lively discussion began, in which every member of the Issyk-kul forum took part. Gorbachev replied to each intervention, often breaking into the discussion and expecting us to do likewise. The discussion was free and informal. He was relaxed and displayed no self-importance, nor was anyone fawning or striving to be particularly tactful. One could not imagine such a meeting with Thatcher or Mitterand. He appeared well informed about global problems. At one stage in the discussion he turned to me and said: 'Your Club must tackle one of the most pressing of contemporary problems, world-wide unemployment that will continue to grow and become increasingly intractable as technology advances. I look to the Club to suggest humanitarian solutions'.

We had exceeded our hour; indeed we had spent three and a quarter hours with Gorbachev to his obvious pleasure. His last words to me as we were leaving were that he would always be accessible to Club of Rome approaches.

We were next taken to a conference hall in a large exhibition hall where a noisy, exuberant youth festival was in progress. The few of our group who still remained were deposited directly on the platform. Aitmatov explained to the 1,000 or more young people what our meetings was about. Then a couple of my colleagues and I were asked to speak. I don't remember exactly what I said, but despite the consecutive interpretation it went down well, judged by the quality of the questions that followed. Tired and hungry we were whisked off to the airport and I had a pleasant flight home to Paris in the agreeable company of Peter Ustinov.

I must break off from my Russian escapades here to mention

some closely related meetings in Poland. I had been in Poland twice, but it was only after my involvement with the Club of Rome and my friendship with its Polish member, Adam Schaff, that I began to understand its politics. Adam is an exceptional, and disappointed, man. He is the foremost social philosopher of Marxism and widely respected. Adam had a difficult life. He had been incarcerated in both Russian and Nazi concentration camps and is no friend of the Soviet Union. Now, in post-war Poland, he had fallen out with the dogmatic faithful of the Communist Party as a result of some heretical views; in fact he had been discarded. However, such was his fame that he appeared to be free to travel in or out of the country at will, was awarded Poland's highest honour and had the ear of the Polish Premier. Schaff was the first Marxist to be invited to membership of the Club. He became enthusiastic and was brave enough not to conceal it. On one occasion in the royal palace at Madrid, King Juan Carlos questioned him as to how a communist could feel comfortable in the Club of Rome. He replied passionately and at length how he saw no incompatibility. He had created a Polish Association for the Club of Rome even before the end of the Cold War.

When I became President, Schaff invited me to Warsaw to talk to his group. I was invited to pay a visit to General Jaruselski, then PM and soon to proclaim himself President. At that time he was the most detested political leader in Europe because of suppression of Solidarity, and I declined the invitation.

By the time I next visited Warsaw, the Cold War was collapsing although the satellite countries had not yet broken free from 'Mother Russia'. I agreed to let Adam take me to see Jaruselski. At first sight he seemed the most sinister man I had ever seen, with his large blackened glasses and rigid stance. After a few minutes conversation, however, my reaction towards him veered 90 degrees. He was relaxed, non-ideological and quite witty. Apparently when he was a young officer, an accidental explosion blinded him and caused back injuries that left him imprisoned in an iron corset, thus his posture. He talked about his difficulties trying to hold the country together, He hated having to play the dictator in suppressing Solidarity and diminishing the personal freedom of citizens, but felt the only alternative was to accept Soviet occupation, with dire consequences. He was certain the

Russians would have invaded within 12 hours had he not acted. They had already placed their commissars in all key ministries. He still felt he had taken the right, if detestable, action. I have no way of assessing the accuracy of this, but he seemed sincere. It is interesting to recall that he was the only head of state of the communist satellites to steer his country through the transition to democracy. He remained in power throughout the process until a new President was elected by popular vote.

We then had a substantial talk about the Club of Rome. Jaruselski listened carefully and understood what we were after. He suggested there would be opportunities to work together in the not-too-distant future. By the time I next visited Warsaw the situation was radically different. There was now nothing to fear from Russia. Jaruselski was still premier and on friendly terms with Gorbachev; he had just returned from Moscow that morning. I went again with Schaff to visit him in the parliament building where he greeted me as an old friend. We discussed the changes and uncertainties of the emerging new Europe, then he broached an idea of holding an informal summit with three or four Eastern European heads of state who would meet a similar number from the West, possibly with a number of experts present, to review the rapidly changing political map and search for counter-intuitive consequences of recent happenings. Such a meeting would not put forward proposals; it would be an educational forum. Gorbachev had agreed to participate and the meeting would take place by invitation of the Club of Rome. This idea was eventually abandoned.

General discussion was more illuminating. He recounted his talk with Gorbachev. They were analysing the reasons why the communist system had fallen so far behind the West in economic performance. Both agreed a fundamental reason was overmanning and lack of labour mobility. This led to low labour productivity compared with the West, low morale and lack of incentive to work hard. Jaruselski caricatured the situation by two stories they had laughed at. He had visited an important construction site outside Warsaw and saw two labourers pushing a wheelbarrow, one on each handle. Aghast, he exclaimed, 'why two on the same wheelbarrow?' The reply was, 'unfortunately the third is on sick leave.' Then, further on his way, he came across an angry labourer who had stepped on a snail and squashed it. His mates were upbraiding him for his cruelty. He responded by saying, ' I know it was wrong, but I was exasperated because it has been following me about all day!'

Jaruselski showed me the reality of this situation. A factory would in a successful year recruit extra staff to cope with increased orders. A few years later demand for its products might have fallen, and thus labour requirements would have reduced. But the state guarantee of job security meant nobody could be dismissed and so hordes of workers would still turn up daily, necessary to pick up their wages. They would hang about in the plant playing cards and drinking coffee (or vodka) and getting in the way of active workers This would result in inefficient and careless work as well as a general malaise. It was extraordinary to hear two communist leaders talking so candidly about the weakness of their system.

During the early days of Perestroika I was in Russia quite frequently. The then Council of the Club of Rome had a meeting there to advise Bertrand Schneider and me on the book we were writing, which appeared later with the title *The First Global Revolution*. We were housed in a quiet retreat for party notables outside Moscow, and had the opportunity for discussion with some of the new party leaders. In particular I remember a useful session with Foreign Minister, Eduard Shevardnadze. He was warm, friendly and well informed. Looking back, my impression was that he was among the real human beings I met in the world of international politics. He felt the quickly disintegrating Soviet Union was being unwisely pressurised by the West to move too rapidly towards a market economy. The magnitude of the transition with its social and economic consequences was so tremendous that more time for reflection was necessary. At his suggestion I gave a talk on Moscow TV, warning that a move to a market economy, although desirable and inevitable, should not be seen as a panacea; market forces were no sure guide for long-term policies or for environmental and social care.

Another interesting visit was the inauguration of the Moscow International Energy Club, designed on the Club of Rome model, on which I had been giving advice. The new body was non-political, aimed at exploring the global energy situation, its trends and socio-political significance, as well as assessing the possibilities of new energy sources. The inaugural ceremony attracted an impressive array of personalities. It was very solemn, idealistic and philosophical: very Russian. As founding members, each of us had a large red scarf laid on the right shoulder. We looked like a meeting of an Orange Order in Ulster, except for an absence of bowler hats. After the meeting Sally and I joined a few colleagues on an immensely interesting tourist expedition to Tashkent and Samarkand.

It was on this expedition that I literally bumped into Yeltsin. Primakov and Aitmatov had arranged an informal meeting with some Russian friends to discuss Club of Rome affairs. This was held in Primakov's room in the Kremlin. I announced myself at the entrance, Primakov asked me to take the lift and he would meet me on the second floor. I stepped into the lift and was about to close its gate when a large ungainly man rushed up, signalled me to hold it for him and barged in, precipitating me to the back of the lift. I wondered who this rather bloated personality could be. Arriving on the second floor, Primakov was there and introduced him as Boris Yeltsin, who I never met again. The meeting with my colleagues was friendly and useful. Aitmatov had just become Soviet Ambassador to Luxembourg. I teased him for having succumbed to Gorbachev's inducements to enter political life instead of remaining among the Kyrgyzian mountains writing his enchanting stories. Soon after this meeting Gorbachev appointed Primakov as head of the KGB to cleanse it of its infamous reputation. This meant an instant change from the eminent academic to what the tabloids termed 'that notorious spymaster'. We asked him to withdraw from the Club and he exited gracefully.

Before closing this account of my experiences with the Russians, I must mention one feature of their behaviour that endears them to me. This is the extraordinary affection that generates when small groups meet around a dining table. Undoubtedly this vibrant manifestation of an apparent collective *love of kind* is greatly encouraged by the apparently limitless vodka that is present on such occasions. Indeed, I first assumed it was a kind of alcoholic stupor acting on the inherent sentimentality that is the other side to Russian brutality. But after a few experiences I concluded it was more fundamental, derivative of the Russian personality if not the soul. I felt the engendered friendships survived the occasion.

The first time I observed this phenomenon was at the Georgian dinner in Moscow when I had taken Helgard Wienert for the preparation of the review of Soviet science. The whole party had seen us back to the hotel arm in arm and an embarrassing situation had been broken. Another occasion was when Sam Nilsson and I were entertained by the professor of physical chemistry at Moscow University in his apartment on the Lenin (formerly Sparrow) Hills. It was a moment of acute food shortages and the dinner was not lavish, but any lack was compensated by a plentiful flow of vodka. There were about ten other guests, mostly academics. The dinner began with the first toast proposed by the host, which was 'Damnation to politics and everything that goes with it'. Then each of the guests had their turn and some of the toasts were long and revealing. We had reached toast number 18 before I lost count. I calculated later that a little less than a litre of vodka per person had been consumed. The atmosphere of friendship was terrific. The rector of Tbilisi University drove us, slowly, back to our hotel. I had no hangover the next morning, but I knew a lot more about how Russians regarded life.

A third occasion was when Sally and I were in Uzbekistan. We had a magnificent dinner in a hotel in the hills above Samarkand. There were six or seven courses with the usual flow of vodka toasts – the final one to the cook. He was summoned from the kitchen, eulogised and drunk to. He returned a few minutes later with a parting sweet he had tossed up for the company. Not so very long ago I was in Kiev and was taken on a day cruise on the River Dneiper. A lengthy lunch occupied much of the day; excellent food, many courses and the usual vodka. As before, the ship's cook was called in, praised and toasted. Once again he brought in the special parting dish. I was glad to know that the custom persists.

My last visit to Russia was in response to an invitation from my now old friend Jermen Gvishiani who had become leader of the Russian Association for the Club of Rome. I was to talk to an international meeting of young people he was organising, to be held at Akademgorsk, the research and postgraduate group of institutes in Siberia. I agreed to go and set off from Moscow on the long journey to Novosibirsk lying on rugs in the belly of a seatless Aeroflot plane. Eventually we reached the science centre by bus across the flat Siberian steppes. Akademgorsk had a certain beauty under a clear blue sky with pine trees covered by thick burdens of newly fallen snow. Accommodation was miserable. The only hotel provided no service and the village shop did not even have soap. What was worst was the absence of anything to drink; no beer, vodka or soft drinks other than a mineral water that tasted strongly of iodine.

The conference was splendid – for my ego. There were about 500 students, mainly postgraduate; the great majority were Russian, but there were a few from Western European countries including a couple from Britain. I had been doing a good deal of lecturing to students in the late 1980s and knew pretty well how to secure their interest in long-term world problems, so my talk went down well. The students decided to elect me as their honorary grandfather. I was amazed to find that one of their main concerns, and indeed bewilderment, was religion. Russia's own Orthodox Church did not interest them; they were looking beyond. Two of the most successful lecturers were Protestant academics from a university in Northern Ireland.

I don't expect I shall be able to visit Russia again. I accept it as right and proper that my long relationship with that country should have led at the end to Siberia. I was sent there, not to a labour camp, but to bring good will and, I hope, some useful experience to the local people!

28

The Limits to Growth

By the end of 1969 membership of the Club had grown to about 30 and we felt it necessary to call a meeting to deepen their understanding of the three points we had identified in Rome as our initial foci of interest. This meeting took place in May 1970 in Bern, on the invitation of the Swiss government. As the Geneva talks proceeded, we had become ever-more convinced that promulgation of the concept of the *world problématique* was the most useful immediate service we could offer to the world of politics, but securing general recognition of its significance was unlikely unless we could suggest some way of identifying the interaction of its elements. The addition of Ozbekhan to our group raised the temperature of debates. Jantsch, in his paper presented at the Academia in Rome, had advanced quite sophisticated ideas on how the problem might be approached, and Ozbekhan elaborated these. He spent some months working on a general proposal concerning the *problématique* and how it might be formulated systematically, as well as a research outline indicating the steps to be taken, the time necessary for its completion and some estimate of the funding required. This work, titled *Quest for Structured Responses to Growing World-Wide Complexities and Uncertainties* was debated by the Geneva Group whose members appreciated its elegance, but doubted if it could lead to any real guidance in policy-making since it still lacked a conceptual framework. Furthermore, useful methodologies were largely undeveloped and would require lengthy research before the project could become feasible. Such a large, ambitious and open-ended project would be difficult to attract funding. We came to Bern, therefore, quite uncertain that we had anything concrete to propose in addition to our general advocacy of the need to study the complexities and uncertainties of the global situation.

Ozbekhan's proposal was received with considerable sympathy by the Bern audience, but with scepticism. Then Jay Forrester spoke up and in a short, but brilliant exposé sketched out the means whereby his technique of *Systems Dynamics* could construct a world model to meet the Club's requirements – and probably in a short time. By means of quickly sketched flow diagrams he showed how population, industry and resources could interact and thus enable a quantitative model to be made with at least some of the main elements of the *problématique*. An executive committee was formed and after a brief debate opted for giving Forrester's approach a try, as the methodology had proved successful in unravelling the interactions within highly complex urban and industrial systems and thus required little preliminary systems research. On his flight back to Boston, Forrester sketched the proposed model, interacting five parameters: population, food, industrial production, resources and pollution, that were to become the basic elements in the model structure of *Limits to Growth*. This he termed *World I*.

Meanwhile Pestel had persuaded the Volkswagen Foundation to provide modest funds for a feasibility assessment. A few weeks later the members of the Executive Committee (other than me, I had obligations in Europe) visited Forrester for five days in Cambridge to learn about his methods, discuss assumptions, the availability of data, and decide whether to go ahead. Jantsch dismissed the method as 'pedestrian' and walked out after the first day. Ozbekhan was critical but seemed inclined to co-operate, the others wanted to proceed. However Aurelio, and to some extent the rest were worried the operation was too mechanistic, taking little account of social and human factors, which were much better incorporated in Ozbekhan's approach. I was briefed about the discussions daily by telephone and finally agreed with the majority that we should go ahead immediately. Aurelio, who had a high regard for Ozbekhan's conceptual thinking, tried to persuade him to co-operate with Forrester, but he refused. This intellectual (and temperamental?) hostility displayed at Bellagio allowed no compromise. Ozbekhan dismissed the proposal as 'a lot of crap', quit the undertaking and resigned from the Club.

We invited Jay Forrester to lead the newly agreed project, but he declined, insisting his role was essentially conceptual in constructing the basic framework of the world model and he did not wish to be involved in its execution. He did propose that one of his assistants, Dennis Meadows, assemble a team of bright young scientists from his MIT laboratory to undertake the project. This we accepted and, the Volkswagen support being confirmed, work commenced. Forrester, anxious to cement his role in establishing the methodology of the first world model, began to construct his own model, similar to ours. It was published under the title *World Dynamics*, before our *Limits to Growth* appeared, but did not attract much attention.

Meadows assembled a team of 16 highly motivated research scientists, mainly post-doctoral systems scientists, economists and natural scientists working at MIT. They immediately started building a world model, using the Forrester systems dynamics methodology, to interrelate the five parameters of significance.

It is not possible to describe the model in any detail here. Suffice it to say that it uses the growth data of all five parameters from 1910 to 1970, integrating them by positive and negative feedback loops. The whole model is thus interlinked and any change of data of any one of the parameters influences the behaviour of all the others. The figures for each are extrapolated until 2100. Most of the five are seen to grow simply by accumulation or depletion, while population growth is exponential. A large number of computer runs were made, each using different assumptions; in all cases they indicate overshoot and collapse by the middle of the twenty-first century. Even with the most optimistic estimates of resource availability, agricultural growth and production, the influence of other factors, particularly population and pollution, lead to global catastrophe. It has to be remembered however, that this is a 'passive' model in that it assumes no important economic, social, technological, or political changes will have taken place during the target years to sufficiently alter the interactions. Important changes that cannot be foreseen, will, of course, take place. The model's extrapolation from the present is thus far from being a prophecy, it is simply a warning that policy changes are imperative if contemporary society is to persist. The model is theoretical also in that it is highly aggregated and takes no cognisance of the huge disparities between major factors in different parts of the world, notably with regard to North/South differences, meaning its generalised statements are inapplicable to any particular region.

I regard it as a prophylactic model, a warning that the system demands new policies. It is in no way a prophesy of coming doom, but a wake up call concerning dangerous long-term consequences of existing trends and attitudes so that these might be corrected and dangers avoided. We have frequently been dismissed as a covey of doom-makers; on the contrary I felt we were doom-breakers. Cassandra was an early prototype of the doom-breaker, her name is the personification of the prophet of doom. She announced Troy would be captured and razed to the ground. She had an elevated position in the hierarchy, so her forecast was heard, but establishment complacency was such that no action was taken. The wooden horse was allowed to enter the city and her prophesy fulfilled. Cassandra was an unsuccessful policy adviser because she had not won the confidence of the decision-makers; she was heard but not listened to.

The work of the MIT group proceeded and the shape of its findings soon began to emerge. Meadows discussed its significance at a few meetings, sent mimeographed copies of technical reports on progress to several hundred scholars in America, Europe and Japan, and eventually a preliminary version of the report, by that time titled *The Limits to Growth*. There was very little critical reaction to all this information – strange in view of the storm of interest and abuse about to break.

By the middle of 1971 there was a good deal of rumour, speculation and partial information about the Club of Rome and its MIT project, brought to a head by two journalists working separately in the Netherlands. One immediately began preparing a series of television programmes and the other published a series of highly controversial articles insinuating that the MIT study was advocating a policy of zero economic growth. All the influential newspapers of the Netherlands took up the theme with headlines such as 'Zero Growth Necessary to Avoid World Collapse'.

Shortly afterwards two peak-hour television features were broadcast as 'the Limits for our Globe', seen by seven million

Dutch viewers. The *Limits to Growth* debate had been well and truly launched despite publication of the book being still six months in the future and the research not yet completed.

The public debate in the Netherlands soon became a political battle when Sicco Mansholt sent an open letter to the President of the EEC, Enrico Malfatti, who he was soon to replace. The eight-page letter was crafted on his conviction that the current international economic order as derived from the Bretton Woods monetary system had outgrown its usefulness and drastic reforms were now needed. He examined each of the MIT study's five parameters and homed in on the Malthusian population-food relation as the main point of constraint; rapid population growth of poor countries being the villain of the piece. This led him to put forward to the EC a number of suggestions for very radical reforms. These involved abandoning economic growth as a central economic goal and reduction of *per capita* consumption of goods and materials in the high consumption countries of Europe if the needs of underdeveloped nations were to be met. This was virtually a plea for zero growth, and hence exceedingly controversial. The debate in the Netherlands was long and bitter and spread rapidly to the other countries of the European Common Market.

All this publicity was a shock. The members of the Club had not yet been officially informed of what had transpired and there had been as yet no discussion. It must be stressed here that the subject of economic growth had never been raised in any of our discussions and we were not envisaging any political repercussions from the MIT study. We just wanted to demonstrate *one* approach to investigating how the *problématique* might be elucidated. Aurelio and I realised we had to react to the Mansholt letter. One thing we agreed was that the Club of Rome must not be linked with zero growth, but it was difficult to dismiss the extraordinary publicity that was putting the Club on the map. In April 1972 Aurelio and I sent a letter to the EC, distancing ourselves from the Mansholt letter. We argued that the limits to growth as now being discussed would never be reached. We made it clear that Mansholt's interpretation was far removed from what the Club stood for. Long before physical limits were reached, the global system would encounter serious social, political and cultural problems. Before non-renewable resources became critically scarce, the world would experience great problems from population pressures and environmental disasters. The debate on zero growth focused attention on a single facet of the much more complex situation of the contemporary world, to the understanding of which the Club hoped to contribute.

We were overwhelmed by all this publicity, much of it hostile and largely misinterpreting the findings of a study not yet available, but already generally misunderstood. It was high time that we had a text to present to the world, despite the fact that the research was not yet complete. It was decided therefore to write a short, clearly written book describing the preliminary results and conclusions: *The Limits to Growth*, written by Donella Meadows, was a brilliant, easy-to-read description of the research for the non-expert that no doubt greatly contributed to the publicity. It has sold between 10 and 12 million copies in more than 20 languages. The full report of the MIT team appeared a year later under the title *The Dynamics of Growth in a Finite World*. It received neither the acclaim nor the notoriety of its popular precursor.

For most of those already engaged in the zero growth debate, *Limits* had few surprises, following quite closely the already divulged consequences to be expected if existing policies and trends were allowed to continue. It did, however, present the case simply and elegantly to millions of concerned people throughout the world. Donella summarised the conclusions succinctly in her introduction:

1 If the present growth trends in world population, industrialisation, pollution, food production and resource depletion continue unchanged, the limits to growth on this planet will be reached sometime within the next 100 years. The most probable result would be a rather sudden and uncontrollable decline in both population and industrialisation.
2 It is possible to alter these trends and to establish a condition of ecological and economic stability that is sustainable far into the future. The state of global equilibrium could be designed so that the basic material needs of each person on earth are satisfied and each person has an equal opportunity to realise their

individual human potential.

3 If the world's people decide to strive for this second outcome rather than the first, the sooner they begin working to attain it, the greater will be their chances of success.

The debate on the book and its message raged on for years and it is doubtful the dust has yet settled. When the Club had the opportunity to discuss *Limits*, it was clear that many of the members were unconvinced by its conclusions on matters that they had never before discussed. Indeed, I felt the majority did not accept it. *Limits* had a distinctly Neo-Malthusian flavour, clearly unpalatable to those amongst us who were technological optimists and we all regretted that the work had paid insufficient attention to the great human and social issues. However, it was a report *to* the Club and not *by* it, so they could all welcome it as igniting a world-wide debate that was patently necessary. Their attitude towards the huge publicity that it provoked was distinctly ambivalent. Aurelio welcomed all publicity, good or bad, as useful in spreading recognition of the Club's existence; most of us accepted it a bit shamefaced.

Criticisms of the model came from all sides, with economists the most vociferous. They felt insulted: their arrogance had been touched by our impudence in promulgating a world model constructed on a material rather than a monetary framework. They had a basic and very reasonable complaint that it had ignored the influence of market forces on the rate of depletion of materials and energy resources.

Technologists and industrialists made the very real criticism that the model had taken no account of the expectedly massive influence of technological development. Business leaders tended to regard us as class traitors casting doubts on the stability of the *status quo*, while environmentalists would say, 'of course the model is fundamentally correct. We have made similar forecasts for years. It didn't take a computer to say so.' People from less-developed countries, feeling that their materials and labour had been exploited, assumed that the rich nations, having messed up the planet with their wastes were calling for a kind of stagnation that would prohibit the development of the poor.

As the unexpectedly large sales of this book testify, there was wide public interest in the uncertainties of world development. However, dramatic headlines in the newspapers illuminated only the first of Donella's three conclusions, ignoring the two constructive ones. Thus the public impression of the book was one of zero growth and doom. The model had many shortcomings; the imminence of materials shortage was overstated, while the magnitude of dangers from pollution could not be fully taken into account until many years later when global environmental effects like global warming and depletion of the ozone layer began to be understood. The criticisms concerning lack of appreciation of technological change, the power of the market and the degree of aggregation were, of course, valid. Meadows repeated the computer runs 20 years later, using more refined data and taking account of some of the criticisms. He could find no reason to alter the general conclusions of the original version.

During the period of controversy Aurelio and Dennis shouldered the main burden of debate; nevertheless, I was involved almost daily in some aspect or altercation. Some of these contacts were quite touching. I shall mention just a few. Arriving from Paris one day, I took a taxi from the airport into Washington. As is normal, I struck up a conversation with the driver. He began to talk about some book he had just read, that greatly influenced him and that he insisted I must read. He became very excited, pulled into parking beside the Potomac Ponds, and began to read paragraphs from *The Limits to Growth*. One of my many invitations to talk about *Limits* was to debate with a professor of economics at the University of Strasbourg. Arriving there, I was ushered into a small theatre to enjoy a performance of a ballet named 'The Limits to Growth'. The students of the drama school had created the choreography and rehearsed the dancers, each one representing one of the five parameters of the computer model. In the dance they weaved in and out, approaching each other in calculated, apparently haphazard contacts, providing a visual representation of the world *problématique*. Aurelio and I also participated in a large gathering in Rotterdam to talk about *Limits*. This was attended by Queen Juliana, who presented a copy of our book to each of the high schools in her country. In our conversation after the meeting she evidenced a profound understanding of the significance of the debate.

Looking back on these events of nearly 30 years ago, my sense of the influence of *Limits* remains undiminished, although I fully admit that most of this influence was indirect. First, it was a brave pioneering attempt to use the computer to construct the first world model. Certainly it had many inadequacies, but it stimulated many to build not only different but often more sophisticated world models. After all, one can hardly expect the first initiative in a new field to achieve perfection. As far as the Club of Rome was concerned, it achieved exactly what we had planned – a convincing demonstration of the *problématique* in operation – how changes in a particular sector had a quantitative influence on the others. It suggested the need to consider the effects that policy changes might have on various sectors so that counter-intuitive surprises could be avoided. A Canadian minister said to me in 1973, 'After *Limits* it will never again be possible to have policy reforms without looking around to see what they will do to other sectors.'

More generally, the debates raised the understanding of political leaders and the general public that many important problems of the future will be of planetary scope and beyond the capacity of individual nations to cope. The debate inflamed public concern regarding each of the five parameters, especially the population explosion and environmental deterioration. In summary, this little book despite its materialistic framework, through the debates it provoked opened the eyes of people to glimpse a new phase of the history of humankind into which they were about to step. It was one of the seminal books of the century.

29

Rest Cure in a California Think Tank

My last years before retirement from OECD were strangely calm. Ever since its foundation in 1968, I had made no attempts to conceal my association with the Club of Rome. Everything was in the open. Yet my position was obviously not politically correct and I feared it would not be easy to make the cat turn round. When *The Limits to Growth* appeared in 1972, with its implicit criticism of OECD policy, my position became precarious, especially as the Council was normally severe on any apparent irregularities. To my relief and surprise the matter never surfaced; I was treated like an elder statesman. The budget committee, uncharacteristically kind, gave me all I asked for, and several of the ambassadors paid me visits to ask my views on what was happening in the world. When I was on the point of departure, the Council invited me to talk about my work and I told them about the prime-ministerial meeting I had recently helped to organise in Salzburg. The cat had spontaneously stretched out the right way and was purring.

A year earlier I had been surprised to receive an invitation from Robert Hutchins, director of the Center for the Study of Democratic Institutions (CSDI} of Santa Barbara, to join a small team to advise on the future of his institution. The invitation intrigued me and, since I learned my friend Bertrand de Jouvenel, the director of Futuribles in Paris was to be a member of the team, I decided to take a week's leave and fly to California. I flew to Santa Barbara, a beautiful and serene little city situated by the ocean with the high Sierras as dramatic backdrop. The CSDI is situated on the top of a green mound in a quiet area just south of the city. It consists mainly of a series of study rooms of

American-Spanish architecture around a large pool in the middle of a patio, I was enchanted. It was a beautiful Californian day, with the patio discreetly shaded from direct sunlight and uncannily quiet. I presumed the Fellows were there, silent and deep in thought in their various cells, immune from all external influences. I couldn't help wondering if this sybaritic tranquillity could really be conducive to creative exploration of the deepest dilemmas of the human spirit. I must admit this thought remained when I departed a few days later.

The Center was a brainchild of the Ford Foundation, being funded through its Fund for the Republic. Its Senior Fellow and animator was Robert M. Hutchins formerly President of the University of Chicago. In the decade before the outbreak of war, he had been a whiz kid amongst American academics. He was a handsome, dignified man with a laid-back manner and sardonic sense of humour. I was given no background information about the organisation, its history, objectives or achievements, so all I relate here was picked up during four days of intense discussion and thus is incomplete.

The functions of the Center were apparently to consider both the burning questions of the day and the basic issues of human behaviour and motivation as well as to attempt to relate them. The method of accomplishment was *the dialogue* – frequent discussions round the table between Fellows, at times with the participation of invited thinkers from the world outside this small circle. This process was and is intended as a rerun of the Platonic Dialogues in the current moral environment. This solemn and somewhat arcane process (if not ritual) had been happening for several years with little emerging in the way of illuminating ideas; the discussion was increasingly seen to be going round in circles. Some of the debates on 'burning issues' had undoubtedly been useful and provocative and the Center had earned a reputation for courage and integrity by standing up to Senator McCarthy's fanatical attacks on the infiltration of communists into American society that he suspected everywhere, to which so many individuals and institutions had succumbed. Obviously some reorientation was needed and Bob Hutchins had asked for advice.

I met up with the other advisers: in addition to de Jouvenel were Arthur Lewis, President of the University of the West Indies, and Paul Erlich the biochemist soon to become the celebrated population and environment activist. We joined Hutchins and the other Fellows in the *dialogue* room and spent the next three days listening to accounts of their work and questioning them in great detail on their intentions for the future. We found the totality of their labours interesting but hardly distinguished. Very little was likely to have an impact on national policies or behaviour. One exception was the dynamic effort of Elizabeth Mann Borgese, daughter of Thomas Mann, to understand and make widely known the significance of the oceans. Her intention was to ensure their potentialities would be made available to all nations, and not left to the most highly industrialised to plunder. Her series of conferences titled *Pachem in Maribus* was the start of a long and successful campaign that continues to this day. But more of this anon. Finally, the advisers discussed the situation between themselves, reported to Bob Hutchins, hoping to have left him with some constructive ideas and departed. I returned to Paris with Bertrand de Jouvenel

A year later as I was thinking about possible activities after retiring from OECD, I received a visit from Norton Ginsburg, a Senior Fellow of the Center, bearing an invitation to spend half a year there as an Associate Fellow. I jumped at the idea. 'What could be better', I thought, 'than to relax for a few months in beautiful surroundings with congenial companions in order to blow 30 years of bureaucracy out of my hair and reflect on future moves.' In the spring of 1974, having relinquished my formal responsibilities at OECD, I led a team of 'examiners' to review the science and technology policies of Australia and then proceeded with Sally to Santa Barbara

The *mis-en-scène* of the Center had changed considerably since my previous visit. Bob Hutchins had retired and Malcolm Moos was now President; I knew nothing about his background, but I heard he had been the speechwriter of President Eisenhower's famous warning about the military/industrial complex. I had little communication with him during my months there; indeed, I had many more conversations with Bob who was still living nearby and frequently accessible in the corridors. I very soon sensed the lack of his unifying influence on the Center. There were many new faces amongst the Fellows. To my great delight, I immediately noticed Lord Ritchie Calder; he had let me down rather badly over his failure to deliver the popular version of the report on OR (Chapter 17), but we had maintained good relations over the years and I was happy to have an ally in this strange environment. Then there was Norton Ginsburg, Dean of the Fellows, who had sought me out in Paris; Harry Ashmore, a Pullitzer Prize journalist; Rex Tugwell, who I had already met several times; Elizabeth Mann Borgese; Harvey Wheeler; John Wilkinson; and Alex Comfort, the English anarchist who was to become famous as the author of *The Joy of Sex*. In addition to this unlikely team, were a few resident visitors like me. I knew two of these: Swedish economist Gunnar Murdal and his wife Alva who had been Director for Social Sciences in UNESCO. Gunnar was uncharacteristically low-key while at Santa Barbara, but for some unknown reason he took a strong dislike to an attractive Chinese/Canadian scholar who he attacked with barbed insults whenever their paths crossed.

I settled down quickly to the routine, sharing one of the poolside studies with Ritchie, participating in the dialogues and introducing one to discuss my Club of Rome ideology. Face-to-face talks with the individual Fellows were particularly valuable. Among others, I had the privilege of hearing Rex Tugwell describing his several attempts at rewriting the American Constitution. He had no expectation that it would be accepted, but it had a deep impact on American political thinking. Day visitors were frequent and interesting. One day for instance, we received a visit from a high-up member of the IRA who described its ideology and attitudes towards various nations and factions. I felt completely relaxed during this idle interlude, one of the happiest periods of my life. I had two short absences during my months there. I accompanied Elizabeth to Caracas to support her at the large formal conference on the Law of the Seas where her ideas were fertilising international policies. Then I went to Moscow for a couple of days to meet Aurelio and Jermen Gvishiani.

The social life of this little community was lively and in the main exceedingly friendly with much château visiting in the evenings; however, Wheeler, Wilkinson and Comfort seemed to have formed a little clique of their own and kept their distance from the rest who remained solidly loyal to Bob Hutchins. I sensed there was some mysterious personality history as well as intellectual differences between the two groups, but I did not pry into what was clearly not my business.

Our domestic life was brilliantly happy in Santa Barbara. With my frequent travels, Sally and I had, for years, enjoyed too few quiet periods undisturbed so this exotic sojourn was a veritable second honeymoon. We lived in a gorgeous house a little distance inland, with the city hidden by a ridge of low hills and directly facing the high Sierras. The owners were on a sabbatical leave in Europe that coincided exactly with our time in California. There was a large, well cared for garden dominated by a grove of tall eucalyptus trees. We breakfasted on a balcony in the cool mornings and could pluck ripe clementines from branches of a tree within reach, with humming birds hovering above our plates. Nearby was an avocado tree, with eight different varieties of the fruit grafted to it. This was the most beautiful residence I have ever been privileged to inhabit. Tragically it and its garden were destroyed in a forest fire a year after we departed.

In this rich and spacious environment, life would not have been possible without a car. So we leased a dignified Buick for the duration of our stay to give us mobility. I would leave the Center at four each afternoon, pick up Sally at our house and speed away, exploring the countryside. Some days, we would mount the summit of the Sierras and walk along the ridge; at other times we would follow steep and secret paths on the sides of mountains, encountering large green harmless snakes; or yet again we might walk along the ocean edge for miles. Occasionally we ventured further afield; on one occasion we drove north to spend a weekend with friends in Carmel, returning by the coastal road via Big Sur. I was amazed to discover that there were miles of unspoiled coastline in central California. We felt it was much as the French Riviera must have looked a few centuries ago, before Queen Victoria had unwittingly opened the doors to the crowds, the hotels and faceless apartments, the hot dog stands and topless bathers. Another weekend we spent with friends in Santa Monica, and on another we flew up to San Francisco, drove up and down the impossibly steep streets and across Golden Gate to meditate under the giant redwoods of Muir Woods. These noble sequoias induce an intensely religious feeling in those who shelter under their branches, not unlike the awe felt in some of the great Gothic cathedrals. It is not simply a matter of their vastness and the dim, soft light they permit to enter; the perfumed air they exude seems to convey the good and evil of the long years of their existence.

But sightseeing and meditation were only one part of the Santa Barbara experience. I managed to complete one piece of solid work. During the 1970s the UN had organised a series of conferences on world issues. Those on the environment, food, population and human settlements (habitat) were successful and it was now proposed to organise one on science and technology for development. Hardly had we arrived in Santa Barbara than I received an invitation from the UN to write a detailed paper suggesting what the substance of the meeting might be and proposing how it might be organised. I had done a good deal of thinking on this topic and decided to accept the challenge. Able to spend time reading and reflecting at the Center during the day, I soon produced a substantial paper of some 65 pages. I described in numerical detail how the disparity in science and technology between the industrialised and less-developed countries was even greater than their differences in economic performance. The primary need was to build up indigenous capacities for science and technology in the poorer nations, in the absence of which development could not take place, nor could the transfer of technology from outside succeed. I outlined a number of burning issues on which scientific intervention was necessary, but argued that these could only be tackled if the national capacities existed. No single conference, however well prepared, could hope to provoke the policy changes the situation necessitated. I therefore suggested that a *process* of activities was needed rather than a once and for all conference. This would start by discussion of the issues in each one of the 'developing' nations, followed by some regional pilot meetings, and culminating in a high-level conference to agree policy changes. My paper was mailed to New York as I was about to depart from Santa Barbara; its subsequent fate I relate later.

At this moment also I had a visit from Joe Slater of the Ford Foundation and the Aspen Institute, as an emissary from the recently established International Federation of Institutes of Advanced Study (IFIAS), inviting me to be its Chairman.

We were very sad to say farewell to the Center. It had been a marvellous and timely experience. I was now unwound and champing at the bit to be galloping off to a new chapter of my career. I retain a vivid affection for the Center and its denizens, but I could never have settled down to work there. Life was too easy, it was too lacking in daily problems and irritations, its pace too slow, without challenge for a workaholic like me.

30

Call Girl or World Citizen?

Before I retired from OECD I had been mildly apprehensive I would not find sufficient interesting work to keep me usefully occupied. Immediately after my return from Santa Barbara the reverse was true, my fear now was that there would be too many demands on my time. My problem had become the selection of those to which I could best contribute and that would interest me most. The next 20 years turned out to be the most satisfying of my career. I still felt young and energetic, I was utterly free of administrative responsibilities, my pension rid me of the necessity of earning money from new activities, and I had friends and contacts everywhere. For the next two decades, then, I roamed the world uncontrolled – lecturing, teaching, consulting, planning and organising.

My paramount preoccupation throughout this period was the Club of Rome, but as long as Aurelio Peccei was active, this was demanding, but by no means full time. I therefore accepted the chairmanship of IFIAS, which I held for ten years. IFIAS was in gestation while the Club of Rome was created and there were several similar international initiatives.

The International Institute of Applied Systems Analysis (IIASA)

One of these was explicitly political in its objective: US President Johnson felt it would be useful to invite co-operation with the Soviets on scientific activity that was not politically sensitive. He asked McGeorge Bundy, President of the Ford Foundation, to identify a suitable topic and find out if the Russians would participate. A small international group, including Aurelio Peccei, Jermen Gvishiani and Solly Zuckerman was set up to explore the possibilities. Aurelio was enthusiastic and naturally made the others aware of his new Club so that the tasks of the two new bodies should be clearly distinguishable. Zuckerman wrote him a rather petulant letter demanding he desist from his Club construction that he feared would confuse the issue.

The group agreed to form IIASA, but could not agree on its site. The French proposed Fontainebleau or Strasbourg, while the British wanted Sunbury on Thames. I had no direct role in the exercise, but was *au fait* with the negotiations and thought it better to situate a new East/West institute nearer to the Iron Curtain, say in Vienna or Helsinki. I mentioned this to the Austrian Ambassador at OECD and very soon had a call from their Foreign Office saying they hoped to put in a bid to house the new institute, had located a site and invited me to inspect Schloss Luxemburg near Vienna to assess its suitability. I did so and reported that it seemed just right for the purpose. I asked them not to mention my part in the choice. Had my name appeared, Solly would have seen to it that Schloss Laxemburg would never have welcomed IIASA. Today, some 30 years later, the organisation is still enjoying the hospitality of the beautiful Austrian castle.

The International Federation of Institutes of Advanced Study (IFIAS)

IFIAS was the brainchild of Arne Tiselius who had been Chairman of the Nobel Foundation and was at the time chairman of its chemistry commission as well as being a Nobel laureate himself. He felt the need for researchers in the different sciences to be more aware of each other's fields and for stronger scientific input into solving contemporary problems. Together with colleagues from the other Nobel committees, he organised a symposium in Stockholm in September 1969 entitled *The Place of Value in a World of Facts*. This was a year of student unrest all over Europe; it also saw the conception of the Club of Rome, IIASA and IFIAS, all with related objectives.

The symposium proved exceptional. It was the first occasion on which all Nobel Prize Committees – physics, chemistry, medicine, literature and peace – acted in concert, demonstrating the value of the multidisciplinary approach. The list of distinguished participants included eight existing or pending Nobel laureates. The debates were serious and at times exciting. Some of those present impressed their personalities vividly on the proceedings: Jacques Monod of the Pasteur Institute repeatedly reminded us to pay more attention to the values aspects of our work; Margaret Mead was ever provocative, but always amusing and relevant. Before each morning's session W. H. Auden intoned a poem on the previous day's discussions – an interesting innovation in conference drama.

One participant, Arthur Koestler, was untypically reticent. Each evening he seemed rather truculent and I had a feeling he didn't much like our meeting or us. A couple of years later his book, *The Call Girls* was published. It was an all too obvious satire on our Nobel Symposium. The Call Girls are people such as me, scientists who fly from conference to conferences indulging in arcane ideas and discussing impracticable schemes. His book shifted the scene of our symposium from Stockholm to Alpbach in the Tirol – another familiar setting. Several of our participants, highly caricatured if lightly disguised, are easy to discern; the portrayal of Margaret Mead is wickedly funny.

At the end of the meeting, Jacques Monod, on behalf of all the participants, invited Tiselius to distil something useful from all the 'wise words' uttered during the meeting and had a positive response. After a year of reflection and consultations he presented his ideas to the Nobel and Rockefeller Foundations. His main contention was that the many research institutes of international reputation were concerned with single disciplines and uninterested in, and incapable of, tackling alone the multifaceted problems that face society. The two foundations agreed to sponsor a workshop to test out the idea. At this workshop in the beautiful Rockefeller villa at Bellagio on Lake Como, Tiselius elaborated his vision of creating a non-governmental, non-profit-making, flexible network of existing first-class national research institutes, rather than yet another large international institution with its own in-house research. This was the blueprint that determined the structure of IFIAS. Sam Nilsson, the young engineer who had successfully organised the Stockholm symposium, was invited to undertake a feasibility study for which he travelled the world, interviewing directors of institutes considered potential members of the proposed new Federation.

Shortly after the Bellagio workshop, Tiselius died suddenly of a heart attack. This was a sad shock in both practical and emotional terms, but it was quickly agreed to carry on with the construction, with Nils Ståhle, executive director of the Nobel Foundation chairing the preparatory group. Statutes were drawn up, a sufficient number of institutes agreed to join and initial funding was assured to enable the founding meeting of IFIAS to take place in October 1972 in the Castello di Duino, on the Adriatic near Trieste. Nils Ståhle was elected Chairman of the Board and Sam Nilsson appointed Executive Director. They were both very active people and very soon IFIAS was in full operation. Nils felt that in line with the Tiselius leadership, the chairman should be someone with a substantial scientific background, which is why he relinquished the post to me two years later to become chairman of a distinguished panel of special advisers, a function he retained until 1989. As a resident of Stockholm, it was easy for the secretariat to profit from his vast experience and wisdom.

I took up my IFIAS responsibilities in the autumn of 1974. It was understood I should be more than a titular chairman and would work with Sam and his deputy, Per Lindblom, in programme planning and implementation as well as fund-raising and contact with the member institutes. I continued to live in Paris to be close to the Club of Rome office, but was in constant touch with IFIAS by telephone and visited Stockholm frequently. The Federation started off in two small offices in the Nobel Building. However, shortly after I arrived, the Swedish King Karl Gustaf gave us the use of one of the Royal Palaces, Ulriksdals Slott, on the shores of the Baltic, 20 minutes from the centre of Stockholm. In a speech at the small reception he gave to welcome us to his castle, he expressed a real interest in the concept of IFIAS and added, 'When I realised that your new Chairman was also a King, I felt that I had to put one of my palaces at his disposal.' Ulriksdal proved to be an ideal centre for our network. I had a bedroom in the palace for use on my frequent sojourns in Stockholm.

My relations with Sam and Per were exceedingly cordial. Sam was an optimist with boundless energy and imagination. During my decade with IFIAS we travelled together to all corners of the globe, visiting research institutes and exploring new ideas and attempts to analyse the elements of the world *problématique*. Initially some 18 institutes joined IFIAS, all selected in accordance with strict criteria, but we soon realised that they differed greatly in their commitment to our concepts of multidisciplinary co-operation, mainly as to the extent of their director's enthusiasm. Walter Orr Roberts of the University Corporation of Atmospheric Research at Boulder, Colorado; Paul Fye of the Woods Hole Oceanographic Institute; Victor Urquidi of El Colegio di Mexico; Adbus Salam of the International Center for Theoretical Physics; and Carl-Gõran Hedén of the Karolinska Institute of Stockholm were particularly supportive of the programmes of IFIAS over the years.

The IFIAS programme built up gradually, comprising seminars, individual studies, training schemes, specialist workshops for the elaboration of specific experimental projects and, most importantly, the projects themselves. These were selected carefully in accordance with strictly defined criteria. They were invariably of long-term significance and with 'direct impact on man and values'. Throughout my decade with the Federation some 14 projects were completed, many of which split into several distinct sub-projects in the course of their evolution. It is not possible to describe these adequately here. I can only give an indication of the flavour of some of the more significant.

Walter Roberts initiated our work on the impact of climate change on the character and quality of human life. By the early 1970s meteorologists were beginning to discuss the probability of a measurable change in the climate of our planet, whether cyclical, human-induced or a mixture of the two. In particular they were anxious to find out how accurate predictions of change could be using the new, complex meteorological models made possible by the computer. However, there was little concern about human and policy implications. This incited Roberts to organise a project design workshop to suggest how IFIAS might contribute to understanding the issue. The workshop, held in Bonn, did indeed formulate the IFIAS projects described below, but it also provided a timely opportunity for the climatologists to

The King of Sweden offers the use of one of his palaces to another 'King'

look at the evidence for trends of change and whether possible changes might be due to human impact on the environment. Thus the significance of phenomena such as that known (erroneously) as the 'greenhouse effect' was debated, without any consensus being reached as to whether it could be a menace to humanity. This situation of uncertainty raised the issue of how to alert decision-makers about the necessity of taking seriously evidence of probable threats to society and environment, such as global warming, ocean-level rise and ozone-layer depletion, that might not become evident for some decades. The impact of such problems for mankind was so great that politicians simply did not dare to take them seriously. IFIAS decided to take a first step in this direction by wide dissemination of its 'Bonn Statement'.

This proved to be highly contentious; many of the climatologists misread it as a prediction of coming, man-provoked climate change, when it was meant merely to initiate an informed debate about whether this was actually starting. This situation was identical to the misunderstanding that followed publication of *The Limits to Growth*. In both cases a red danger warning was mistaken for a prophecy of doom. Interestingly enough the 'Statement' was better appreciated by decision-makers and planners who set up commissions in many countries to investigate climate change. The Statement was a historical document, leading the way to today, when these matters are taken seriously by both the general public and politicians.

In 1972 television had stunned viewers all over the world with its pictures of the ghastly suffering of the populations of the Sahel and Ethiopia as the result of a terrible drought in much of East Africa. Generous food and other aid poured into the region from all over the world and there was a determination that such catastrophes should not be allowed to happen again. IFIAS launched a study to compare two completely different regions to indicate how useful long-term 'reliable and credible' climate forecasting might be in reducing the impact of drought and famine. The regions selected were some of the arid Sahel states and wheat-growing Saskatchewan in Canada. Evidence soon accumulated to convince the project leader, Micky Glantz, that such a forecast would be of immense value to the individual farmer in Canada, while in Africa it would have no practical impact at all.

He concluded that human disaster is inbuilt in the social system of most of the Sahel countries and hence the well-known recipes for famine relief cannot touch the long-term situation. What is needed is change in their entire socio-political and economic structure.

This introductory study stimulated IFIAS to deepen its concern with human involvement in climate change and its consequences. This led to one of its biggest projects, 'Drought and Man'. Walter Roberts found Rolando Garcia, a distinguished Argentinian Marxist, climatologist and philosopher of science, to lead it. With his structuralist approach within a holistic vision, he possessed just the right mixture of qualities needed for this complex task. Garcia established a small core group at the Graduate Institute for International Studies in Geneva and some dozen contact points in Africa, Asia and Latin America.

The investigation had several objectives. First, it sought to elucidate the causes and consequences of the response to the drought of 1972 and from these to provide suggestions on how future drought situations could be more effectively faced. Then it analysed the changes that had happened in many parts of the world in connection with the apparent climatic anomalies of that year. A large number of case studies were carried out around the world, including a special study of climate change in the Soviet Union. From the rich data that accumulated, Garcia had to conclude, much as the initial probe by Micky Glantz had suggested, that the catastrophe's roots lay, not in the triggering drought, but in the vulnerable socio-economic systems of famine-prone areas. The exceptional severity of the drought, excessive population increase and overgrazing undoubtedly contributed, but were not the fundamental cause. Emergency provision of aid or short-term technical fixes, while essential to alleviate immediate suffering, could do nothing to resolve the problems of socio-economic vulnerability or prevent recurrences. The results of these studies were published in three concise volumes; the first had the provocative title *Nature Pleads Not Guilty*. The series dealt with every aspect of the complex system including agriculture, malnutrition, health and education. It proffered many recommendations to IFIAS, national governments and UN Agencies such as FAO and WHO. It had a profound influence on development strategies.

1973 had seen the world oil crisis, the publication of *The Limits to Growth* and the creation of IFIAS. It was not surprising the new Federation should look at energy problems as a priority. Energy is the fundamental element transmutable to matter, in accordance with the Einstein equation. The Universe is Energy. On the microscale of our planet, energy is likewise fundamental. Money is but an uncertain and primitive surrogate for energy. As the world population grows and material wants increase, the consumption of energy rises. The direct capture of solar radiation by hydroelectricity and other renewables provides some of this, but by far the larger amount consumed by humanity comes from the combustion of fossil fuels; petroleum, coal and natural gas. These reserves of stored solar energy, accumulated over millions of years are being depleted rapidly and cause pollution.

The oil crisis forced attention not only on temporary disruption of supply, but on the world energy problem as a whole and highlighted the vulnerability of the main energy consuming countries to disruption from political and other events beyond their control in distant lands. Denmark, as a country without fossil energy resources and hence dependent on oil imports, was deeply disturbed by the world crisis and in political turmoil as a consequence. IFIAS invited its Danish member, the Niels Bohr Institute of Physics, to convene a workshop to consider what research it could contribute in the situation. This meeting of energy experts and policy-makers surveyed the world energy situation broadly. They discussed the extent of fossil fuel reserves, the need to slow down the rate of world consumption, fuel conservation and increased efficiency of energy generation, transmission and conversion, nuclear energy and its dangers and 'soft' energy options such as solar.

IFIAS decided to take up two of the suggestions of this group. The first project, energy analysis, was led by Malcolm Slesser, a chemical engineer from the University of Strathclyde who had pioneered the subject of energy accounting, a method of determining the energy requirements of goods and services. Planning depends heavily on monetary cost/benefit considerations, but with the new concern for energy economy, awareness of energy costs became necessary in addition to monetary evaluations. For example, the net energy yield of a new power production unit will be influenced by the amount of energy used in its construction. In an extreme case of an attempt to harness non-traditional energy sources, the energy used in manufacturing its equipment may approach its yield. A balanced approach to energy use by society requires a detailed knowledge of the energy used, not only by sector, but also in the large number of goods and services provided within each.

Slesser undertook a world tour for IFIAS, visiting individuals concerned with such problems. On his return, key experts were convened to a workshop at which the foundations, standards and methods of application of what was seen as a new discipline of energy analysis were formulated. The report of this meeting became an immediate best-seller and undoubtedly had real influence on policy-making.

The second component of our programme on energy and quality of life, was a case study of the energy economy of Denmark, undertaken by the Niels Bohr Institute in 1974. Thorkil Kristensen, now retired from OECD, acted as general supervisor of the Danish team that went painstakingly into the major sectors of the economy and in 1976 presented a plan for future energy use. This indicated how the energy demands of the country could be reduced considerably, not counting technological breakthroughs, by action over the period 1990–2005 without reducing lifestyles or impairing economic growth. The report provoked much political debate and its findings were largely accepted. In fact, the Danish Minister of Trade took on the Nils Bohr team as energy advisers. Ten years later it was agreed that the investments stemming from the report had been well justified in savings on imported oil. Although this study was aimed to alleviate the situation of a single country, its approach and methods proved useful to others such as Japan, heavily dependent on imported oil.

One of our most influential programmes was on the socio-economic implications of enzyme engineering led by Carl Héden and concerned with energy technology in the food industry, in medicine and in the monitoring of environmental pollution. One of its elements achieved close co-operation with the main institute engaged in theoretical and practical aspects of the development of a biological fuel cell. Significant, too, were the advances made in enzyme labelling in the diagnosis of tropical diseases. This wide-ranging programme gave an important impetus to biotechnology, the programme involving 200 scientists, 36 research institutes in 15 countries, 5 international organisations and 7 governments.

The potentialities of biotechnology arising from the enzyme project together with insights from energy studies suggested an integrated approach to Third World development at the village level. An Indian scientist, Dr Seshadri, had reported on his work in assisting the village of Injambakkam, south of Madras, to approach self-sufficiency in food and energy. He joined forces with Malcolm Slesser to develop this project on an interestingly quantitative basis. Seshadri measured the inputs of 1,000 villagers, while Malcolm, with his computer in Glasgow, was able to incorporate the data in a simulated energy model of the village. His 'Model for Energy and Resource Development Analysis' was capable of taking account of a very large number of variables: crop production, human population, cultivated land, numbers of animals, cow dung availability, fertiliser, weedkiller and insecticide import, biogas generation, etc., making it possible to simulate various development scenarios from the data.

Seshadri faced the sensitive task of winning acceptance and willing co-operation from the villagers in furnishing the detailed inputs the Slesser model required. This was not easy in the complicated social system of India, exacerbated by superstition and suspicion from some inhabitants. But he was largely successful and the interest of the villagers was aroused as they began to see practical results. A number of local youths were recruited to assist the team in introducing and explaining the significance of new technologies such as solar cookers and dryers, windmills for pumping water, biogas plants and aquaculture. The Glasgow team was not merely concerned with its theoretical work: several of its members spent prolonged periods in the field. The outputs from Slesser's model were impressive, indicating that with the optimum mix of bio and solar technologies, substantial increases can be gained in energy and wealth of the community. This pioneering experiment blending theoretical modelling with practical fieldwork was welcomed by both developers and authorities in the Third World and was emulated widely. However, as a highly disaggregated approach to development requiring exceptional devotion, skill and sensitivity from project leaders, it is not easy to envisage its generalisation.

These examples give some indication of the IFIAS methods of work. I shall conclude by just mentioning some of our other projects. In the early 1970s there was much concern with the loss of topsoil world-wide, stimulating IFIAS to mount a project in association with FAO and UNEP named 'Save our Soils' (SoS) that developed well in several regions including the Soviet Union. It also spawned an interesting study in Kenya of the population capacity of the country.

A further group of projects fell within the ABC (Analysing Biospheric Change) programme of approach to environmental problems, with special reference to socio-economic and structural elements. These included investigations on coastal waters pollution, rapid agricultural change and urban development, particularly in the underdeveloped countries. These projects, executed in the most diverse areas, involved co-operation between large numbers of institutes. Finally, I must mention projects entitled 'Science Technology and Development – the Role of the Transnational Corporations', 'Adapting Economic Theory to Changing Global Conditions' and' Science and Technology in Islam and the West'.

When IFIAS was created it was decided that at the end of five years, a panel of independent assessors should review its work. This was duly done by a group of undoubtedly illustrious men, who duly pronounced themselves satisfied, strongly recommended that IFIAS continue, and provided some constructive criticism of methods. After a further five years a different review panel repeated the exercise with the same outcome. In 1984 I relinquished the chairmanship and was replaced by Sir Herman Bondi, the cosmologist. It had been an exciting decade of innovation. 'IFIAS was a process rather than a structure', as one of our members described it. It devised new procedures of networking between research groups internationally that will assuredly be adopted by other organisations in the future. While its research product justified the resources deployed, its greatest success was the way it engendered the need to consider the individual human and social consequences of their work throughout the scientific community.

Throughout the period IFIAS suffered from lack of secure

funding. It was relatively easy to obtain resources for interesting, well-prepared research plans, but much more difficult to find money to support the central secretariat that had to carry out these plans, an expensive task that involved much consultation and travel. In spite of this, the overheads were surprisingly modest. Sam had to spend much too high a proportion of his time in fund-raising to the detriment of more intellectual work. I found my association with IFIAS extremely rewarding. It brought me into contact with many world leaders in science, and forced my concern on detailed elements of world problems. I remember with much gratitude and affection, the kindness of Nils Ståhle and the indefatigable camaraderie of Sam Nilsson.

The Foundation for International Training (FIT)

One day early in 1973, Aurelio Peccei and I were in Toronto for a Club of Rome meeting. As it finished, a young Canadian introduced himself and expounded his ideas about the development needs of Third World countries. He was Ranjit Kumar, at the time a lecturer in a large technological college, from which he had initiated some interesting projects overseas. We were impressed by his ideas and visited him the next day at Seneca College to see him at work. He discussed his ideas on development with great enthusiasm, convinced that a much more useful and lasting transfer of Western skills and methods could be achieved at grass roots level than was the case with most existing aid measures. He argued that development projects undertaken in particular towns or villages, however successful, were of quite local significance because their improvements would spread slowly if at all to neighbouring communities. Replication of successful projects to the thousands of villages of the region would be logistically impossible. It was thus necessary to devise schemes that would have a multiplier effect, and that could only be achieved by training indigenous people to take responsibility for introducing elements of change themselves. This suggested priority should be given to training and especially to training trainers. This could be done either by supporting and bringing new skills to such training centres as already existed or by stimulating the creation of new ones.

Ranjit was convinced that in Canada, as in many other countries, a large reservoir of skilled people existed who would be willing to help the development process by undertaking missions abroad to share their experience and techniques. There would be many people in universities and technical colleges, and probably in industry who, through concern, idealism or a sense of adventure would relish the opportunity of spending months or sabbatical years in Third World countries, taking part in well-designed aid schemes, their travel and living costs being covered plus perhaps a modest remuneration. Ranjit described a few initial experiments of this type that he had organised from his college with limited means.

Aurelio and I felt that his initiative was sufficiently promising as to warrant support, at least on a pilot basis and agreed to talk to the Canadian aid agency to find support for it. Our approach was successful and it was agreed that Ranjit, while remaining on the college staff, should devote half his time to operating an independent aid programme of the type he had suggested. This duly happened, was successful and quickly led to the creation of FIT, with Ranjit as its full-time executive director. An international Board was assembled to guide the work of the new organisation, and I became chairman, remaining in the post for eight years. I finally retired from the Board in 1994.

FIT is a splendid organisation. Its training projects have been mounted all over the developing world, and the subjects treated are extremely diverse, ranging from management of small-scale enterprises, low-cost housing, export promotion, credit systems for women, community education, agricultural co-operatives, tourism to biogas generation – to mention only a few. All this has been done at very low cost but even so it has never been easy to attract funds, which in later years became constrained by national self-interest. Annual levels of activity have grown regularly if somewhat erratically to their present high point.

Ranjit proved to be an excellent director. He had the rare quality of being at once a first-class meticulous manager and also a leader of imagination, ever exploring new openings for FIT in the light of a profound understanding of the ever-changing world theatre. He has managed to carry his Board with him. Its annual meetings have always been interesting and often entertaining. After effective decisions on financial and other issues and discussing the progress of current projects, time was always reserved for exposition and debate of novel ideas. This contrasts with many other Boards and committees on which I have served, which became so bogged down by administrative minutiae that no time remained for mature reflection and the generation of ideas for future development. Over the years I developed a deep affection for FIT and its devoted and enthusiastic staff; its successes may not appear dramatic, but they are very real.

Having described this Toronto-based activity, I must just mention another Canadian function of these years. On the very day that Aurelio and I first met Ranjit, I was invited to act as an adviser on technological development to the Provincial Government of Ontario. This proved to be an agreeable and undemanding task, involving occasional short visits to Toronto that usually began with a dinner discussion with the deputy ministers (i.e., permanent secretaries) of the various departments of government. I was expected to give a survey of significant trends in world technological development with potential physical, social and political developments. This was followed by a long and serious debate. On occasion discussion would focus on a single complex issue; for example, we discussed the significance of global warming to Canada, to its climate, agriculture and the transmission of its grain exports from the areas of the Great Lakes. On this occasion, the chief meteorologist of the federal government was invited to deepen the quality of speculation. The next day was devoted to talks with individual officials.

Science and Technology for Development – the UN Conference and Yet One More

In the last chapter I described how I devised a scenario for the conference on science and technology on the invitation of the UN. I had been rather pleased with my proposals that I felt converted the conference from a single static event to a process engaging all UN member states. The time has now come for me to relate what happened to it. It was well received by the UN Secretariat and even praised by an advisory committee. Then the organising machinery sprang into motion. The conference organiser was selected. He was a diplomat, ambassador to

UNESCO of one of the Latin American countries, who immediately entered into violent altercation with the head of the Science Secretariat. I doubt if he ever read my plan. At any rate, no trace of it appeared in his own structuring. That is not quite fair; his proposal accepted as axiomatic that science and technology could only be effective in a developing country if it possessed an indigenous capacity of skill and training. That was a big step forward.

It soon got abroad that the conference was to concentrate not on science, but on the funding of science. This alarmed the scientific community, which knew all too well that new funds were essential. But there was an even greater need to determine how best science and technology could function before finance could be deployed usefully. Maurice Strong, then head of the UN environment programme, spoke out on this theme and the international scientific bodies growled. I was naturally indignant. I gate-crashed a meeting of the International Council of Scientific Unions (ICSU) being held in Budapest. I argued that if we had no power to remould the official conference, we should at least mount an unofficial counterpart to expose the most advanced thinking on how science could revivify the development process. ICSU agreed and in the end I was invited to organise an unofficial world science conference on science and technology for development a few months before the UN meeting to be held in Vienna. The results of the informal meeting could be reported to the officials.

Time was short, but I managed to persuade Maurice Goldsmith of the Science Policy Foundation in London to cooperate in making the exercise possible. The conference duly took place in Singapore in January 1979. It achieved its aim. Nearly every prominent thinker in science and technology for development was present and the Third World representation was impressive. Discussion was frank and there was consensus that a new development strategy must be constructed, based on a better understanding of the needs and strengths of developing countries, and giving prime importance to their social and economic aspirations. Existing aid tended to be automatically patterned on the lines of development that the West had evolved slowly over two centuries on a top-down basis, and there was little sign of trickle-down effect. Discussion at Singapore of sectoral issues, such as agriculture, water resources, education etc., was rich and new insights gained. All in all, I felt this was one of the most constructive meetings I had attended.

A report on the Singapore findings was forwarded to the UN, but it did not surface at the official conference in August that year. The official conference was much as I had envisaged. The national delegates were mainly diplomats, not of the highest rank, although there were many science advisers around. The conference considered financial needs usefully, but pledges by the countries were hardly adequate to meet them. Negotiations continued seriously for years after the meeting. Its real value lay in the preparatory meetings in countries and regions that had deepened appreciation of the promise science offered them and some vision of the means required to actualise this promise.

Picnics in the Groves of Academe

Even in the OECD days, my contacts with universities were frequent, but when age cast me off from official activities they became considerably more intense. I gave lectures at universities on all five continents, attended meetings on university campuses and, as the years passed, made replies on the reception of honorary degrees. Then in 1987 I felt the urge to return to my origins and see if I could still face up to undergraduates, avid for learning, passive or rebellious. I took the plunge and accepted an invitation to be a visiting professor at Brandeis University in Massachusetts. This was a suitable location for us to try our little experiment. Brandeis was just a few miles west of Cambridge where our friends Sylvia and Derrick Denny-Brown lived. She was a school friend of Sally and he a professor at Harvard Medical School, a prominent neurologist. We had been in touch regularly since before the war and they had spent a holiday with us in France.

Our arrival in America was not a happy one. I had been stressed trying to finish a job before leaving and was very tired. Our plane was delayed and we sat in de Gaulle airport for six hours waiting for it. By the time we landed at Boston I had a high temperature and felt like death. Sylvia and Derrick were there to meet us and drove us to their home for the night. In the morning he took one look at me and drove me straight to the hospital, where I stayed a week with pneumonia. I could not have had a better impression of American hospitals, but I was worried I would miss the beginning of term and after a week's convalescence I feared I would be dropping into a new drama without having the chance to learn my part.

Providentially nature at her most savage came to my rescue. Just as I came out of hospital and entered our rented flat, a monstrous snow storm descended – streets were blocked, parked cars submerged, the ocean overflowed and there was an electricity black out. The whole city came to a standstill. A state of emergency was declared. It was nearly a week before the army came in with its heavy machinery and dug us out, so I could relax in my snow cocoon without anxiety. We had a beautiful apartment facing the Charles River that belonged to a Harvard professor on sabbatical leave in Europe.

Brandeis is an elite university of New York Jewish intellectuals, but it has a wide diversity of race and religion among its students. It is, in fact, extremely ecumenical and has three chapels; for Jews, Catholics and Protestants. Likewise, it offers a wide range of studies. I discovered that there was even a course on 'the English crossword puzzle', a subject to which several members of the faculty were passionately addicted. My chair was in the social science faculty and I was designated as a professor of political science, a rare appointment for a physical chemist. It was announced that I was to give a course of 25 lectures on 'the State of the Planet', the title of my book for IFIAS written a decade earlier and doomed to obscurity by the wiles of Bob Maxwell. To my horror more than 500 students inscribed, an embarrassing situation as the average number of students taking social science was 12.

Being completely ignorant of American student behaviour I was apprehensive about the impact I would make on this mob. But this proved groundless; I quickly established an easy relationship with these young Americans who were if anything too docile, and as many turned up to my last lecture as to the first. Their questions were on the whole apposite. I even brought in a guest lecturer for one occasion. This was a friend, Father Enrico Cantore, a Jesuit priest and celebrated luminary of the history of science. In addition to the lectures, each had to be followed by a series of tutorials, interesting conversations with smaller numbers. With the large numbers it was not possible for me to do this solo, so the university provided me with four postgraduate assistants to share the task. My real difficulties came with the onset of examinations. The threat of having to correct 500 papers and decide on grades was daunting, but here again the assistants shared the burden. I was surprised how many students questioned the grades assigned to them and sought interviews to argue for reassessment.

I enjoyed this interlude enormously and it gave a welcome serenity to our personal lives. Sally was delighted to be with Sylvia and made several new friends. We spent a wonderful week in early summer with the Denny Browns at their house in Maine. The academic picnic had been a huge success.

The next year I plunged again into a visiting professorship at the Université de Montréal. This was for only a few months, in a small graduate school on the history of science, where I took a series of seminars on the evolution of science policy since the war. It was less of a challenge since it involved no undergraduate teaching, but was intellectually stimulating and interesting to see something of the workings of a French Canadian school. Sally and I revelled in the experience of a Canadian winter.

These 20 years of so-called retirement activity which terminated only when I suffered a stroke in 1994 were the most hectic of my life. I was deeply convinced that human exploitation of the biosphere was nearing the point of no return. Thus the threats to the planet and to society obsessed me so that I had a compulsive urge to spread some understanding of the situation to the public. Never before had I experienced the zeal of the evangelist, but I now felt that I was a finger in the dyke. I was certainly dépaysé, a free radical in the world and, in the last years, even to some extent disembodied. I would be conscious of myself as if from outside, standing on a platform giving a talk, in eye contact with the audience and perfectly at ease with my subject.

I was still living in Paris, or rather working from Paris. These were difficult years for Sally. She often came with me on longer visits abroad; for instance she came with me on the Concorde to Rio and we had memorable days together after my meetings. Then there was a miraculous two weeks in Japan and China, visits to Kenya, Ethiopia and Cairo. She even came with me to

Amsterdam in 1987 when I was awarded the Erasmus Prize 'for services to Europe'. Nevertheless, she had many dreary weeks when I was away on short visits to Dusseldorf, Stockholm or New York. I was aware I was neglecting her scandalously, but felt little guilt. Later, after my stroke made me realise I had robbed Sally of the happy post-retirement years we had hoped to live together, a deep remorse descended on me and is ever present. If my time permits, I will write in expiation, a second autobiography about our love story of nearly 70 years and our spiritual strivings.

31
Two Controversial Asians

Rioyichi Sasakawa

In 1979 when IFIAS was preparing its programme *Analysing Biospheric Change,* Sam Nilsson met in Tokyo with some 400 members of the Japan Science Society to explore Japanese interest and possible co-operation. Even more significant for IFIAS' future was his meeting with Rioyichi Sasakawa, Founder and Chairman of the Japan Shipbuilding Industry Foundation, Japan's largest private foundation. He was extremely concerned with world environmental problems and with global issues generally, and had generously supported activities in this connection through his foundation. He had been till then, the largest private contributor to UN programmes.

We applied to his foundation for a grant in support of the ABC project, receiving a very generous response. A design workshop was held at Ulriksdals Castle on World Environment Day the next year. The Swedish king participated, as did Mr Sasakawa who expounded his life philosophy of universal brotherhood and sisterhood. He accepted the honorary chairmanship of the ABC programme and, in an atmosphere of euphoria, planted two Japanese cherry trees in the castle grounds in the presence of the king. The relationship with Sasakawa was long lasting; during my IFIAS years our work benefited by $2.8 million from him, making possible valuable activities that otherwise would have been difficult to finance. He was an agile elderly man, short in stature and long and emphatic in speech, but he spoke only Japanese, so that the pauses necessary for interpretation somehow dimmed and erased the emotional nuances of his words. On initial contact he transmitted an image of arrogance and autocracy. There was something of the Inscrutable Oriental about him, yet he was in some ways gregarious and even convivial. He enjoyed proposing toasts and his kanpai 'cheers' boomed out deafeningly on each occasion.

At the outset we knew little about him apart from his world concerns and great wealth. We sensed a certain ambivalence towards him on the part of Japanese officials; they neither encouraged our approaches to him, nor warned us against him. Gradually we began to know something of his history, which made him more of an enigma than ever. The following details are largely hearsay, but I believe them to be essentially true. Before the war Sasakawa had been a rich and powerful man, extremely nationalistic and fanatically devoted to the emperor. When the war came, his chauvinism drove him to become a kamikaze, a solo suicide candidate pledged to dive his plane or torpedo on to the enemy target. Presumably he had not yet become operational before the war ended. Be that as it may, after the defeat of Japan he found himself a war criminal, imprisoned by the Americans. At that point he experienced an epiphany like St Paul on the road to Damascus and realised that all men and women were brothers and sisters. This at once became his leitmotiv and preoccupation. He enjoyed displaying his vitality in relation to his great age. For instance, his early morning joggings around the streets of Tokyo were well known. I remember reading in a Paris newspaper about ex-President Jimmy Carter accompanying him on one such expedition. He frequently announced he would live to be 120 – a prophesy that was alas not fulfilled.

Once released from gaol, he determined to act on his revelation. As the war had depleted his fortune, his first task was to create a new one to finance major activities all over the world to alleviate human distress. I have no indication of how he provided capital to launch his principal money-making enterprise. At any rate he succeeded. In view of the scarcity and high cost of exploitable land on the Japanese Islands, he decided to operate on the sea. Apparently he established a system of speed boat racing, establishing centres in many locations around the coast where he could organise meetings, built factories to manufacture the boats, and controlled the betting. Money flowed in and was transferred to the charity foundation created to receive it.

I established a comparatively easy relationship with Sasakawa, especially during one of my visits to Tokyo. This was surprising in view of the language barrier, but was made possible by his brilliant and beautiful lady interpreter. He received me in his headquarters, a large and flamboyant palace in the centre of the city. His office was monstrously large, easily outstripping Mussolini's, and the ostentation was overwhelming. I was ushered into the 'presence' by an obsequious attendant and advanced to his distant desk, avoiding the hunting trophies, mainly tigers, that carpeted the floor. He talked at length about the IFIAS work and about the world's problems, assuring me that I also would live to be 120! He proposed taking me to visit the emperor, but we soon found out that he was away on a state visit so, much to my relief, my reception was postponed till another time. It never happened. Nevertheless, he insisted on dragging me off to one of the royal palaces to visit one of the princes, with whom he was friendly. We had tea with the prince and his wife, pleasant, relaxed people, who spoke sophisticated English.

After I withdrew from IFIAS, I continued to have frequent contacts with Sasakawa and his eldest son who later succeeded him. They were very disturbed by the graphic images of human suffering in Ethiopia broadcast throughout the world in October 1984 at the climax of the Sahel drought and famine. He felt it his duty to make a substantial contribution towards the alleviation of this wave of human misery gripping Africa and invited me to advise him on how much to donate. He similarly approached a few of my colleagues and his friend Jimmy Carter. I suggested he should not give a vast sum to the emergency relief effort already in train. Public response had already been enormous and generous, yet, essential as it was, it could only be but a Band-Aid attempt to staunch a running sore. Would it not be better to plan for an attack on the fundamentals of the problem in the hope of subduing the intolerable recurrence of drought and famine in Africa?

A few weeks later a small meeting took place in the Carter Center in Atlanta, between Carter, Sasakawa's son, myself and Norman Borlaug, the pioneer plant breeder who had created the 'green revolution' in India. There was consensus that a new initiative in African agriculture was called for in view of the series of persistent droughts and famines and that as a start, Rioyichi Sasakawa might consider calling a meeting of key people concerned with agriculture and development in Africa. In due time this meeting was held in Geneva, chaired by President Carter. The meeting listened to a broad sweep of ideas on how African agriculture might be made more resilient to its unreliable climatic swings. Then Borlaug gave a convincing outline of his approach in modifying the green revolution methods to the small farmer pattern of Africa. No detailed plan was presented nor was agreement sought, but the general attitude of the group favoured this approach.

The day after the Geneva meeting there was an interesting lunch party in a sun-swept Alpine meadow near Annecy, at which Carter was able to explain the detailed strategy to Sasakawa, who then promised to provide the funds. I spent most of the lunch in conversation with Rosalind, the President's wife while we intermittently witnessed exciting hang-gliding from a neighbouring Alpine peak.

Events moved quickly. Carter borrowed a plane and, making use of contacts from his presidential years, visited many of the African capitals in an attempt to select a few possible cases where political stability and agricultural potential would seem to proffer success. Finally, Ghana was chosen for the first experiments and Zambia followed. The basic plan was to provide each cooperating farmer with a packet containing an appropriate amount of low-interest credit, the new short stock seeds, the minimum necessary amount of pesticides and weed-killers and technical advice. A surprising number of farmers accepted the first year and many more the second. A couple of years later we gathered with Carter in Nairobi. The experiment had been a brilliant success; crop yields were much higher and more than 90 per cent of the loans had been repaid after the first year. Sasakawa then established an ongoing programme for Africa.

The Reverend Sun Meung Moon

My contact with the Moonies was unsought and inadvertent. One day in Paris, a visitor announced himself, hoping to talk to me about the unity of science. The concept of the unity of knowledge through the interaction of its elements had long interested me. Was it not akin to my obsession with the *Problématique*? I agreed to talk to him. He was a rather suave man, with an English accent. I don't remember his name and I never met him again, but the conversation interested me. He concluded with an invitation to take part in an international conference on the unity of science (ICUS) to be held two months later in New York. This intrigued me. The sponsor was the International Cultural Foundation, a name that sounded respectable if a bit vague. I had, in any case, to be in America about the time of the meeting, so accepted.

A pleasant young man met me at the airport and drove me to the Waldorf Astoria where the conference was to be held and the delegates housed. In the foyer I encountered my friend Pierre Piganiol, who knew as little as I what to expect. To our astonishment the conference was opened by the Rev. Moon. It became clear immediately that he was the sponsor; the ICF and ICUS were, as the Unification Church itself, part of a web of organisations owned or controlled by him. At that time I was hardly aware of the existences of the Moonies; I had only a slight impression of them from the press as a somewhat disreputable cult.

Moon read his opening homily with assurance in rather troubled English. It was a strictly philosophical exposition on the unity of all knowledge and the universality of values. No mention of the fundamentals of belief or the doctrines of his Unification Church, no hint of political ideology. Then, having wished the meeting success, he withdrew from the hall and did not appear again until the closing session three days later. After some general announcement of the themes to be discussed, the meeting split up into six separate groups, each on one of the six chosen topics. There must have been between 80 and 100 participants, many of them scientists of great distinction, coming from some 50 countries including the communist bloc, Israel and Japan; here were several Nobel laureates. Excellent meals were provided in a large private room, with no assigned seating. We were joined by the Moonie assistants, well-dressed and well-spoken relaxed young men and women who joined in conversation with the scientists freely. Over the three days of the meeting one inevitably made random contact with a large number of these young people

The organisation of the conference was impeccable. Topics for discussion had been cleverly selected with excellent introductory papers for each. Discussion was free and at times exciting, as was to be expected from the quality of the participants. No shadow of Moonie thinking was cast on the meeting. Its intellectual content was no different from that of any scientific meeting. The only difference was it was more professionally run. To my mind it was thoroughly justified by the interdisciplinary and transdisciplinary qualities it generated. The Rev Moon appeared again at the end of the final session. He made a speech rather similar to that given at the opening and invited the participants to a farewell banquet. This was a formal occasion with all the Moonie boys and girls dressed in their best. There was a high table where Dr Moon officiated, flanked by a couple of civic notables, some of the conference chairmen and a protestant clergyman who pronounced grace. The host was in a more jocular mood now, and there were some mild jokes. Then followed an entertainment. First, we listened to some Mozart played by a symphony orchestra which Moon supported in New York and then saw a brilliant performance by the Korean National Ballet which Moon apparently owned. Then he made a friendly farewell speech and wished us safe return to our various countries

This had been a rather impressive occasion. I had enjoyed the meeting, had many interesting conversations and a tentative invitation to a repeat performance a year later. But what was it all about? Why had he spent enormous sums of money in bringing all these scientists to New York and feasting them there unless he expected some reward; yet there had been absolutely no hint of proselytising. I did attend the ICUS meeting the next year; then one the year after, it had become an annual event. One of them took place in Korea, the others in large American cities. The pattern was always the same – topping and tailing by the Rev. Moon, uninhibited discussion in between. Gradually, in company with a few of the other independent scientists I got sucked into the preparation of these meetings, the selection of topics, discussants, etc. I met many of the Moon people including some of the lesser reverents. They never talked about their beliefs, nor did they attempt to influence our choices. I didn't ask questions about their theology, but putting two and two together, I gradually gathered a hazy and possibly false view of an alien belief. It made no sense or emotional impact.

I met the Reverend Moon himself on a number of occasions. He was an alert, short figure with a friendly approach. If he is an inscrutable oriental, his inscrutability is utterly different from that of Sasakawa and masked by a seemingly approachable man-of-the-world image. What struck me most about the man is the diversity of his interests and action, together with competence and professionalism in everything he instigates, whether it is ICUS, his orchestras and ballets or the *Washington Times*, his right-wing national newspaper. I know nothing about his industrial enterprises in Latin America or his Atlantic fisheries. The Unification Church, whatever its value and significance, is undoubtedly effectively organised.

I had, of course, read about the suspicion in which Moon and the Unification movement was held, the rumours that he kidnapped children from the streets, brainwashed them and purloined their wealth. Having talked over the years with perhaps 100 Moonies at the ICUS meetings, I had a very different impression. They were all relaxed, seemingly normal young people with easy, worldly conversation. Nothing about them betrayed hints of brainwashing. I remember a dinner conversation with a Canadian Minister of Justice in which the Moonies came up. He remarked, 'I thank God every day for the Rev. Moon who helps me by removing distressed young people from the streets thus making my job easier.' On my last appearance at ICUS, I met a young Scottish girl married to another Moonie, a Dane. She was just completing a degree course at a well-known American university, supported by Dr Moon's church. She was in obviously good relations with her parents who had come out from Dundee to visit her. I became well acquainted with the young American responsible for organising the conferences. He had been brought up within the Unification Movement, had been to university, risen to positions of authority and was soon to be a 'reverend' with his own parish. He was in an apparently happy marriage with a charming Korean girl and they had a child. He was a well-informed person of considerable cultural breadth and his views on world events had obviously not been constrained by cult prejudices. He could well have become a friend.

The American authorities were not comfortable with the Rev. Moon and his activities and looked for some way of trapping him. In the end they charged him on tax evasion by the Unification Church and he was sentenced to a short period in prison; I think it was one year. However, it proved to be a troublesome judgement as it seemed that others, including the main respectable churches, were equally vulnerable, thus setting a cat amongst the clerical pigeons. When the Rev. Moon entered prison he was accompanied as far as the gates by a small band of leaders of the main religions as a gesture of solidarity. Apparently he was very popular with his fellow prisoners as he helped in the kitchen and produced flavoursome Korean dishes. I was invited to visit him in gaol, but declined. Good behaviour assured his release after a half of his sentence had been served and as he passed through the prison gates the same band of eminent colleagues were waiting to welcome him home.

This was the extent of my connection with the Moonies. I found them pleasant, straightforward and co-operative with none of the holier than thou attributes that characterise 'true believers' of most other cults and sects. I hold them in great respect, but I still wonder what it is all about.

32

The Club of Rome

It is inappropriate to sketch here the history of the Club in detail. Peter Moll has provided a scholarly and fair account of its development in *From Scarcity to Sustainability* (Peter Lang: Frankfurt 1991), so I shall present only a personalised and highly subjective account of some of the events.

The controversy over *The Limits to Growth* was welcomed by the members of the Club as initiating a highly significant debate world-wide. All the same, we were highly embarrassed by misunderstandings and misinterpretations that presented a distorted image of our message and generated hostility toward us in some sectors. The image created by *Limits* was so strong it has haunted us ever since.

What to Do Next?

The Executive Committee (Excom) met in The Hague in 1973 to discuss the way ahead. We were struggling to establish our identity and there were heated arguments about what we were. The Club was not a research organisation and could claim no professional eminence as a group for the subjects of concern, despite the exceptional competence of members in their diverse fields. Likewise, we were not a political club seeking power. But we often aspired to be a catalyst in the evolution of policies and had even enjoyed modest success. We were a group of activists without vested interests, surveying the trends and discontinuities of society, attempting to foresee situations and problems and advocating modifications of policy and approach that might confront them.

We saw in the publicity generated by *Limits,* the opportunity for raising understanding of the significance to governments and the public of impending global problems. We were convinced it was urgent to explore the *problématique* and press for longer-term thinking throughout the political process. To this end we would:

- Organise conferences throughout the world.
- Confer with government, business and union leaders.
- Study specific aspects of the *problématique.*

Excom invited Eduard Pestel to construct a second world model disaggregated sufficiently to illuminate meaningfully the problems and trends of different regions. I was asked to write a detailed account of the present stance of the Club. This appeared as a substantial paper, *The Club of Rome – The New Threshold*, and circulated widely. The Club was up and running.

Peccei's Work Methods

It may be of interest to see how a structureless body such as this actually operated. In its early years the Club essentially *was* Aurelio Peccei. He possessed so much natural authority, was so outgoing, warm, genial and persuasive that he was able to charm associates into accepting all his propositions. Excom had some influence on him. It met irregularly, often at the end of one of our own or somebody else's conference and sometimes in my apartment in Paris. At these meetings we discussed world events or personal encounters relevant to the Club, and at times novel ideas were debated. The level of discussion was profound and I looked forward to them. Coherent decisions or formal agreements were seldom reached or sought at these meetings, but I know Aurelio appreciated them and incorporated many of our ideas in his thinking. I had frequent meetings with Aurelio, travelled often with him and we spoke on the telephone nearly every day. Our relationship was so close and relaxed that he took no umbrage even at my sternest criticisms. Indeed, many of his public pronouncements were developed jointly. I was usually in agreement – but not always. Aurelio had a deep belief in the

'Monarch of all he surveys'

basic goodness of people that I found a bit naïve and sentimental; I was more cynical.

Formal consultation with members of the Club outside this inner circle was perfunctory and mainly limited to one short 'members only' session during our annual conferences. Even then they were often squeezed out by the excitement and overflowing of the plenary sessions. But he saw most of the members individually from time to time. All seemed happy with what was going on. There was little sign of democracy! I remember only one incident of revolt against Aurelio: at one Excom meeting in Tokyo he advocated acceptance of a manuscript submitted for publication as a 'Report to the Club of Rome'. The members rejected it unanimously as of insufficient quality. Nothing more was said, but the printed report was on sale a few months later.

The Secretariat was Peccei's office in Italconsult, of which he was still president and for which he retained some executive responsibility. I had the impression that as time passed, he became so engrossed with his Club activities that his Italconsult interests gradually faded. He had two loyal and efficient secretaries able to work in several languages. They soon became as devoted to the Club as Aurelio. During the first decade of its existence, members were not expected to pay annual dues, but these were introduced later. Foundations provided funding for projects, and local sponsors paid for conferences. These were local foundations, development banks and sometimes government agencies. No government or corporate funds were sought or received for day-to-day activities.

Reports and Conferences

The main products of the Club were its conferences as well as books or reports *to* it, based on studies it commissioned. In a number of cases, discussion of a recently-issued report was the focus of a conference. During Aurelio's presidency, some 17 books or reports were published. Space will permit only a few of these to be mentioned – together with a few personal anecdotes.

The annual conferences had two objectives: first, to convey the thinking and messages of the Club to intellectuals and decision-makers of the region in which it was held; second, to provide a forum for the members. Thus, over the years they took place on all the continents and in many countries as large and often brilliant gatherings of club members with invited internationally known guests and speakers, together with an equal number of local influential personalities. Some of us felt that these conferences were unnecessarily grandiose and referred to them flippantly as Aurelio's 'Three Ring Circuses'.

Pestel, who had set up his own institute in Hanover and, by the time the *New Threshold* had appeared, was already planning his disaggregated world model, joined forces with Mihajlo Mesarovic, professor of systems science at Case Western Reserve University, Ohio, who had recently developed a theory of multilevel hierarchical systems. This, together with aspects of the *problématique* rationale formed the methodology on which the Pestel/Mesarovic model was based. It distinguished between ten different world regions and was exceedingly complex, involving some 200,000 equations. It was able to process social and technical data. The designers of the model clearly hoped it would be sufficiently refined to be of use to politicians in their decision-making. However, despite its technical superiority, size, complexity and handling strategies made it difficult for specialists, let alone politicians, to use. I was present at one experimental session where a number of less than senior political leaders were able to question the model on energy policy options. The process was long and laborious and their response mixed. Some thought it could be valuable for testing new policies to detect any counter-intuitive consequences before implementation.

The non-technical version of the study was presented as a report to the Club with the title *Mankind at the Turning Point*. The presentation lacked the elegance of *Limits* but gave a much less technocratic image. A preliminary report was presented at the Club's annual conference in Tokyo in 1973, a high-profile occasion at which various other approaches to global planning were discussed. The completed report was discussed at the corresponding meeting in Berlin the next year.

The New International Economic Order

1974 was a confused year in international politics: the monetary arrangements agreed at Breton Woods had all but broken down; Third World countries demanding a bigger say in their own and world affairs had joined together in a group of 77 to enhance their influence; and a petroleum crisis arose resulting from OPEC action taken in October of that year. The creaking world system clearly demanded an overhaul, if not radical restructuring. This was raised at the UN with a proposal by the less-developed countries for a New International Economic Order (NIEO), which led to contentious debate in the General Assembly and the agencies, lasting from 1974–76.

We realised the new phase of the Club's work would have to be relevant to these shared concerns. In any case the North/South issue had arisen forcibly during the *Limits* debate and appealed to the humanistic instincts of Members. Obviously our *problématique* leitmotif would appear prominently. A small group was set up to investigate some aspects of *problématique* interconnections, led by Dennis Gabor, the Nobel laureate who invented holography. Its particular concern was with the mutual influences between population, food, energy and technological change with reference to Earth's capacity to feed the very large numbers of people expected in the next century. The resulting report, *Beyond the Age of Waste*, reflected a moderate technological optimism – with strong reservations.

But much larger game was in view. Aurelio and I had often talked about the understanding that might be generated if leaders of countries could meet occasionally for private discussions without the media, not to negotiate but merely to talk informally about their problems and attitudes. This we termed grandiosely, *The World Forum*. We realised that informality and real conversations could never take place in a large assembly open to all nations. We thought rather of smaller groupings of a dozen heads of state or PMs, not on a regional basis or sequentially with the same individuals. We decided to experiment by trying to organise a meeting between a few leaders of smaller Western countries with Third World counterparts and some of our colleagues. My OECD work had brought me into cordial relations with Bruno Kreisky, the Austrian Chancellor. I found him very open to my concerns and thought he might be persuaded to act as host to the kind of meeting we envisaged. Aurelio, the charismatic persuader had no difficulty in so doing. He spent the next months in hectic visits to explain our intentions to some 20 national leaders in the hope of identifying a smaller number able and willing to meet together. This was no easy task. Quite apart from the difficulty of inducing leaders to spend a couple of days abroad without any concrete outcome that might appeal to their voters, there was the difficulty of finding days when these busy people would all be free. However, Aurelio did succeed and a meeting was scheduled for 4–5 February 1974 in a picturesque eighteenth-century castle outside Salzburg.

The plan was for the meeting to take place behind closed doors with the media and ministers' civil servants excluded, with a press conference at the end. Club of Rome participation was to be restricted to the ten members of Excom, and Aurelio would be chairman to avoid problems of protocol *vis-à-vis* ministers. Aurelio had deliberately not invited leaders from the larger European countries and particularly America and the Soviet Union to lessen the danger of descent into banal political bickering. Thus the invitees were the premiers of smaller European and Third World countries – Leopold Senghor of Senegal, Luis Echeverria of Mexico, Olof Palme of Sweden, Pierre Trudeau of Canada, Joop den Uyl of the Netherlands, Nello Celio, former President of Switzerland, and Bruno Kreisky, host of the meeting. In addition, Presidents Boumedienne of Algeria, Bhutto of Pakistan and Cosgrave of Ireland sent senior ministers to represent them.

Important tasks were to increase public interest in the activities of the Club and to persuade the ministers to take account of the interactions of the *problématique* in the formulation of policies. Five two-page papers had been prepared in advance to initiate discussion on five major problem areas. Each of these was introduced by a member of Excom, after which it was left to the ministers to debate, orchestrated by Aurelio. This proved to be a good procedure; discussions were lively, hard hitting and, in the main, managed to keep to global issues rather than individual nations.

The agreement to resist from publicity until the end of the meeting was blatantly ignored by Echeverria who brought with him a couple of television crews and a flock of journalists. He repaired to Salzburg both evenings to broadcast about the meeting, including his own speeches for home consumption. The Club members produced a 'Salzburg Statement of the Club of Rome' giving a balanced account of the meeting with some useful if somewhat anodyne recommendations to governments. This was the basis for the press conference that was attended by some 300 journalists and lasted several hours. Ministers responded to questions vigorously and, in the case of Trudeau and Palme, delightfully contentiously. In retrospect, I feel that the Salzburg meeting served its purpose. It was a learning occasion for many of the ministers; its political effect could only be indirect but it contributed to the UN debate on the NIEO. I think it may have influenced the thinking of some of the participants. For instance, the exposure of Echeverria to a calm discussion on world demographic prospects probably helped to modify Mexico's population policy. For the Club, the publicity was welcome in adding a more humanistic and global dimension to the *Limits to Growth.*

My happiest memory from this rather flamboyant gathering was a whispered conversation during one of the sessions. I had just introduced the two-page paper on energy and had returned to my place at the table between Senghor and Trudeau. The latter bent over and whispered, 'What poet were you quoting?' 'I am unaware that I quoted anyone let alone a poet', I replied. 'You said *Had we but world enough and time.*' I thought for a moment and whispered back, 'Yes, it must have slipped out, it's from *To His Coy Mistress,* by Andrew Marvell – you know:

Then worms shall try thy long preserved virginity
And thy quaint honour turn to dust
And into ashes all my lust.

Trudeau chimed in:

Thou by the Indian Ganges' side
Shouldst rubies find: I by the tide
Of Humber would complain

We spent most of the session trying to reconstruct the poem and in the end completed most of it. I found it touching that a French Canadian Prime Minister had such a feeling for a seventeenth-century English poem that it resided so long in his memory.

President Echeverria invited us to hold another meeting at prime ministerial level the next year in Mexico. However, remembering the weight of effort that Aurelio had exerted to achieve Salzburg, we knew it was unrealistic to expect them to come again so soon. Nevertheless, we compromised and invited premiers to designate senior ministers to represent them in Mexico. The meeting eventually took place in Guanajuato, a delightful colonial-style city. A somewhat larger and more diverse number of nations took part than at Salzburg with high-level participants. There were many interesting discussions, but they tended to be more politically charged and unfocused. In retrospect the Guanajuato encounter was the least successful of our major meetings. It was smothered by the generosity of our Mexican hosts; there seemed to be bands playing on every possible occasion, meals were too rich and too long, prolonged by still more music and songs.

Here I must add another personal anecdote. Travelling from Paris, I sat in the plane beside one of our most revered members, Cardinal Fritz König, Archbishop of Vienna, a wise, open-minded man whose conversation made for a memorable journey. Mexico, as a secular country, does not permit clerical dress on the streets, so the Cardinal was in a business suit like the rest of us. Once arrived in Mexico City Aurelio and I had to pay courtesy calls on the Prime Minister and Foreign Secretary. On entering the office of the latter, the Chef de Cabinet, introducing us in Spanish, confused König and King and announced me as the Cardinal. Much to my alarm, the Foreign Minister knelt on the floor searching for my non-existent ring to kiss. The whole scene dissipated in laughter.

We had another of these high-level meetings in Stockholm in 1977 with leaders of the Nordic Countries plus others, after which the idea of a world forum was quietly forgotten. The NIEO debate was still in full swing at the UN and we felt that, as a group of independents, we still had something to contribute. Immediately after the Salzburg meeting therefore, Aurelio approached Jan Tinbergen, Nobel Prize economist, to undertake a study of what kind of new international order should be recommended to politicians in order to meet the urgent needs of present and future world populations. Tinbergen was one of the few economists who had received *Limits* sympathetically and had already devoted much study to the North/South disparity issue. He accepted the challenge and immediately assembled a group of scholars on aspects of global development and food provision. These were mainly from the First and Third Worlds. A serious attempt was made to attract the co-operation of Soviet and East European scholars, but was largely a failure.

The work soon became known as the RIO report (Restructuring the International Order). Working sessions took place frequently during the next 18 months in Rotterdam. The work started with the selection of a number of topics; trade, financial policies, agriculture, energy and technology, and inviting one member from a developing and one from a developed country to combine in writing a paper on the global situation. For example, I wrote the one on technology with the brilliant Ethiopian scientist Aklilu Lemma. These papers were discussed at a working session and, modified, formed the material from which Tinbergen later wrote the RIO Report. However, the group did not wait until its work was finished to make known the trends of its thinking. They sent reports to some 300 institutes or individuals and used responses received for later discussion. 5000 copies of a paper commenting on the discussions were distributed at the seventh special session of the UN in September 1975 in the hope it would contribute to the NIEO debate.

The report began by comparing the disparity between the richest and poorest countries. The ratio of 13:1 seemed intolerable both as a reflection of world poverty and as a destabilising element in the global economy. In the absence of reform in the system, the disparity would certainly grow because of the differential advantages the industrialised countries possess, social and political disruption could follow. The obvious conclusion lay in massive redistribution of wealth, necessarily a long process. There is certainly need to increase development aid to the Third World, but more important, the RIO group concluded, was the task of restructuring the global economic and trade systems, seen as grossly unfair to developing countries. The report included an important series of recommendations for action. It was well documented and argued, but suffered, in common with reports written by committees, from efforts to reflect the views of all its members to achieve some semblance of unanimity which led to a low-key boring presentation.

Aurelio, who had taken little part in these discussions, persuaded President Boumedienne of Algeria to host a meeting to launch RIO. This took place in Algiers in October 1976 and turned out to be a very large, noisy affair attended by ministers and development pundits of many countries. It abounded in expressions of conflicting political opinion. During one of the sessions Aurelio whispered to me, 'This is more like a meeting of the UN than one of our own.' Some of our members were disappointed, feeling that the report was too unadventurous, insufficiently radical. Towards the end of the meeting our stalwart Indian member Romesh Thapar announced forcefully, 'I am a WOG, a Westernised Oriental Gentleman. Even if you do achieve your New Order, it will only create more WOGs like me: wealth simply does not trickle down to the poor.'

The RIO report was a substantial, wise and academically respectable piece of work, but its lack of impact was discouraging. It had appeared just too late to have much effect on the NIEO debate. Times had changed since it was conceived. By the time it appeared, the consequences of the oil crisis had been more or less coped with. The Western powers were no longer jittery about their vulnerability to events in distant countries and hence were less willing to look with sympathy on the complaints of the underdeveloped world. But RIO had been a brave try.

The Post-RIO Phase

During the next decade a number of reports were issued – always *to* and not *by* the Club. Some of these were the work of individual members, notably *Road Maps to the Future* by Bohdan Hawrylyshyn, *Dialogue on Wealth and Welfare* by Orio Giarini and *The Future of the Oceans* by Elisabeth Mann Borgese.

Others were the work of small groups assembled for the study of a particular element of the *problématique*. The first of these, on education, was written by Jim Botkin (USA), Mircea Malitza

(Romania) and Mahdi Elmandjra (Morocco) with assistance from Ricardo Diez Hochleitner (Spain), representing four very different systems. Its report entitled *No Limits to Learning* stressed the importance of the broader concept of learning rather than the more formal process of education. Two of its themes were participation and anticipation, the latter calling on the educators to prepare the young for the emerging new type of society, rather than devoting undue attention to fading objectives of present education systems. This report, presented at a conference in Salzburg, was useful in debating a number of principles to guide the future development of education, but clearly a follow-up study was needed to concretise its proposals. The Club has not yet found the courage to do so.

The Report *Microelectronics and Society* likewise resulted from the work of a group of experts, mainly from outside the membership, under the leadership of Adam Schaff and Gunter Friedrichs, a West German trade unionist. Each member wrote a chapter on a specific aspect of the coming IT revolution. For instance, there was a deeply disturbing account of the impact it was having on military defence and aggression. I contributed an essay on the potential problems from incorporating IT at low cost to practically any device used by humans, while Adam followed this up with a strong final chapter on its inevitable influence on the nature of work, employment, leisure and education. This work had little impact on the English-speaking world as a result of Maxwell's failure to distribute it. It was presented at yet another meeting in Salzburg, from which two of the participants took follow-up action: Cardinal König introduced it to the Vatican, where its messages were incorporated into its youth education plans, while Jermen Gvishiani took it to the Kremlin. The Club again missed an opportunity to follow up the new social thinking by not issuing a hard-hitting warning about impending structural unemployment.

Our report *Africa beyond Famine* was the product of many years of discussion of the problems of Africa that had virtually deleted the continent from the geopolitical map. The very low GNP of countries South of the Sahara has made development a prime objective of economic aid programmes. We had long been aware of the gravity and diversity of Africa's difficulties. Consolidation of decolonisation was, and still is, incomplete:

disparities of wealth, infrastructure, industrialisation, capacity for science and technology and most other factors, were intolerably large. Thus programmes of development aid were obviously necessary if the continent was to find its way in the modern world. It was difficult to decide how the Club should respond. I had my other channel of concern to African issues through the IFIAS climate work. Widespread concern about droughts and famines in Africa suggested we should enter the *problématique* from there. Aklilu Lemma, my Ethiopian partner in the technology chapter of RIO teamed up with Pentti Malaska of Finland to assemble a team of African and European scholars who undertook a lengthy study of the many factors. Their papers were discussed at a number of meetings. It became clearer that the famine problem with its desertification element, tribal rivalries, political chaos, civil war and the whole gamut of poverty and underdevelopment issues, could not be tackled in isolation. So in the end *Africa beyond Famine* became a treatise on development with many useful recommendations. Prior to publication, a final meeting of the study group, together with other African scholars, took place in Lusaka under the patronage of Kenneth Kaunda, President of Zambia.

Around this time we organised other meetings, smaller than those of our annual conferences, in S. Korea, Australia and the UN. One of the most interesting was in Seoul, S. Korea by invitation of the Chancellor of Kyung Hee University. This was more of an occasion for reflection than promoting action. It included people from Europe, America and Japan, and an equal number of Koreans. This was a quiet, friendly meeting, our first venture into Asia other than Japan, which had a troubled start. Aurelio and I flew out together from Los Angeles. When we transited in Hawaii, the news came through that President Park of Korea had been assassinated. Martial law had been declared, and it was doubtful if we would be allowed to disembark; all public assemblies were prohibited. After some negotiation we were allowed to land at Seoul and were driven under guard to our hotel. Aurelio had to visit military headquarters to plead for our conference to take place, while I had to go immediately to the palace to pay homage to the dead dictator whose body lay in state. The next day all seemed calm. A feared *coup d'état* had not happened and we were able to confer peacefully.

President Kenneth Kaunda offers King a cup of coffee

Mayor Chirac, soon to become President of the Republic, presents King with the Great Medal of Paris

The National Associations

It was to be expected that the early, enthusiastic members of the Club would seek to discuss our ideas in writing and speeches. This happened quickly in the case of Spain. I had known our first Spanish member, Ricardo Diez Hochleitner, in his days at UNESCO. Ricardo had become deputy Education Minister during the last years of the Franco regime and, when the dictator died, was anxious to assist in a safe transition to democracy. He invited Aurelio, me and a few other members of the Club to visit Madrid to meet with some influential Spaniards from industry, economics, science and politics. This was an important event in the return of Spain to normality. After the meeting we had a serious discussion with King Juan Carlos and Queen Sophia on world affairs and our new approach. The widespread interest in the Club thus generated, encouraged Ricardo to form a Spanish Association for it, which he presided over until 1990 when he relinquished it to become President of the International Club itself. This Association had an active programme for many years. Its success induced other countries to emulate it, and numbers gradually crept up to over 20. Some members were suspicious of this development, fearing self-proclaimed associations might all too easily misrepresent the Club's ideas and detract from its global mission or damage its reputation. Indeed, they might. One enterprising person advertised for people who wished to become members of the 'Australian Club of Rome' on payment of a sizeable sum. A number of people responded and the entrepreneur was never heard of again

Particularly significant was Schaff's creation of the Polish Association for the Club of Rome, the first in a communist country, providing a meeting ground for people from the Communist Party, Solidarity and the Church. President Jaruselski expressed hope that the new body would prove to be a point of national reconciliation. In March 1987 a meeting of all national associations was held in Warsaw to foster communication between them, to agree their functions and establish their relationship and status *vis-à-vis* the Club itself. As a result a charter was drawn up and agreed by all.

The national associations vary greatly in membership and

significance. A few are very active, organising meetings and studies regularly; others are quite small and have little impact.

Forum Humanum

In 1979 Aurelio launched the last great initiative of his life. He had always stressed the importance of involving the young in Club concerns about the future. We had all agreed, but never found an effective way of doing so. Then he became passionate about the need to give the young a place in shaping a future in which they would be actors. In many of his speeches he introduced this theme in a highly romanticised vision: 'The youth of today are so good, so clean, so much better than was my generation.' He decided to build an international network of young people in First, Second and Third World countries to consider the pressing problems of the world. This he termed *Forum Humanum*. He discussed his proposal with Excom. For once he met with a lukewarm response. We all appreciated his objectives, but were not eager to embark on a new, complicated, costly and tangential adventure. Some also felt Aurelio's richly sentimental exposition was at odds with the Club's objective stance. He therefore decided to go it alone and from 1974–79 conducted the most intensive campaign of his life, setting up groups of young people and seeking funds to support their studies. The campaign was only modestly successful. Groups of about 15 people were set up in Rome, Madrid, Geneva, Rio de Janeiro and Buenos Aires, while others were in formation when he died. Donors were reluctant to provide funds. I am sure much of the funding came from his own pocket. Some of the groups presented sensible and constructive reports, but they did not long survive his death. This was a sad ending to a noble initiative. He was sad that his project didn't have even the moral support of members of his own Club.

Involvement of the young in constructive criticism of society and planning the future is important and difficult. I had never found an effective way of encouraging it, but during my years as President I came across a student movement that was seriously discussing global problems. This was AIESEC, an international association of students of economics and business with branches in many schools and countries and a solid base in Brussels. We invited their representatives as observers to many of our meetings and I lectured at their meetings in many European cities even after I relinquished the presidency.

Enemies in the Night

Throughout its short history the Club of Rome had its detractors and evoked hostility in some quarters: from economists furious about the *Limits of Growth*; businessmen denouncing us as class traitors; people in the Third World suspicious we were a hidden hand of the European industrial Mafia; or others simply disliking us as elitists. Most of this was understandable and without malice. However, in the mid-1970s a more sinister form of opposition appeared. Small groups of people and at times mobs, tried to disrupt our meetings in at least a dozen countries. These people were followers of the American Lyndon LaRouche, a right-wing extremist and marginal presidential candidate on several occasions. I came across groups of them at many US air terminals – well-groomed boys and girls, selling some of the many LaRouche publications, particularly one aimed at business people, *Executive Intelligence Review*. A friend, noticing my name in a copy he had bought at Kennedy airport, lent it to me. There it was; an article reporting the opening of a La Rouche branch in Bangkok. I quote 'The audience gasped when Tnapura read a quote from Club of Rome founder, Alexander King, stating his commitment to murdering the brown, yellow and black populations throughout the underdeveloped sector.' They appeared to operate by sleaze and slander: Kissinger was vilified; Queen Elizabeth was the centre of the world drug trade; the Club of Rome, in unlikely association with the Vatican and Muslim Brotherhood, had planned the downfall of the Shah of Iran – and so on. They informed me I was to be tried for genocide by their 'people's court'; they telephoned my wife and daughter with threats several times in the middle of the night. The Club's US Association commissioned a lawyer to demand they desist from persecuting us. LaRouche replied with a curt refusal to do so, adding it was his duty to attack us as the Club of Rome was more evil than Hitler. I have no idea what, if anything, this cult stands for except for their declaration that nuclear fusion would solve all the world's problems.

The King Administration

Aurelio Peccei died suddenly in March 1984. The loss of a leader so highly regarded, inevitably left the Club in disarray. The instinctive reaction of most members was that Aurelio's demise meant the end of the Club, but after a few days of reflection some of us decided the global *problématique* was still insufficiently understood and it might be our destiny, or duty, to soldier on. Preparations were well advanced for the annual conference in Helsinki a few months ahead, so why not continue until then before coming to a decision. I managed to round up a few members of Excom, who met with me in Paris to examine the options. The outlook was bleak. Not only had we lost a leader, but also the Secretariat had disappeared with him and we had no resources. We hoped to identify a member who, without remuneration, would be willing to act as Secretary-General and cope with the day-to-day affairs. Bertrand Schneider, a Parisian management consultant, rashly volunteered to do so. Schneider, a former diplomat had been close to Aurelio and fully identified with the Club.

The next step was a formal meeting of the committee in Toronto which a few other dedicated Club members attended. This was an important occasion of brutally frank discussion of the pros and cons of continuing. In the end there was consensus that the Club should be shaken up and restructured to be more democratic without sacrificing the flexibility and quick response of the Peccei regime. Could it continue as a non-organisation? There was then much discussion of issues and priorities, and determination there would be much more active participation in particular projects by members. I was appointed President, subject to ratification at the Helsinki meeting, and invited to write a paper concerning the future of the Club for presentation to that meeting and for wide distribution thereafter. This duly appeared with the title *The Club of Rome – Reaffirmation of a Mission*.

At Helsinki, a closed meeting of the membership brought us back to first base. They attacked the 'to be or not to be' question ferociously, but gradually came round to just about the position we reached in Toronto. The ensuing discussion of structure and programme was equally fierce without a clear conclusion emerging. Bertrand and I were left to carry on without any clear guidance. The conference itself was essentially a memorial to Aurelio. But after the eulogies had been pronounced, it settled down in the presence of the President and Prime Minister of Finland to serious discussions of pre-documented themes. I mention only one of these. Aurelio and I had long felt that the Club ought to have something useful to say about the impact of the arms race on development. When in Tokyo, we animated a useful seminar on the subject at the UN University, the new Rector of which was one of our wisest members, Sudjatmoko of Indonesia. Then I invited all our members to express their views in writing, an exercise I continued after Aurelio's death. This evoked many interesting reactions that I harmonised into a possible report to the Club, but I decided not to present it, feeling it was too conventional and short term. Nevertheless, I wrote a paper, *The Arms Race and Development – Waste and Want*, to give a dimension to the debate that followed.

Back in Paris Bertrand set up headquarters in his own office in Neuilly, and with the new open approach, it soon became a busy port of call for members and colleagues, so much so that he soon had to find larger accommodation. I had retired from IFIAS just before Aurelio's death, so was able to devote all my time to the Club's affairs. I badly needed moral support, so invited Ricardo Diez-Hochleitner and Kurt Furgler to be my Vice-Presidents. Under Aurelio, Excom, despite its formal mandate, did not attempt to make decisions; its functions were simply advisory. My democratisation made it the true power centre of the Club. Members adapted themselves to the new situation, buried themselves in detail and took an unconscionably long time in doing so. It has always amazed me how otherwise wise people become fascinated by administrative trivia about which they know nothing and profess to despise. My naivety was punctured by the realisation that democracy begets bureaucracy. So I had to trim my principles in order to achieve anything, resorting to Aurelio's methods as far as I dared. In practice Bertrand and I made the decisions. We discussed Club problems almost every day. A few minor reforms were introduced. I revived the procedure of inviting members who had become inactive to withdraw to make room for younger people, especially women. Nearly all did so gracefully. Then I instituted a new category of *Honorary Members*, prominent individuals who shared our ideas but were debarred by their offices from becoming active members. Initially these included the King of Spain, Queen Beatrix of the Netherlands, Jimmy Carter, Crown Prince Hassan of Jordan, Cardinal König and Abdus Salam, the Pakistani physicist. Most of these have given us valuable advice and encouragement over the years.

The Activities

I started by attempting to secure the participation of members in a project on *Governability and the Capacity to Govern*, a topic proposed at Helsinki. The world, its structures, ideologies and problems had changed unrecognisably during the previous century, yet the machinery of government, its role, structures, methods and political conventions remained much as they had been created to meet the needs of simpler times. Complexity and uncertainty were straining the capacities of archaic institutions. Here was a place for analysis and innovation by a politically independent college of people of good will. I received a number of interesting responses and these served as the basis of debate at our next major conference held at Santander, Spain, in 1985. This was an extremely difficult meeting with strong arguments for and against our continuing involvement in the subject. Some felt it was a subject so vast and open to partisan pressures from all sides that it would inevitably draw us into political dialectics. In the end we invited Yehezkel Dror, of the Hebrew University of Jerusalem, an outstanding thinker and writer on the topic, to do an in-depth study for us. This eventually appeared as a brilliant exposition of the inherent problems, needs and underlying philosophical principles, but provided little practical advice for political innovators.

Ever since the RIO exercise, we had been preoccupied by the special difficulties facing Africa. In 1985 we published Bertrand Schneider's study of the grassroots approach to development,

The Barefoot Revolution, an account of the thousands of self-help groups that had appeared spontaneously all over the Third World. Work had already begun on the study of Africa's famines, which appeared later as *Africa beyond Famine*. However, we felt the need to take a more comprehensive view of Africa's problems and planned to hold a conference there with the help of our member, Edem Kodjo, recently retired Secretary-General of the Organisation of African Unity (OAU).

On the invitation of President Biya of Cameroun we held a meeting in Yaoundé, endorsed by OAU at the end of 1986. We brought together 30 members of the Club with an equal number of African development experts and political leaders. The group had a broad and frank discussion on economic planning and development in Africa, including the controversial matter of Third World debt. Much concern was expressed about the continent's environmental problems, raised by awareness of rapid desertification in the Sahel and Southern Africa.

On the margin of the conference were a number of interesting situations. In front of the typical mid-twentieth century 'Palais des Congrés' was a large parade ground. When we arrived, large numbers had already occupied this, cheek by jowl with a dozen or so music or dance troupes. The President arrived and we were introduced. The first act was for he and I to inspect sizeable contingents from the army, navy and air-force, a ludicrous situation for someone of my background. Then, after the end of the conference, President Biya presented in our honour a performance of a play written by one of his ministers, at which the government elite and diplomatic community were expected to attend. The play, *The Old Nigger and the Medal*, was a wickedly sweet satire on the stupidity of both illiterate peasants and their former colonial masters, skilfully fashioned and elegantly presented. After the applause had faded, the Director of Protocol advanced to the front of the stage and announced the order of dispersal, 'Firstly the President of the Republic will leave the theatre accompanied by the President of the Club of Rome. A few moments later Ministers may withdraw together with the Corps Diplomatique. Thereafter the audience may disperse.' I rose and walked out with due dignity beside President Biya, barely able to suppress my giggles.

We continued working with OAU to arouse interest in

mobilising a meaningful attack on halting the southern advance of the Sahara. Bertrand and I visited Brazzaville, invited by President Nguesso of the Congo to discuss follow up from the Yaoundé talks and I pressurised the Secretary-General of the UN, Pérez de Cuéllar, to make a major *demarche* on world desertification. The Club lacked the means to launch operational attacks directly, but we were reconciled that increased action by UN agencies had resulted partially from our efforts.

During this period, rather than promoting planetary studies such as *Limits to Growth,* we were endeavouring to widely promote Club ideas, especially to encourage long term thinking and understanding of the *problématique.* This we did through conferences and more generally by the personal diplomacy of members. Some examples of my own personal diplomacy were described in Chapter 27. One of the more formal approaches was on European-Arab relations. Prince Hassan of Jordan had created a Club of Aqaba (later renamed the *Arab Thought Forum)* modelled on the Club of Rome. The secretariats of the two Clubs established fraternal contact, Prince Hassan becoming a member of our Club, while Bertrand and I participated in a meeting of theirs in Amman. This led to a joint meeting in Carmona (near Seville) in the presence of the king. It faced up to serious long-term Middle East issues, particularly those of water, food and agricultural deficits.

A quite different series of meetings was planned in a few key locations, on the theme *Thinking Globally and Acting Locally.* Cities noted for their concern with regional planning and preferably distant from national or federal capitals were selected. Two successful conferences of this type took place in Denver Colorado and Fukuoka in Southern Japan.

Then in 1988 came the twentieth anniversary of the founding of the Club. The celebrations, held in Paris, were probably the most flamboyant of our history. The meetings represented two decades of work and were well attended. The annual meeting of 1989 in Hanover was more significant. Its theme was *World Industrialisation – Panacea or Nightmare?* We planned to lighten debate on this rather grim topic and humanise the image of the Club by interjecting presentations of some very different views of the world and humanity. One of these, an angry, passionate denunciation of global injustice was delivered by the German novelist Günther Grass; another was an evocation of life in the vast, empty plains of Central Asia by Chengiz Aitmatov, while a Native American poet gave a wonderful sense of the serenity and primeval faith of living with the spirits of ancestors on the prairies and mountains.

The meeting highlighted environmental constraints on industrial growth, questioned the wisdom of pressurising underdeveloped countries to seek advancement by imitating Western patterns of industrialisation, and stressed the role of energy in future world development. This complex of interdependent issues gave a vivid picture of a zone of the *problématique* particularly critical for the next stage of world development. The participants, including a number of industrial invitees, were so impressed by the gravity of the situation that the Club decided to spend 1990 re-examining the world situation and reassessing its own mission in the context of the turbulent world that followed the end of the Cold War. A council of 14 of the most active members was elected and invited to spend a year considering the *problématique*, followed by a presentation of their views on areas most menacing for the future of humanity. I produced a paper for preliminary discussion that was subjected to intensive debate by the Council at a meeting on the outskirts of Moscow, arranged by Gvishiani. A draft report was then circulated for comment before it was again debated in depth in Santander, Spain. These occasions appealed to me as high points in the history of the Club, provoking fertile, sustained intellectual struggles between members. There was unanimous agreement that the Santander conclusions provided the basis for a statement of the Club's credo that would be, for the first time, a report *by* and not *to* the Club of Rome. Bertrand Schneider and I then wrote it up in a more or less popular form and published it as *The First Global Revolution*. This eventually appeared in 24 languages. The book is in two parts; the first being a comprehensive overview of the danger spots of the *problématique* – population, environment, urban decay, global economy, poverty, technological development, etc. But unlike our earlier reports, it ranges over social and behavioural zones in addition to the more material problems looming on the horizon. For the second part we coined the phase *The Résolutique* to complement *The Problématique*. By this we meant strategies for the resolution of problems in their place within the complex of the *problématique*. We did not presume to propose concrete solutions; rather we suggested lines of action to follow in pursuit of them.

On 1 January 1991, Ricardo Diez Hochleitner replaced me as President of the Club a few months before the book's appearance. I remained an active member, attending meetings and indulging in my personal diplomacy until my stroke in 1994. I shall mention a few of these, mainly for their anecdotal value. They all concern East European countries, selected, no doubt, by nostalgia for my youthful globetrotting adventures. I had crossed from Finland with Sally, spending joyful days in Tallinn when Estonia was what the Russians referred to as 'a bourgeois republic'. Years later I attended a Soviet environmental meeting there in an unrecognisable dark city of suspicion and smouldering discontent. Now I returned to the old Tallin, light and cautiously hopeful, its charming old buildings having somehow emerged from the high-rise Soviet flats that had overshadowed them. I lectured and discussed with the new cohort of intellectuals and had interviews with the President and Prime Minister.

Prague was much the same. I had visited in 1929 by canoe, again during its dark, Stalinist days and then during its 'Springtime'. Now I went to meet its new poet President Vaclav Havel and recruit him to the Club of Rome. As we drove slowly up the hill to his office in the historic castle on the hill that dominates the city, I suddenly recognised the small dark house that I remembered was Kafka's home. I felt sick with horror. The castle I was approaching was Kafka's Castle. How could I ever enter that symbol of unreason, contradiction and confusion! The mood vanished as I entered a bright welcoming room and the presence of Havel. He was a pleasant quiet-spoken man, looking a little tired. Conversation was easy enough. We discovered we had one thing in common; we were both Erasmus laureates, he in 1986, I in 1987. I told him about the ideas and actions of the Club of Rome and he offered to be honorary president of a Czechoslovak national association. I promised to visit Prague once more when this had begun functioning. During this visit I met several Czech Ministers and had a talk with Dubček, the hero of the 'Prague Springtime', a simple, cheerful and seemingly guileless man.

I returned six months later and attended a meeting of the

Association that Havel chaired. I stayed in the Academy's guest-house and spent a few happy days meeting friends and giving interviews. On my last morning I was sitting at breakfast when the Minister of Trade and Industry telephoned. Prague was hosting an international conference of industrial leaders, which was very important for them. Not only was it the first major international meeting since liberation, but it repeated a corresponding event of the same group that US President Harding had sponsored just after World War I and the creation of Czechoslovakia, hence its highly symbolic significance: 'Would I please come and speak to the delegates that day'. I refused, explaining that I had to return to Paris immediately and was, in fact, waiting for a car to the airport. He insisted, saying the Congress Hall was *en route* to the airport, so I could stop and address the meeting for 20 minutes. They would delay the departure of my plane until I arrived. I protested, but to no avail. I was more or less kidnapped and delivered to the podium in front of perhaps 1,000 industrialists. I was given no theme or indication about what I was expected to say; nevertheless, I managed to speak for half an hour and it seemed to go down reasonably well. I was accompanied to the airport by the Industry and Foreign Ministers. We dashed across the tarmac to the steps of the waiting plane and as I said farewell, the Foreign Minister shouted, against the noise of the jet, 'Your visit has been more useful to us than that of Mrs Thatcher last month.' I boarded the plane and staggered to my economy seat at the back amidst the hostile scowls of delayed passengers.

My last vignette is of another descent to Eastern Europe – Romania this time, a land even more embedded in memories of my youth. Sally was with me this time. We landed at Bucharest and I spent the morning signing copies of the Romanian edition of *The Global Revolution* and responding to the award of an honorary doctorate at the university. Then in the evening my friend Mircea Malitza drove us up to Sinaia high in the Carpathians where Dan and I had camped in 1928. The landscape was every bit as grandly romantic as I remembered, the primeval forest with its gigantic trees supporting long horizontal branches that barred all light from penetrating to the forest floor. The castle, summer palace of the Romanian kings, was very eighteenth-century. We were installed in the Royal Suite that offered a different species of comfort from that of my camp site 70 years earlier.

The meeting that began the next morning had been designed a few months earlier at Budapest. It brought together people from most Central and Eastern European countries to explore means to attack the many problems they faced. I felt strongly that accumulated experience of inter-country co-operation in the rebuilding of Western Europe could be modified to meet the new needs of the East. In chairing the meeting I had the greatest difficulty in establishing a real discussion. As I had already found in Russia, committees there function very differently from those of Western Europe, America and Japan. Participants tend to give lengthy descriptions of their own organisations or of particular schemes with which they were identified. So it was in Sinaia where the delegates seemed to go out of their way to avoid mentioning practical action. This was obviously the result of years of the communist system when it was too dangerous to do so.

President Ionescu drove up from Bucharest, joined us for lunch and joined actively in the afternoon discussion. He appeared interested in our ideas and even suggested to Malitza afterwards that he would gladly accept an honorary presidency of the Romanian Association. Whether he took this seriously or not, I do not know. It didn't matter anyhow because soon afterwards he was voted out of office. On the second day, with my relaxed leadership, the group began to discover meaningful dialogue, and I concluded the meeting had been a softening up success. At breakfast on the morning of our departure, the palace was bathed in warm sunshine and some light seemed even to have penetrated the canopy of trees. A small brown bear with two appealing cubs came to the door begging for bread, fittingly rounding off my relationship with Sinaia.

Sombre Conclusions

Thirty years of activity of the type I have just recounted and innumerable debates had clearly modified my view of the world and society. I had attempted hesitatingly to express this in a *Call for Solidarity* on the last pages of *The First Global Revolution*, but I must do so more boldly here, stressing that my thoughts are strictly personal and not those of the Club. My firm conviction is that human intervention in the complex mechanisms of the biosphere has already reached a level that threatens the survival of civilised society and possibly of humanity itself. The explosion of world population, destruction of the environment, misuse of technology, especially in armed conflict, and impending energy shortage seem to have compounded to a disruptive synergy that imperils the planet and challenges our capacity to manage the ever-more vulnerable world society.

There is much we could do to stabilise the situation and slow down further deterioration. Contemporary wisdom suggests that, through political negotiation, technological fixes and social manipulation, many catastrophes might be avoided. However, most of the remedies would result in considerable deterioration of lifestyle in rich societies. In democratic countries it would be well-nigh impossible to obtain majority agreement to the painful measures necessary to save the planet unless a majority were to understand the nature of the problems ahead and of the apocalyptic consequences of not facing up to them. The only hope is to begin the construction of a World Solidarity based on the collective will and consciousness of humanity – no easy task to persuade the many societies, ideologies, cultures and religions of our world to work together. This has to be seen as a matter of enlightened self-interest, a form of collective self-interest. It could only happen if people were convinced of the fatal consequences of *laissez faire*.

The initial step, then, must be to inculcate an understanding of the gravity of the perils facing humanity, amongst all peoples. At the same time they must be offered hope that by making immediate sacrifices, the way would open for a peaceful and fairer world of modest prosperity for all that their children and grandchildren could enjoy. Realisation that our current exploitation of the planet is a time bomb for our eventual destruction has grown considerably in recent years. The Club of Rome has helped to shake apathy and complacency from the thinking of a small proportion of humanity, but a quantum leap of awareness has to be energised before coherent action can be taken. Yet even if that were to succeed, it would be but a holding position.

On environmental issues, *The First Global Revolution* concluded, 'the common enemy of humanity is man.' I am ever-more convinced that this applies far beyond environmental deterioration to a vast number of ills from which we and our societies suffer. If this is true, the fundamental difficulty lies within the very nature of man. Any durable solution to his external and internal problems can only come from evolution beyond the egoism that motivates every individual. The wisdom we desperately require can only come through inner transformation. All religions have, in their purest aspirations, attempted to induce such a change, with very little success. Much as we need a miracle, we can hardly rely on one appearing. Nevertheless, we should strive, through deliberate efforts of inner development and new insights into consciousness and the working of the mind, to cultivate an enlightened communal sense. This seems the only hope of breaking the impasse of both great danger and great promise, in which mankind is presently constrained.

33

The Scheme of Things (*Curiously Planned*)

Ah Love! Could thou and I with Fate conspire,
To grasp this sorry scheme of things entire,
Would we not shatter it to bits and then
Remould it nearer to the Heart's Desire.

OMAR KAYAM FROM THE *RUBAIYAT*, TWELFTH CENTURY

Ah Ladies; it was ever thus,
Men are unwise and curiously planned
They have their dreams and do not think of us.
We take the Golden Road to Samarkand.

JAMES ELROY FLECKER FROM *HASSAN*, A POPULAR PLAY OF THE EARLY TWENTIETH CENTURY

I am approaching the end of my story. The rhythm of my busy life came to a sudden halt in January 1994 when I suffered a stroke paralysing the left side of my body, but miraculously sparing mind and memory. I recovered quickly but was left with walking difficulties. We felt vulnerable, living in Paris, so moved to a small flat in London to be nearer to our children. I compensated for my lack of mobility by rushing about the streets of London on an electric scooter. We still maintained a small *pied à terre* in Paris and our house in the Midi; it had become a family summer home, much used by children, grandchildren – and now by six great-grandchildren.

Sally looked after me well as I recovered from the stroke in London and we settled down to the quiet life together that we had long looked forward to, but which I had delayed until then by my frenzied activities. But it was not to be. It soon became apparent Sally was losing her memory and sinking slowly into dementia. So the situation was inverted and for the next six years I cared for her every need. These were sad and often agonising times, especially as she was aware of her condition. 'It is terrible', she said, 'to realise that I am dying by inches.' Right up to the end she recognised our daughters and me and from time to time she still manifested aspects of her Valkyrie character.

During these years I was virtually a prisoner in our home. On a very few occasions I did manage to go abroad on Club of Rome business and to give a couple of lectures in London. But I had ample time for reflection, began to assess my past and sketched out a framework for my autobiography. This final chapter is more or less an addendum to the book, a post-Club of Rome *cri de coeur* evoked by a sudden realisation of the extent of society's descent into decadence.

I recognise that my life has undoubtedly been of a piece, with a persistent ego running lines of continuity throughout. Nevertheless, I have experienced many unforeseen discontinuities in my intellectual, emotional and geographical environment that diverted me into a series of disparate small worlds, each with its own horizon, against which my life flowed. I have witnessed most of the twentieth century and have marvelled at the extent of history that has rolled out in front of my eyes, the unprecedented enormity of change within a single lifespan.

I have lived through two world wars and the Cold War, during which the Damocles sword of the nuclear deterrent hung over humanity – as it still does. I have witnessed the rise, devastation and collapse of Italian Fascism, German Nazism and Soviet Communism, the more or less peaceful dissolution of the British Empire, the decolonisation of the Third World and the emergence of American hegemony. I have enjoyed, or suffered, existentialism, psychoanalysis, the sexual revolution, women's emancipation, feminism, post-modernism, the death of God and the end of history, as well as a plethora of cults and movements too numerous to mention. On a personal level, I have memories of travelling to the railway station in a hansom cab and riding in a horse-drawn tram, yet have lived to fly supersonic in Concorde. I cannot believe any other epoch could have seen so much history packed into a single lifetime, or that its momentous changes should have taken place simultaneously in so many elements in the life of individuals and society. Our information and knowledge have grown astoundingly and we have technical and economic instruments in place that are capable of shaping an equitable, just and modestly prosperous world. Despite this, we have lived through the bloodiest century in human history. I recognise few signs that human wisdom has developed in the last 5,000 years.

I began my career with a passionate conviction that science would create knowledge, making vast improvements possible in the human condition. Now, more than half a century later, I still believe this to be true; but my optimism is tempered by the painful recognition that science can also serve exploitation, tyranny and destruction of life and the planet. Whether the application of a break-through discovery will be used to the benefit of humanity or its detriment depends on the vision, (doubtful?) wisdom or greed of decision-makers. To my aged eyes, technological innovations, despite their enormous contributions to human wellbeing have caused a substantial if unforeseen leakage of quality of life.

When I first sketched the outline of this book, its last chapter, to be named 'Terminal Reflections', was planned to express my end of century thoughts and conclusions concerning my lifetime's work, with some gentle criticism of the motivations, judgements and behaviour of myself and others. Having now come to the end of the book and the beginning of a new century, I feel that such an ending would be irresponsible. I am so disturbed by the rank materialism of society and trivialisation of the human spirit that I feel it is my duty to express my views on society by assessing the 'sorry scheme of things entire'.

The last 20 years have seen a great change in the nature of society. Materialism has triumphed; consumerism is the dominant driving force. With the collapse of communism, capitalism has been almost universally accepted as the one undisputedly successful economic system. I have no difficulty with that. However, it is the metamorphosis within capitalism itself that is reshaping the global system and endangering both human and planetary futures.

The relatively benevolent, controlled capitalism that reigned in the West until the end of the 1970s, serving up unparalleled affluence, has gradually shed its controls and emerged as what is politely called the Free Market, but I would term 'enlightened jungle capitalism'. I use the word 'enlightened' cynically to indicate that the system is sufficiently vigilant of its own self-interest to ensure that there is just enough in the way of safety nets to prevent disruption through social unrest. This is not a sudden aberration corresponding to one of the occasional flips in the magnetic axis of the planet, nevertheless in the social realm it may be just as significant and could prove irreversible. The social axis of the world has moved many degrees to the right of its previous pole. Consequently, the objective of societies – as of individuals – is to become richer at all cost. Aspiration to wealth has always been there, but it has been but one of several incentives. Hitherto, leaders of society have probably responded even more to the acquisition of power and fame than to wealth as such. The power incentive in excess led throughout the ages to the emergence of dictators – the Stalins and Hitlers, and multitudes of petty tyrants – to exploitation, loss of freedom and war.

Deregulation, privatisation, technological break-through and globalisation of trade have brought about the emergence of today's jungle-capitalism. It is proffered as the sure basis of a dynamic economy that will generate endless new wealth. This might well be so, but at what cost? It is not simply a matter of freedom to buy or sell; it brings profit into every transaction, private or collective. We see the progressive absorption by the market of so many spheres that until now were not seen in relation to money. Science, like other areas of knowledge, is now regarded as something produced and, as with any other product, sold in the most profitable way. The new brutalist system shows no sign it possesses a heart. It pays no heed to the plight of displaced employees or their impoverished communities, it weakens the authority of governments, accelerates environmental destruction and denigrates the human spirit.

It is necessary to make a few remarks about 'globalisation' as now understood. With my international concerns, I have always been interested in proposals and dreams for a single united world in which the diversity of races and cultures could relate in harmony. But most of the suggested schemes, such as world federation, seemed too closely reliant on old political ideas already obsolete. Thus I am naturally curious about the implications of trade globalisation and critical of its lack of ideology beyond that of wealth maximisation. It seems to me obvious that the production of riches without consideration of how they will be distributed and for what purposes they will be employed would merely raise existing human issues to a higher level of magnitude and make their resolution still more difficult.

The process of globalisation has highly contradictory implications for the future of existing nation-states. On the one hand it is highly integrative and intensifies global political, economic, military, technological, scientific and cultural interconnections. This might generate a more tolerant, multicultural and universalist atmosphere for international relations – or it might do just the opposite. On the other side of the coin, globalisation is disruptive. It decomposes communities, destroys jobs, weakens state structures, sharpens conflicts of interest and exacerbates social stratification. Globalisation embraces the intolerable dichotomy of closer partnership and social disruption.

After my long years of struggle to improve productivity in industry and agriculture and speed up the rebuilding of war-flattened Europe, I was horrified by what I saw as a deteriorating situation. We had achieved gains mainly by generating shared motivation between partners in production, persuading them that results would be shared equitably between them. The productivity movement thus had positive social as well as economic values and goals. By 1980 productivity growth had become important once more, but in a quite different social and economic setting. With cut-throat competition, the only way a firm can establish an edge over rivals is to maximise output per worker. Improvement in personal efficiency can only be marginal, so the only way to achieve higher productivity is by upgrading the technology of the process. In effect this means replacing people with machines. When this proves successful, wealth is created, jobs are lost and lives ruined. There has been far too little discussion of the human dimensions of such changes. Job losses are engulfed within the general unemployment statistics while increased profits flow quietly to management and shareholders. In my youth I was taught that economics was for people. I now wonder.

In 1968, distressed by the 'growth for growth's sake' syndrome, I had posed the question, what is growth finally in aid of? I have received as yet no coherent answer. The question is more urgently cogent today with the globalisation being pressed on us being utterly and exclusively concerned with money. It is only logical to enquire, what should happen to the money? For stability any government must come to an understanding with its citizens on the just distribution of wealth.

The practice of the jungle-economy was worked out in America, although it can be argued that Thatcher invented it simultaneously. It is spreading rapidly in European countries, with some opposition from France, and is beginning to penetrate elsewhere. It has become virtually impossible for any nation to opt out of the brutal marathon of the free market for fear of being left behind with a fading economy in a world of growing profits.

There is much vocal opposition to jungle-capitalism, manifested at WTO meetings, but this comes from many uncoordinated groups with widely different, sometimes contradictory, reasons for their objections. This opposition is bound to continue, with most intellectuals and artists horrified by the blanket of materialism that has descended on them. However, the majority of US and UK citizens accept it uncritically, unaware of its implications. Decades of economic growth and increasing affluence have raised expectations of ever-more material prosperity, now seen by many as a right. This fuels the engine of consumerism and encourages economic irresponsibility in individuals.

It is said that shopping has become the national hobby of the British! The temptation to spend more than one earns is great. Personal consumer credit is easy to obtain; indeed, its sales pressure is so insistent it is difficult to resist. In America combined personal debt has already reached 90 per cent of annual income, it is hardly coincidence that America, with its regular high foreign trade deficits, has a mammoth external debt. In poor countries people get into debt to avoid starvation; in the rich, credit is sought, not as in the past to redeem debt or for essentials, but to purchase luxuries and passing fashions. In such countries, living in permanent debt is the norm. How different from my youth when living on credit was thought disgraceful and had to be concealed at all cost. The credit card is marvellously convenient, but a trap.

This situation of living on future earnings rather than providing for the future, inevitably influences the level of national

Alexander King aged 87

savings. In some East Asian countries people save a large proportion of their income; in Japan about one-third. Continental Europeans put aside about a fifth of their incomes, while British savings have sunk to a tenth and American to one-twentieth and still falling. In Britain there is a curious ambiguity; on the one hand policies are designed for growth through consumption, while at the same time, government urges more savings to provide pensions for the increasing and longer-living retirees of the future.

Under otherwise benevolent controlled capitalism, European countries accumulated a heavy burden of unemployment. This was partly due to very heavy social changes reducing the competitivity of their industries. In America, high economic growth and the onset of jungle-capitalism reduced unemployment to a very low level, although the official figures conceal the real situation. Many new jobs offer low wages, are temporary and insecure. To be truly fair and comparable, international unemployment statistics need to be weighted to take account of the quality of employment. The British level is somewhere between that of America and continental Europe, due to a little clearing in the jungle – the political need to retain at least some fiction of the Welfare State.

Structural unemployment and particularly its quality elements are likely to persist within the new system. This is due partly to replacement of people by smart technology, partly by multinational firms moving production to countries where labour costs are low. It is frequently argued that lost manufacturing jobs in developed countries will be compensated by new service sector jobs. This is questionable; it is in banking, insurance and financial services where large-scale dismissals are most evident. Furthermore, a high proportion of employment in service industries is poorly paid, seasonal, part time and insecure. The elimination of meaningful, secure jobs and the shortening of hours of work would inevitably cause enormous social problems and fundamental restructuring of the patterns of society. Jungle-capitalism is hardly conducive to social justice.

Insecurity is a main characteristic of contemporary society. Employers are free to hire and fire in the name of efficiency while concepts of loyalty, care for the fate of families and communities have been abandoned. Trade Unions have few legal instruments to protect their members. They still organise protests and strikes; but these are less frequent and less effective. The business world is floundering in insecurity. The jungle is full of predators large and small. Successful small or medium firms are easily swallowed up by rapacious entrepreneurs. Many very large enterprises are assailed by hostile, difficult to resist take-over bids. Management cannot expect loyalty from shareholders as the majority holders are generally institutional investors whose sole objective is better yields or capital growth. Such mergers often result in dramatic losses of jobs. Some of those discharged are later re-engaged on less favourable terms.

This brings me to the difficult questions of the very large differences in wealth between countries and within national societies. The greatest, most shameful of these disparities is between rich industrial countries of the North and poor underdeveloped nations of the South. This has long been admitted by the industrialised nations, with aid programmes mounted to combat poverty, hunger and disease, but these efforts are too feeble and too often distorted by the unsuitable development models of donors.

The poverty problem has been with us for so long, the suffering is so distant from the experience of citizens of rich countries and the numbers involved so great that they tend to be regarded merely in terms of statistics. It is only when television shocks the public with pictures of emaciated children or earthquake-devastated villages that governments and people begin to react generously. Indeed, the spontaneous reaction of nations and people to catastrophes, acts of war and terrorism and abuse of human rights is a happy indication that caring still flourishes in today's jungle. Care for people in danger in distant lands, however fleeting, contrasts vividly to the situation a century or more ago when the despair of victims of the Irish potato famine found little echo in actions of sympathy and relief by the English.

Growing disparity in the distribution of wealth in society seems to be a feature of the new economy. The rich become richer and the poor, poorer. In Britain the best-off 50 per cent of the population own 93 per cent of the national income, in contrast to 7 per cent earned by the other half. This is surely intolerable! The last two decades have seen a veritable epidemic in the numbers of very rich. Some of them have been catapulted to superwealth in frenzied Internet speculation, others through corporation mergers and upgrading of entrepreneurial success. Football millionaires and popstars are not grudged their wealth; the lottery creates three new millionaires per week.

However, the most bizarre situation is America, where the market reaches its ultimate of freedom. Some 350,000 US households are credited as each possessing more than $10 million, while Forbes magazine in its annual list of the 400 richest Americans notes that 267 had a net worth of more than $1 billion each. The absurdity is signalled by realisation that a single rich American can possess more wealth than several countries in Africa. At the same time US urban poverty is extensive. The new wealth is supposed to trickle down eventually to the poor, but the system seems unwilling to create a mechanism for doing so.

The mighty dollar has always been venerated; today its worship is the secular religion, even incorporating within it some evangelical Christian sects whose members support the political right. The financial mushroom associated with world telecommunications and computer development throws a spotlight onto the behaviour of the jungle-economy. To a naive person like me, it seems extraordinary how a firm like AOL, with its entrepreneurial panache and assertiveness, is able to attract enormous capital and merge with an established major corporation, even before it has made a profit. It makes one doubt the reality of these billions being tossed around and generating more and more (paper?) billionaires. Is there a bubble here, expanding to burst?

What, then, is the nature of this urge to become rich and what are its rewards in human terms? I would start by looking at the primitive biological instinct of all life, that resides within each one of us, the spark that distinguishes the animate from the inanimate in Nature. The Life Force (a quaint Victorian term) appears to be possessed in equal measure by all species from a blade of grass to an elephant and including us. It is the compulsion to metabolise and survive the attack of predators; to persist. It is the universal ego and has three associated urges – to search for food, to reproduce and to defend or extend territory. In man and many other species, the drive to survive has persuaded individuals to work together in defence against enemies, evolving the family, the community and eventually the nation. In this way social behaviour gradually developed. Protection of territory and aggression to extend it led to war, chauvinism and lust for power.

These primeval urges still remain and the universal ego activates all of us to a greater or lesser extent, stimulating us to feats of creativity and innovation. It also produces destructive aspects like selfishness, domination and exploitation. An element of this is self-aggrandisement, through which we seek to raise our esteem above that of others including colleagues and rivals. Politics is the constant striving to maintain a safe, constructive equilibrium between positive and negative aspects of ego.

We do not seem to have been endowed with any special mechanism to force us to struggle for wealth as such – the life force could hardly have envisaged the concept of money. It presumably arose as an offshoot of the need to survive. It exists to an extent in most people in most cultures, excessively in a few. In the latter it has a powerfully compulsive character that those addicted are seldom able to break. I recounted in Chapter 19 how Robert Maxwell, already a multimillionaire, said how easy it would be for him to acquire much more wealth, but that he was bored by the 'game' and wanted to be remembered for having done well for humanity. Nevertheless he was obviously unable to resist the compulsion to increase his wealth still further, finally indulged in criminal methods and was found drowned in the sea off Tenerife.

The United States is the most prolific of countries in generating the megarich; it is also exceptional in the esteem that an envious public accords to their Midas-like role models. This seems to be connected somehow with the Protestant origins of American society, traces of whose mores seem to have persisted in the racial melting pot that is America. A capacity for hard work, at times obsessional, together with a firm conviction of the virtue and respectability of success were bequeathed by the Pilgrim Fathers and ensure a favourable entry to the jungle of the new millennium. Generalising, I would argue that very rich Americans differ from the same people in most other countries. Their residual Puritanism makes their image appear dry and grey; workaholism allows only a sinuous contact with society; as with other Americans they allow themselves only two weeks vacation a year; they are neither exuberant nor flamboyant and no hint of hedonism is allowed to cloud their image. As was the case in the last century with the Rockefellers, Carnegies, Mellons and Fords, they tend to leave their vast fortunes to charitable foundations that become memorials to their financial prowess. Although it is not to my taste, I can understand the desire of individuals to acquire great luxury, to possess cars, yachts and aeroplanes, leaving sufficient capital to guarantee security for life. However, beyond a certain limit of extravagance, the drive to acquire more and more wealth seems pointless and pathological. It is a form of collection mania, no more admirable and much more harmful than collecting postage stamps. It is difficult to see the monstrous effort of self-enrichment as a version of the traditional 'pursuit of happiness'.

In Britain the enormous earnings and flashy lifestyles of footballers, models and rock stars are approved and marvelled at. The ostentatious lifestyles of the non-entertaining very rich are tolerated, except occasionally at board meetings where it may be revealed that the earnings of chief executives are as much as 400 times that of workers.

Disparities between rich and poor are likely to increase if globalisation advances unchecked and the issues of poverty are not faced. Will this eventually lead to social disruption? Not necessarily so. Throughout history the differences in conditions of life between aristocrats and peasants has always been vast, yet public discipline has been surprisingly strong. Grinding poverty in the face of splendour was accepted as the ordained state of existence. Throughout the centuries hope of paradise and fear of hell, with occasional helpings of bread and circuses by the state, usually managed to keep populations reasonably docile. The Marxists denounced religion as 'the opiate of the people'; how right they were to recognise it and how wrong to denounce rather than exploit it. Even in the muddy streets of the villages of Russia today, the colourful vestments of the priests, the gilded ornamentation and the power of sonorous bass voices give hope and colour to otherwise drab lives.

However, in today's material society, loss of religious faith and lack of respect for authority and contempt for politicians signal the evaporation of the former opiate, yet, remarkably, without noticeable loss of coherence in society. A new and effective opiate had to be found and is now freely available, ENTERTAINMENT. It comes in many flavours – radio and TV, videos, the Internet, electronic games, discos, beer binges and rave festivals, to which we must add spectator sports, especially football which has rapidly become a secular religion. These largely passive pastimes really do indeed pass time in a big way. They leave little space for conversation, reading books, interest in politics, active sports, playing with one's children, contemplating the stars or meditating on the aim and objective of existence. The 'masses' are engrossed in entertainment, with much passion and at times hysteria. Serious subjects such as educational topics have to be dressed up as entertainment to secure attention on television. The entertainment industry, with its momentum and profits maintained by new fads and successive waves of new technology, is highly commercialised. Look, for example, at Manchester United Football Club, valued on the Stock Exchange at £1 billion. Entertainment is thus itself an important part of the consumerism that drives the economy.

It is time to bring some of these matters together in proposing a model of what the world economy is likely to be a few decades hence when the jungle-economy has penetrated more deeply into the many countries only now beginning to espouse its virtues. Let us call this New World Order 'The Sorry Scheme'. Within this world-wide scheme, I see creation of wealth as being the unique and all commanding objective. Globalisation of trade will be fully operative and the jungle-economy highly successful. Countries we now classify as developed will be considerably richer than today. Some will be actually wallowing in wealth, yet still with high numbers of the insecure, poorly paid and unemployed. Disparity between rich and poor will thus be greater than ever. The Entertainment industry will have grown rapidly, keeping up with the economy, so that the masses are likely to remain politically supine. Their only function will be as consumers.

Most industry will have migrated to poorer countries, with the exception of agriculture, highly mechanised and operated by huge corporations competing for the most fertile regions to exploit. The existing nation-states will still exist, if only to preserve popular pictures of their former traditions for packaged tours. The living standards of poor countries will have risen considerably as a consequence of the success of the jungle-economy world-wide. In addition, having realised the need for ever-more consumption, the increase of purchasing power of underdeveloped countries will be prioritised. At the same time the traditions of indigenous societies will be corroded and corrupted significantly. To minimise the perennial outbreaks of civil and tribal wars, the opiate of entertainment will have to be administered lavishly. The world corporations, to propagate the gospels of consumption and entertainment and to sell hard and software devices, will send 'Missionaries' to keep people noisily happy. The equipment will be relatively inexpensive as a result of the enormous markets and will no doubt include plenty of those effective opiates, 'adult' videotapes.

The rich and the monstrously rich will mainly live in the great cities of former developed countries, with several secondary residences in climatically attractive regions overseas. They will still be hell-bent on adding still more to their senselessly large fortunes. Hence their essential activity will be maintaining or increasing the market share of their corporations' products and devising new methods of financial and technological innovation, including the essential task of sustaining the momentum of the entertainment industry. The world corporations will have a near monopoly in the support of scientific research, including artificial intelligence, genetic manipulation and nanoscience, that have enormous profit-making potential, but also the power to remake the world. Already these are raising difficult ethical issues. With the overwhelming dominance of the profit motive and the absence of universal control, it is difficult to see how these enormously significant developments can be controlled to safeguard the future of the human race and the planet

With the leakage of sovereignty from nation-states, corporations will take most of the major national decisions, individually or in unison. Politics as we know it will become of little interest and will operate mainly at the city, community or tribal level. Power will be seen in terms of control of the capacity for the production of wealth, so it will be difficult to be sure who is in charge of the actual running of the planet. There will still be speculators, shoving billions of dollars instantaneously from market to market each minute of the day, hoping opportunistically to benefit from small changes as they appear. These anonymous casino-playing 'strangers in the night', as Wouter van Dieren so charmingly calls them, are likely to be the only group feared by the corporate leaders, influencing their decisions and threatening their strategies, not in terms of power seeking or ideology, but simply through fortuitous opportunism. They will possess great power, but will have little interest in exercising it, their only ambition, like that of the tycoons, being to increase their fortunes.

Who then will be in charge of the planet's future? Are we revolving around the universe in a rudderless ship or simply sinking into the bosom of a self-regulating Nature – Creatures of Gaia?

A prophecy is the last thing this pernicious model pretends to be. I intend it as a warning for society to change direction sufficiently to ensure that it does *not* happen. I fully admit that my 'sorry scheme' model is as impossibly simplistic as it is darkly pessimistic. For instance, it avoids any mention of war and peace, nor does it face up to the environmental threats flowing from global warming that, probably by the middle of our new century, obliterate this whole scenario. By that time also, world energy shortage will disturb prospects for further growth. Indeed, it would be prudent to begin consideration of a post-oil-age energy strategy now. After all, we shall have consumed the fossil fuel reserves, built up by nature from the absorbed energy of millions of years' solar radiation, in two short centuries. Energy is the most fundamental factor in the physical equation, money being its surrogate. How will our multibillionaires adjust to the bleak economy as oil and coal disappear forever? It would be salutary if they could be persuaded to work with an economics based on energy instead of money!

In the period following World War II we all realised that a high proportion of the world's population had succumbed to totalitarian regimes with a horrifying history of genocide, oppression, loss of individual freedom and dignity. The ease with which this happened led many to fear that the apparently stable, free countries could, in situations of crisis, just as easily fall under the control of 'Big Brother' dictators. In 1949 George Orwell published *1984*, a nightmare account of totalitarianism with its brutal repression of the human spirit. This warning about the abuse of power had an enormous influence. The year 1984 passed with society unscathed.

I anticipate the appearance of a new Orwell who will write a scary novel, perhaps *2084* or *The Descent into Materialism*, to warn the world about the impoverishment of everything from daily life to spiritual development that would result from the continuation of our adoration of the one God – MONEY. I would love to have a go at it myself. There would be plenty of room for intriguing speculations; for example, would the saturation of poor countries with entertainment extinguish the population explosion and force our successors to breed consumers much as we do now with cattle to satisfy our carnivorous nature? Alas I have insufficient time left and I lack the necessary dazzle of imagination. All I can do is to offer my 'sorry scheme' as a skeleton frame on which that writer might build.

I am confident that when 2084 dawns, the world will have quite other preoccupations; the jungle society will have passed into history, as do all civilisations. Hopefully the eternal qualities of humanity will have survived the perils of ultimate materialism and be searching in new directions. The then current planetary problems will no doubt be every bit as threatening as jungle-ism seems to me today. Despite all my fears for the future of man and society, I find that I am still incurably optimistic. I still believe that homo sapiens has the inner capability to develop to be something greater than itself.

I must rise early tomorrow to reach Waterloo by taxi in time to board the eight o'clock Eurostar for Paris. I have to present a paper at a conference on the influence of science on peace and harmony between nations.

So here I go, *curiously planned* and with Sally in my heart, I still hobble along the golden road to Samarkand.

Acknowledgements

Dr Stevens (I never discovered his first name) a young Cambridge graduate not only 'deepened my knowledge of chemistry but inculcated a love for English literature and European theatre that has persisted all my life.

Professor Baker, my first research supervisor, demolished my respect for formal qualifications when he mounted a campaign against the PhD degree.

An invitation from Sir Henry Tizard abruptly changed my career line shifting my activity from the laboratory to the office of the Minister of Production. He influenced my subsequent work and behaviour.

G M Gurdjieff illuminated for me a different and bizarre way of living.

I was devoted to Aurelio Peccei, my friend and colleague over many years. This splendid man inspired life-long the vision of a better fashion of life for all humanity.

Throughout my work for the Club of Rome, my two daughters encouraged with imagination and subsequently helped me to prepare this book for publication. Marie Jennings, Brian Locke and Mihaela Smith were constant encouragements to me in writing the book, Penny Simmons was the ideal copy-editor and Mike Hauser undertook the design and typesetting. Thanks to *La Vanguardia* for kind permission to reproduce the photograph on page 378.